STRATEGIES FOR TECHNICAL COMMUNICATION

STRATEGIES FOR TECHNICAL COMMUNICATION

Nancy Roundy

with David Mair

LITTLE, BROWN AND COMPANY

Boston   Toronto

Library of Congress Cataloging in Publication Data

Roundy, Nancy L.
  Strategies for technical communication.

  Bibliography: p.
  Includes index.
  I. Technical writing. I. Mair, David. II. Title.
T11.R68 1985   808'.0666   84-26134
  ISBN 0-316-75923-6

Copyright © 1985 by Nancy L. Roundy with David Mair

All rights reserved. No part of this book may be reproduced in any form or by any electronic or mechanical means including information storage and retrieval systems without permission in writing from the publisher, except by a reviewer who may quote brief passages in a review.

Library of Congress Catalog Card Number 84-26134

ISBN 0-316-75923-6

9 8 7 6 5 4 3 2 1

BP

Published simultaneously in Canada
by Little, Brown & Company (Canada) Limited

Printed in the United States of America

Acknowledgments

The authors wish to thank Lee Blyler, Cathy Hamlett, Steven Lubahn, Jackie Brennecke, Duane Stall, and Virginia Quay for the use of their names and ideas.

"Egocentric Organization Chart." The concept in this chart was originated by Dr. Dwight Stevenson and Dr. John Mathes in *Designing Technical Reports,* 1976 (figure 2–3, page 15). Permission granted by the publisher, Bobbs-Merrill Educational Publishing, a subsidiary of ITT.

Page from serials catalogue reprinted by permission of Iowa State University.

Page from *Guide to Reference Books,* ninth edition, page 747 by Eugene P. Sheehy reprinted by permission of the American Library Association. Copyright © 1976 by the American Library Association.

Sample entries from the *New York Times Index,* reprinted by permission of The New York Times Company.

Sample entries from *Readers' Guide to Periodical Literature.* Copyright © 1980, 1981 by The H. W. Wilson Company. *Applied Science & Technology Index.* Copyright © 1980, 1981 by The H. W. Wilson Company. Material reprinted by permission of the publisher.

Sample entries from *Science Citation Index*®. Reprinted by permission of the copyright holder, Institute for Scientific Information, Philadelphia, Pennsylvania.

Sample entry from *Engineering Index*. Copyright © Engineering Information, Inc. Reprinted by permission of the publisher.

Selection from Rudolph Flesch, *The Art of Plain Talk*. Reprinted by permission of Harper & Row, Publishers.

Text from *Optical Fiber Telecommunications* reprinted by permission of Academic Press, Inc.

Excerpts from December 1979 *Bell Laboratories Record,* Copyright 1979, AT&T Bell Laboratories. Reprinted by permission of AT&T *Bell Laboratories Record.*

Text by Bret McCleary used by permission. Copyright © 1985 by Bret McCleary.

Text by Virginia Quay used by permission. Copyright © 1985 by Virginia Quay.

Text by Mark Nelson used by permission. Copyright © 1985 by Mark Nelson.

Text by Dale Erickson used by permission. Copyright © 1985 by Dale Erickson.

Text by Curtis Leonard used by permission. Copyright © 1985 by Curtis Leonard.

Text by Steven Lubahn used by permission. Copyright © 1985 by Steven Lubahn.

Text by G. K. Sammy used by permission. Copyright © 1985 by G. K. Sammy.

Text by James Gaunt used by permission. Copyright © 1985 by James Gaunt.

Text by Karen Christian used by permission. Copyright © 1985 by Karen Christian.

Text by Tonga Noweg, Ismail Subah, and Michael Curley used by permission. Copyright © 1985 by Tonga Noweg, Ismail Subah, and Michael Curley.

Text by David Wells used by permission. Copyright © 1985 by David Wells.

Hierarchy skills chart from *What Color Is Your Parachute?* by Richard Nelson Bolles. 1984 Edition. Copyright 1984 by Richard Bolles. Reprinted by permission. Available from Ten Speed Press, P.O. Box 7123, Berkeley, California 94707.

Text adapted from *Effective Business Communications* by Murphy and Hildebrandt. Reprinted by permission of McGraw-Hill Publishing Company.

Resume by Jean Ann Murphy Langie used by permission. Copyright © 1985 by Jean Ann M. Langie.

Text by Arlen K. Honts (''Arlen Hon'') used by permission. Copyright © 1985 by Arlen K. Honts.

Text by Roger Bagbey used by permission. Copyright © 1985 by Roger Bagbey.

Text by Rick Stebens used by permission. Copyright © 1985 by Rick Stebens.

Excerpt from ''Proposal for Research Studying Immunological Defense against *Cryptococcus neoformans*'' used by permission of Dr. Juneann W. Murphy, University of Oklahoma.

Text by Mark Hermanson used by permission. Copyright © 1985 by Mark Hermanson.

Text by Donald Murry used by permission. Copyright © 1985 by Donald Murry.

Text by Lance Thompson used by permission. Copyright ©1985 by Lance Thompson.

Text by Ann J. Eisenbraun used by permission. Copyright © 1985 by Ann J. Eisenbraun.

Text by Ted Jones used by permission. Copyright © 1985 by Ted Jones.

Text by Andrew Masterpole used by permission. Copyright © 1985 by Andrew Masterpole.

Text by Tonga Noweg used by permission. Copyright © 1985 by Tonga Noweg.

Text by Scott Crosser used by permission. Copyright © 1985 by Scott Crosser.

Text by Rebecca Rose used by permission. Copyright © 1985 by Rebecca Rose.

Troubleshooting table, isometric drawing, exploded view drawing, and cutaway drawing reprinted by permission of Technical Oil Tool Corporation (TOTCO), Norman, Oklahoma.

Text by Gregory Natvig used by permission. Copyright © 1985 by Gregory Natvig.

Text by Cleanweld Turner used by permission. Copyright © 1985 by Cleanweld Turner.

Map of rivers in Iowa and adapted map from 1977 *Statistical Profile of Iowa.* Reprinted by permission of Iowa Development Commission.

Text by John Cheslik used by permission. Copyright © 1985 by John Cheslik.

Preface:
To the Teacher

We have designed this text to assist students with the process of composing various technical documents. Although we also present finished products, our emphasis is on the process by which those products are written rather than simply on the completed documents.

The content and organization of our text reflects this process orientation. Our overview chapter illustrates the composing process with a case study of an experienced writer at work. In Chapters 1 through 6 we then present five strategies necessary for successful writing: analyzing the communication context, gathering information, selecting and arranging content, planning style, and revising the draft. Although the first four strategies are traditionally prewriting tasks and the fifth is traditionally a rewriting task, we suggest introducing students to all five strategies *before* they compose entire documents, because the composing process is not linear: Students may begin to revise before finishing a first draft or may gather additional information and arrange it after completing that draft.

Despite the nonlinearity of the composing process, we also stress the fact that some use of the first four strategies should precede successful composing: Students should analyze the communication context of their documents, gather information, select and arrange that information, and make stylistic plans *before* writing to ensure a well-directed and efficient writing stage.

In Chapters 7 through 14, we apply these five strategies to composing various technical documents. We illustrate the process of composing with case studies: extended examples where we follow writers step-by-step through the operations they perform when producing the documents being discussed. These case studies introduce students to realistic situations when the documents might be written. The case

viii   Preface: To the Teacher

studies also show the dependence of particular writing decisions on context: audience, purpose, and situation. However, the studies do not limit the document being discussed to the particular example or discipline involved: The *process* of composing the document is the same, regardless of situation or discipline. We suggest this fact by using a variety of disciplines in our studies and by including additional samples of documents at the end of each chapter.

In general, the chapters proceed from easier documents to those that are more difficult to write because they are longer and involve more decisions. The chapters also proceed from types of writing that may form parts of documents (e.g., definitions) to those that are whole documents in themselves (e.g., proposals). In each case, however, the emphasis is on the steps by which the document is written rather than simply on the resulting product.

The samples at the end of each chapter continue our process orientation. A brief context analysis precedes each sample, giving the circumstances under which it was composed. In general, one sample is annotated to point out important writing decisions, while another is unannotated. The first sample may be used to extend the material in each chapter, while the second sample provides material for class discussion. Some exercises at the end of chapters, ''Topics for Discussion,'' then encourage class participation and give practice in the individual steps necessary for composing the particular document. Other exercises, ''Topics for Further Practice,'' ask students to produce a finished product.

In Chapters 15 through 17, we discuss three aspects of preparing a document for readers: visual aids, format, and report supplements. Here we indicate the conventions governing document preparation as well as ways to achieve effective visual aids, format, and supplements. We end the text with a Handbook of Style containing the rules of standard written English, for the students' use as a reference.

We hope this text will introduce your students to the complexities of composing. We also hope that our process approach will demystify these complexities and make the difficult act of writing a little easier.

The samples in this text are intended to illustrate various forms of writing rather than to convey information on technical subjects. We have attempted to ensure the accuracy of the source material: However, any inaccuracy would not reflect on the validity of the samples as forms of technical writing. In addition, the names of persons and firms in some of the samples have been changed to protect privacy. In these samples, no resemblance to existing persons or firms is intended.

ACKNOWLEDGMENTS

We would like to thank the following persons whose assistance made this text possible: our reviewers, Virginia A. Book, University of Nebraska; Rebecca E. Carosso, University of Lowell; William O. Coggin, Bowling Green State University; Michael Connaughton, Saint Cloud University; David Fear, Valencia Community College; Dixie Elise Hickman, University of Southern Mississippi; John M. Lannon, Southeastern Massachusetts University; Michael J. Marcuse, University of Maryland;

Preface: To the Teacher

Richard N. Ramsay, Indiana University-Purdue University; Russell Rutter, Illinois State University; and Victoria M. Winkler, University of Minnesota, who patiently read our drafts and directed our revisions; Charlotte Thralls and Richard Freed, colleagues at Iowa State University, who also critiqued the manuscript and, in several cases, field-tested it; Terry Abbott and Rita Newell, whose typing and editorial skills were invaluable in allowing us to meet deadlines; Joseph Opiela, Allison Hoover, and Barbara Breese of Little, Brown and Patricia Torelli, all of whom oversaw and produced the text; our contributors, whose work appears throughout the text; and last, our students, whose response to the material was more useful than they may ever have known.

Preface:
To the Student

We have designed this text to assist you with writing the kinds of documents you will compose on the job. In our overview chapter, we introduce you to the process of writing by means of a case study picturing an experienced writer at work. These case studies continue throughout the text, as illustrations of the ways different documents are composed and as real-life examples of the types of writing you will encounter.

In Chapters 1 through 6, we discuss the steps you must perform in order to write effectively: analyzing the communication context, gathering information, selecting and arranging content, planning style, and revising the draft. In these chapters, we do not show you finished documents. Instead, we illustrate the particular step being discussed and give methods for carrying it out. The exercises in these chapters are also designed to give you practice in these steps.

In Chapters 7 through 15, we lead you through the process of writing various technical documents. The case studies in these chapters show writers composing these documents; the finished products are also presented. Our samples at the end of each chapter then illustrate the range of possibilities for a given technical form. Some exercises following the samples, "Topics for Discussion," encourage class participation or give practice in specific techniques useful for writing the particular document. Other exercises, "Topics for Further Practice," place you in situations where you would compose such a document and provide practice in the steps to follow.

In Chapters 15 through 17, we then discuss preparing documents for readers. Chapter 15 concerns visual aids, Chapter 16 format, and Chapter 17 report supplements. The exercises in these chapters focus on the skills necessary for effective document preparation.

Preface: To the Student

We end the text with a Handbook of Style for your use as a reference tool. This handbook will assist you with the rules governing standard written English: capitalization, grammar, punctuation, spelling, and usage.

In conclusion, we hope you will find our text useful to you now, as a student involved in writing technical documents, and later, as a person who must write on the job. If the book makes the difficult task of composing a little easier, the text will have served its purpose.

Contents

PART I

AN OVERVIEW   1

The Composing Process	3
Summary	5
Exercises	6

PART II

STRATEGIES: PLANNING AND REWRITING THE DOCUMENT   9

1 Analyzing the Communication Context: Audience, Purpose, and Use   11

The Effects of Audience, Purpose, and Use	11
Audience Analysis	12
Purpose-and-Use Analysis	22
Checklist for Analyzing the Communication Context	25
Exercises	26

2 Gathering Information: I. Writer-Directed Techniques — 27

Brainstorming	28
Free Writing	30
Relating Devices	31
Checklists	37
Grids	37
Checklist for Gathering Information: Writer-Directed Techniques	37
Exercises	40

3 Gathering Information: II. Library Research — 42

Library Resources	42
Procedure for Conducting Library Research	55
Checklist for Gathering Information: Library Research	68
Exercises	69

4 Selecting and Arranging Content — 71

Selection	72
Arrangement	72
Checklist for Selecting and Arranging Content	81
Exercises	82

5 Planning Style — 84

Stylistic Plans	85
Checklist for Planning Style	94
Exercises	95

Contents

6 Revising the Draft

A Definition of Revising	98
Techniques for Revising	99
Checklist for Revising the Draft	112
Exercises	113

PART III

**APPLYING THE STRATEGIES
TO TECHNICAL FORMS** 119

7 Definitions

121

The Purpose of Definitions	121
Bret's Definition of Asphaltic Surfaces	123
Checklist for Definitions	129
Sample Definitions	130
Exercises	135

8 Descriptions of Items

137

The Nature of Item Description	137
Mary's Description of a Hand Auger	139
Checklist for Descriptions of Items	150
Sample Descriptions of Items	151
Exercises	157

9 Descriptions of Processes

159

The Nature of Process Description	159
Jim's Instructions for Operating an Atomic Absorption Spectrophotometer	160
Checklist for Sets of Instructions	170

Karen's Process Narrative on a Technique for
Analyzing the Microstructure of Bread 171
Checklist for Process Narratives 180
Sample Process Descriptions 181
Exercises 188

10 Technical Letters and Memorandums 190

The Nature of Technical Letters
and Memorandums 190
Jon's Positive Message:
A Request for Information 190
Checklist for Positive Messages 197
Terry's Negative Message:
A Response to a Request 198
Checklist for Negative Messages 204
Sample Letters and Memorandums 204
Exercises 207

11 Resumes and Letters of Application 209

The Purposes of Resumes and
Letters of Application 209
Dave's Resumes 210
Checklist for Resumes 222
Dave's Letter of Application 223
Checklist for Letters of Application 232
Sample Resumes and Letters of Application 232
Exercises 236

12 Proposals 237

The Nature of Proposals 237
Arlen's Proposal 239
Checklist for Proposals 256
Sample Proposals 257
Exercises 260

Contents   **xvii**

13 Progress Reports   262

The Nature of Progress Reports   262
Don's Progress Report   263
Checklist for Progress Reports   275
Sample Progress Reports   276
Exercises   282

14 Final Reports   284

The Nature of Final Reports   284
Sam's Final Report   285
Sam's Executive Summary   299
Checklist for Final Reports   301
Sample Final Reports   302
Exercises   306

PART IV

PREPARING THE DOCUMENT: VISUAL AIDS, FORMAT, REPORT SUPPLEMENTS   309

15 Visual Aids   311

The Nature of Visual Aids   311
Purposes of Visual Aids   311
Types of Visual Aids   316
Appropriate Visual Aids   340
Integrated Visual Aids   345
Checklist for Visual Aids   347
Exercises   348

16 Format   350

The Importance of Format   350
Elements of Format   350

xviii   Contents

Layout and Standard Parts of Letters and Memorandums	356
Checklist for Format	364
Exercises	365

17 Report Supplements 366

The Nature of Report Supplements	366
Prefatory Material	367
Supplemental Material	383
Checklist for Report Supplements	387
Exercises	388

PART V

HANDBOOK OF STYLE: CAPITALIZATION, GRAMMAR, PUNCTUATION, SPELLING, USAGE 393

Capitalization	395
Grammar	396
Punctuation	398
Spelling	403
Usage	406

Appendix: Selected Bibliography on Technical Writing	411
Index	413

STRATEGIES FOR TECHNICAL COMMUNICATION

PART I

AN OVERVIEW

As a student in a technical discipline, you are familiar with processes — operations that take place, or are carried out, step by step over a period of time. Photosynthesis is one example of a process; distillation is another. However, you may not have thought of writing as a similar procedure, in which you perform a series of activities to arrive at an end.

These activities may be grouped into three divisions: prewriting, writing, and rewriting. Prewriting starts the moment you know you will be writing about a certain subject and involves planning your document. Writing involves composing your drafts. Rewriting includes all the changes you make when composing your drafts and turning the last draft into a finished document.

This idea of writing as process has two important implications. First, if you skip or do not complete an activity in the writing process, you may end up with an inferior product. If, as a biologist, you did not sterilize your equipment before trying to grow a culture of a single organism, you would probably obtain many organisms besides the one you wished to isolate. You would not obtain the desired results. The results of taking shortcuts in the composing process may not be so immediately apparent, but they will be just as damaging. Unless you are lucky, you will produce inadequate and poorly written reports.

Second, strategies or plans for action can assist with the process of composing. For instance, as a biologist, you would follow the scientific method when isolating your organism. This method is a strategy because it directs your experimental procedures. Five strategies direct the writing process: analyzing the communication context (audience, purpose, use), gathering information, selecting and arranging content, planning style, and revising the draft. These strategies help ensure successful reports.

Although composing is a process, it does differ in one important way from other processes with which you may be familiar. In most of these, you do not redo a step once you have completed it unless you have made a mistake or have decided to perform the process again. In writing, however, activities do not usually proceed in a straight line from prewriting through writing to rewriting. Instead, you may find yourself discovering and incorporating new ideas (a prewriting act) or changing sentences or words (a rewriting act) while composing your first draft (a writing act). This alternation between writing, prewriting, and rewriting is very common; the activities in the composing process overlap and blend.

In the next section, we illustrate this discussion of the composing process with a case study of Lee Blyler, a chemical engineer with Bell Laboratories in Murray Hill, New Jersey. Five years after joining Bell Labs, Lee was named supervisor of the applied research, properties, and processing group of the Plastics Research and Development Department, not only because he is a skilled researcher but also because he communicates well. In fact, he writes a great deal, anything from extensive technical papers and journal articles about his experimental work to letters and interoffice memos about his supervisory duties. All this writing, including his current project, is an integral part of his job.

THE COMPOSING PROCESS

Prewriting

Lee's writing project involves his research, a study of the deformation and flow of polymer melts and solids and the processing behavior of molten polymers.[1] Specifically, he is examining organic polymer coatings for fibers used in light-wave systems. These glass fibers carry signals in the form of light pulses. If the fibers are damaged in any way, their transmitting ability is reduced, so Lee has been investigating ways to coat the glass in order to protect it from abrasion and corrosion. He and his research team have developed a uniform coating and devised an application technique using a flexible die or feeding apparatus. Lee has been asked to write a brief report on this research for the *Bell Laboratories Record*, a journal produced and distributed internally.

Lee knows he must approach his communication task systematically, just as he did his technical work. He begins by analyzing the communication context of his report — his audience, and the purpose and use of his document — because he has found he writes documents differently, depending on who will read them and how readers will use them. He then uses this information about audience to select the content for his document, structure that content, and plan his style.

Lee often spends a great deal of time on these prewriting activities, because he has found that thorough planning is efficient. It aids him in the writing stage and reduces the time he must spend composing his first draft. Although he spends more time planning complex documents than brief letters or informal memorandums, he always reflects on audience, purpose and use, content, structure, and style before he writes.

Writing

Lee finds the transition from prewriting to writing the hardest step he takes. At this point, he relies on techniques that help him make this step and that continue to assist him as he writes. Although some of these may not work for you, we do stress one important point: Experienced writers *consciously* use a variety of techniques to help them compose.

First, Lee tries to find circumstances which will help him write. He sits down at his desk as soon as he arrives at work, because he has discovered that composing is easiest when he is fresh. Instead of beginning with the opening of a document, he frequently begins with the easier sections, because he finds that getting some words down helps him start the writing process. Second, Lee tries to write his first draft without extensive rereading or revising. He feels that, at this stage, simply getting his information on paper is the most important consideration. As he goes along, he may cross out words or phrases and substitute better ones that occur to him, for fear

---

[1]Polymers are large molecules made up of long chains of simple chemical units repeated over and over.

of forgetting them, and rearrange small bits of material. However, he does not let these actions interrupt his progress. In addition, he does not reread his entire document, although he may reread brief parts, because he knows he might be tempted to begin revising before he has a conception of the entire draft, perhaps unnecessarily editing material he will later discard. Lee has found that premature revision can cause writer's block, where he might polish and repolish the first sections but never progress beyond them.

Even though Lee tries to write his first draft without a break, we wish to stress the fact that his composing activities, though somewhat distinct, also overlap. For example, he will often discover additional material he wishes to include, or a more effective structure, while writing his document. Either he will immediately make these changes or he will jot a note in the margin of his paper to incorporate them later, if he does not wish to stop a train of thought. Thus, Lee gathers information and revises as he writes his first draft, even though these activities are traditionally prewriting and rewriting tasks.

Although this overlap bothered Lee as an inexperienced writer, he now realizes it is a natural part of the composing process. Writers can never be *completely* sure of what they wish to say before they begin a draft. Moreover, the ideas and structures that occur to them *while* composing will frequently strengthen the final document. As Lee has learned, composing is a dynamic, idea-gathering time, and a time for making additional writing choices.

Rewriting

Lee does alter a draft while writing it. He adds words, phrases, and occasionally entire blocks of material, crosses out words or sentences, and substitutes others. When a better order occurs to him, he rearranges material by drawing arrows to places where he feels the details ought to go. Sometimes, if he is writing a simple letter or memo, this revision while he is writing is sufficient. He simply gives his draft to his secretary at this point. For complex documents, however, he must review the draft to be sure he has communicated fully (content), logically (arrangement), and well (style). Lee has several techniques to accomplish these purposes.

First, unless the document he is composing is short or he is under the pressure of a deadline, Lee sets his first draft aside for a time, because he has found he cannot revise a lengthy manuscript effectively immediately after composing it. Then, when he feels he has gained sufficient objectivity about his writing, he reads the entire document through from start to finish, checking its content and structure. He begins revising at the level of overall content and structure rather than that of individual paragraphs or sentences so that he will not be altering parts he might later omit.

When Lee has finished his first reading, he goes back to revise the document paragraph by paragraph and sentence by sentence. He reads his manuscript aloud to hear the construction and connection of sentences. If he finds grammar, punctuation, spelling, or usage problems, he corrects those as well.

Summary 5

When Lee has time, he tests his manuscript on readers who have the same level of technical expertise and interests as his intended audience. If he cannot test the document on real readers, he gives it to colleagues who can role-play, or put themselves in the place of, this audience by simulating their interests and needs. (When he is under the pressure of deadlines, he role-plays this audience himself.) Even if he thinks he has expressed himself clearly and well, he knows he is too close to his document to be an accurate judge. He must trust his "mock" audience to read actively and critically, indicating where he should make further revisions. He repeats this testing procedure, which is much like the verification of experimental work, until he and his readers are satisfied with the document.

SUMMARY

Prewriting

Prewriting consists of analyzing your communication context, gathering information, selecting and ordering material appropriately, and planning style. These strategies help you make preliminary writing decisions. (We emphasize the word preliminary, since you may alter these decisions as you write.) Though prewriting activities will continue throughout the composing process, this preliminary planning makes composing more efficient.

Writing

You will want to find writing techniques that work for you; we suggest those Lee has used:

1. Find the best circumstances for writing.
2. Compose the document in any order you choose. Try beginning with the easier sections first.
3. Compose the first draft without rereading in entirety or stopping for premature revision.
4. Incorporate writing decisions you make while composing.

Rewriting

Although rewriting consists of the changes you make at any point in the composing process, we suggest the following techniques after you have completed your first draft.

1. Put the draft aside. Read it later with a fresh eye.
2. Revise from larger to smaller units. First, go straight through the draft for major content and structural changes. Then pause after each paragraph and sentence to consider minor changes.

3. Read aloud for sentence construction and connection.
4. Edit for mechanical correctness.
5. Test for audience effectiveness. If you have time, give your draft to readers for critical judgment. If you do not have time, role-play the audience of your document yourself.

We would like to make two points about rewriting. First, as you gain experience, you may find yourself combining all revision with writing. That is, you may make major and minor changes *while* you write, and your first draft may become your last. We suggest, however, that you separate composing and revising somewhat while you are learning, so you will not skip any important steps.

Second, we want to distinguish revising from editing, a specific step in the rewriting process. Revising involves reviewing your drafts and making major or minor changes as you find necessary. Editing involves more mechanical operations: reading for grammatical errors; correcting spelling, punctuation, and usage problems; and proofreading the final manuscript for typographical errors.

Both of these activities — revising and editing — are important in rewriting. We have found, however, that students frequently ignore the first and think that editing is the only rewriting operation they need to perform. Experienced writers know that revision is more complex, as with Lee's rewriting process. He read first for substantive content, structural, and stylistic changes before he corrected mechanical faults.

In addition, we believe that even if you combine rewriting and writing, a form of editing should end the composing process. Although this editing procedure may be as simple as proofreading your typed draft, only mechanically perfect manuscripts ought to leave your desk.

In the remainder of this book, we take you step by step through the process of composing several kinds of documents you may use on the job. We begin by discussing strategies — analyzing your communication context, gathering information, selecting and arranging content, planning style, and revising — in Part II. In Parts III and IV, we apply these strategies to technical writing and discuss preparing your draft. We end with a handbook in Part V.

EXERCISES

Topics for Discussion

1. Consider the following questions as a basis for class discussion:
   A. Prewriting
      1) Do you use all four strategies we have presented when you plan your writing? Discuss your use of each: how do you analyze your audience and purpose and use, gather information. select and arrange it, and plan your style?

Exercises 7

            2) If you omit any strategies, indicate why. Is the omission effective? Why or why not?

      B. Writing
          1) Consider the techniques we have suggested. Which do you use and why?
          2) Do you use any alternative techniques? Name them and discuss their effectiveness.

      C. Rewriting
          1) Consider the techniques we have given. Which do you use and why?
          2) Do you use any alternative techniques? Name them and discuss their effectiveness.

2. Do you experience any difficulties with writing? List each, then reflect on techniques you use to overcome that difficulty. If you do not have a technique, reflect on suggestions from our chapter which you might use.

3. Interview a professional in your field concerning his or her writing process. Gather information on use of the five strategies and any techniques employed to plan, write, and revise the work. Use this information as a basis for class discussion on the composing processes of writers of technical documents.

PART II

STRATEGIES: PLANNING AND REWRITING THE DOCUMENT

1

Analyzing the Communication Context: Audience, Purpose, and Use

THE EFFECTS OF AUDIENCE, PURPOSE, AND USE

If you are to communicate effectively, you must know a great deal about your audiences, your purposes, and your readers' uses of your documents. For example, suppose that as an agronomist you have been investigating various means for controlling agricultural pests. You must now write two documents. The first is a report for a local farmer who has asked for information on available methods of pest control. The second is a technical memorandum to members of your research team, detailing the results of your investigation. Although you are using the same body of information, the content, arrangement, and style of your documents would be very different.

For the local farmer, you would describe the various methods of pest control to be used, selecting only those methods suitable for his conditions. You might arrange your discussion of methods from most to least important in terms of his pest-control needs. You would write simpler, less complex sentences, because you would be dealing with technical material unfamiliar to your reader, and you would use synonyms or define all technical terms. In this way, you would accomplish your primary purpose (to inform the farmer of his options), and he would gain the knowledge he requires.

For members of your research team, you would describe in depth all the methods you had investigated. You might arrange this discussion of methods in several ways, for instance, most to least complex or effective, or by categories (chemical/natural). You would use longer, more complex sentences and technical terms, since your readers would have no difficulty understanding the technical material expressed in complex sentence structures. Moreover, technical language would aid rather than hinder

11

FIGURE 1.1
Possible audiences for Steve's reports

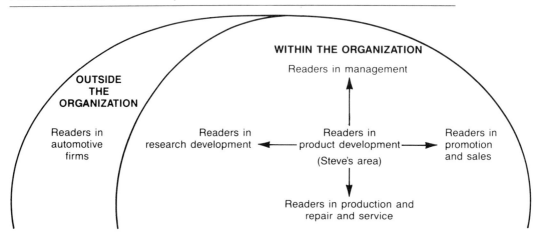

communication. You would then accomplish your primary purpose with this audience (to inform them of your research results), and they would gain the technical information they need.

Because the nature of your readers, your purposes, and the uses your readers will make of your document dictate writing choices, we devote this chapter to analyzing the communication context. We use the writing situations of Cathy Hamlett and Steve Lubahn as examples. Cathy, a self-employed farm-management consultant, must compose a letter asking Strom Computer Corporation for information on their latest microcomputers. She will use this information to help her recommend computer equipment to farmers she advises. Cathy's audience is one person — Doris Wilson, director of sales for Strom.

Steve, an electronics engineer with Acme Systems, a firm specializing in automotive hardware, is working on the design of a digital dashboard for automobiles. His reports may be read by a variety of audiences, as diagrammed in Figure 1.1.

Notice that reports such as Steve's can travel horizontally or vertically within organizations, to readers on the same level, above, or below the writer. Reports can also travel outside the company and may have a "lifetime," if they are kept on file. They may then be read by any number of audiences in the future. In the following sections, we discuss audience analysis and purpose-and-use analysis for limited communication situations such as Cathy's and complex ones such as Steve's.

AUDIENCE ANALYSIS

We first present the elements of any audience analysis, then discuss methods for determining significant audience characteristics.

Audience Analysis   **13**

Elements of an Audience Analysis

Even the briefest audience analysis involves three operations: identifying, classifying, and characterizing readers.[1]

Identifying Audience

You can identify your audience by name, title or job role, affiliation (e.g., company or institution), job description, and relationship to you as writer (e.g., client, supervisor, recipient of information). At times, you may not be able to designate your readers by name. For example, your audience may be an entire group (e.g., Association of Biomedical Engineers) or a person whose name you cannot discover even though you have tried (e.g., director of personnel). In these cases, you might identify only by job role or title and by affiliation. In fact, this information may be more important than names, since people change positions but job roles or titles usually remain.

Identifying your audience serves several purposes. First, identification forces you to pinpoint those to whom you are writing. Pinpointing readers is not a problem when your communication situation is fairly simple: Cathy knows precisely who will receive her letter. However, if your situation is more complex, you might easily mistake the recipients of some of your documents. For example, Steve might think members of his group would be his only readers and therefore might compose a technical memorandum on the digital dashboard. Knowing that readers outside his group will receive a document causes Steve to rethink his report in order to meet these readers' interests.

Second, identifying your audience gives you clues to help you satisfy their interests. For example, the title or job role and the brief job description would tell you how much your readers may know about the material you are presenting and what their particular needs will be. If Steve sends a report to George Brown, supervisor of promotion and sales for Acme Systems, Steve knows from Mr. Brown's job role that he will not necessarily require the technical details of Steve's research. Instead, Mr. Brown will be interested in the marketability of the product.

Third, identifying your audience assists you in considering the tone (approach to the reader) you will use by asking you to define your relationship to your reader. For example, George Brown is not Steve's immediate superior, but George does outrank Steve in the company hierarchy. Therefore, Steve should use a more formal tone than the one he might employ in an informal memo to his immediate colleagues.

Classifying Audience

Audiences may be either primary or secondary. A primary audience acts on or makes decisions on the basis of your document. Because these readers use the document for information or instruction, you direct it to them. A secondary audience does not

---

[1]We are indebted to J. C. Mathes and Dwight W. Stevenson for this information. See *Designing Technical Reports: Writing for Audiences in Organizations* (Indianapolis: Bobbs-Merrill, 1976).

directly act on or make decisions based on your document. However, these readers may use the document following the actions of the primary audience and thus may need the document for background information, or they may simply be interested in what you say.

The primary audience for Cathy Hamlett's letter is Doris Wilson, director of sales at Strom Computer Corporation. Unless Ms. Wilson happens to be out of town, in which case her secretary would read and reroute the letter, Cathy's letter has no secondary audience.

Steve Lubahn's reports may have primary and secondary audiences. For example, he might write a report to upper-level people in management, promotion and sales, and product development of Dennison Motors, an automotive firm that might market cars with the digital dashboard. These readers would be his primary audience because they are decision makers. However, his report might also be distributed to sales personnel as background information for selling the cars with the digital dash. These secondary readers would not make decisions on the basis of the report but would find it of considerable interest.

Classifying readers helps you pinpoint those who will be directly affected by your document so that you can fulfill their information needs. For example, in his report to Dennison Motors, Steve knows he must include information on the unique features of the digital dash, its marketability, the cost of production, and the suggested price because his primary readers would require these details. However, he may omit such details as his research methods and results, facts his primary readers do not need.

Characterizing Audience

You can characterize or define your audience by educational background, technical experience, and reader attributes affecting your document.

Educational Background

Your audience's educational background would include years of schooling and degrees attained, field of study, and additional training. Perhaps your readers have taken refresher courses or in-house seminars on the topic of your document. Perhaps they have attended conferences or workshops on this topic. You should note this informal training in your characterization, as well as the audience's formal schooling.

Technical Experience

Evaluating technical experience asks you to assess your audience's expertise concerning your topic. Sometimes technical experience will be related to educational level, but these may also be unrelated. Regardless of their years of schooling or degrees, your readers will usually be nonexperts when your topic is outside their areas of interest or study.

The following checklist will assist in categorizing your readers' general level of technical experience:

Expertise closest to writer's →

1. Same discipline and specialty
   Example: mechanical engineering, kinetic design ——→mechanical engineering, kinetic design

2. Same discipline, different specialty
   Example: computer science, programming ——→computer science, hardware design

3. Different discipline and specialty, same general concentration
   Example: zoology ——→botany

Expertise furthest from writer's →

4. Different discipline, specialty, and general concentration
   Example: aeronautical engineering ——→fisheries and wildlife biology

Once you have a general idea of your readers' experience, you can diagram their specific expertise with your subject on a grid:

Audience's level of experience			
Experience	*Extensive*	*Moderate*	*Little*
Theory			
Design			
Construction			
Operation			
Maintenance			
Management			

All six categories of experience may not apply in every writing situation. However, the grid provides a systematic means of considering an audience's experience as background for any communication task.

Knowing your audience's educational background and technical experience aids with content selection, arrangement, and style. For example, the primary readers for Steve's report to Dennison Motors have varying educational levels and most of these readers are outside his area of expertise. Therefore, his report will have to be general and nontechnical in order for him to communicate with the least experienced of his readers. However, since readers in product development will require the technical features of the digital dashboard, he may decide to append such material to the end of his document.

Reader Attributes Affecting the Document

Reader attributes affecting a document include personal or professional characteristics of your readers that could influence your writing. Perhaps the audience for your document has preferences that might color its reception. For example, Steve might know that upper management at Dennison Motors is conservative when considering the sale of a new product. This personal characteristic will affect Steve's report. He will have to include sound reasons for marketing the digital dashboard, perhaps using buyer opinions as support. He might structure his report so that his recommendations come last, after his explanations. In this way, he may lead upper management to agree with his conclusions.

Other reader attributes affecting the document might include professional concerns. For example, Steve might know that upper management at Dennison has not been happy with certain new designs in the past because of structural difficulties. This fact means that Steve must persuade upper management of the dashboard's soundness by providing a convincing technical discussion.

Methods for Audience Analysis

Unwritten Analyses

Experienced writers often use unwritten means of audience analysis when their reporting situations are not complex (e.g., when they are writing for one person or for a homogeneous group with similar backgrounds, interests, and expertise), or when they are under the pressure of time. For example, Cathy's audience analysis is unwritten:

Identification: ——▶ I know I'm writing to Doris Wilson, director of sales at Strom Computer
Name, Job role, Affiliation, Brief job Corporation. Since she handles the publicity for the firm, she is my likeliest
description, Relationship to writer source of information. She is my primary and only audience.

Classification I don't know Ms. Wilson's level of education, but I would assume she
Characterization: holds a B.A. Though it may be in business, she will have had much on-the-job
Level of education, experience. Obviously, Ms. Wilson will have more technical know-how about
Technical experience, computers than I do. The only reader attribute affecting my letter is Ms.
Reader attributes Wilson's busy schedule. She won't have a great deal of time to search for
affecting the document answers and reply, so I must make her task as easy as possible.

Even though experienced writers often analyze their audience without writing anything down, we suggest that, as a beginning writer, you use written analyses at first. In this way, you will become accustomed to the procedure. Moreover, you will be able to use these written techniques when your reporting situations are more complex.

Audience Analysis **17**

Written Analyses

Checklists

A checklist is useful when you are addressing a limited audience: one or two persons or homogeneous groups, or a small number of primary and secondary readers. Consider Steve's report to George Brown, supervisor of promotion and sales. This report will have one primary audience (George Brown) and one secondary audience (workers in promotion and sales). The report will then be kept on file and sent to other division supervisors, if requested. Because Steve's audience situation is limited, he uses the checklist we present in Figure 1.2.

FIGURE 1.2
Audience-analysis checklist

Audience analysis

Audience #1

1. Identification
   a. Name: George Brown
   b. Job role or title: Supervisor, promotion and sales
   c. Affiliation: Acme Systems
   d. Job description: Directs sales campaigns for all Acme products
   e. Relationship to writer: Supervisor of related division. Is above me in the company hierarchy

2. Classification
   a. Primary: yes
   b. Secondary:

3. Characterization
   a. Educational background
      1) Years of schooling attained: College + 2
      2) Degrees held/discipline: MBA, advertising
      3) Additional training: Knowledge of electronics as it relates to Acme products, gained on the job
   b. Technical experience
      1) Relationship to writer's area: Different discipline, specialty, and general concentration
      2) Level of experience:

	Extensive	*Moderate*	*Little*
Theory			X
Design			X
Construction			X
Operation			X
Maintenance			X
Management			

18   1│Analyzing the Communication Context

FIGURE 1.2 *continued*

c. Reader attributes affecting the document: George Brown does not market a new product until he is convinced it is in demand and will sell well. Therefore, I must persuade him of the dashboard's viability.

Audience #2

1. Identification
   a. Name: N/A
   b. Job role or title: Workers, promotion and sales
   c. Affiliation: Acme Systems
   d. Job description: Carry out George Brown's promotional campaigns
   e. Relationship to writer: Receivers of information. Are on my level in the company hierarchy

2. Classification
   a. Primary:
   b. Secondary: yes

3. Characterization
   a. Educational background
      1) Years of schooling attained: College +
      2) Degrees held/discipline: B.A., advertising
      3) Additional training: Knowledge of electronics gained from advertising campaigns for other products
   b. Technical experience
      1) Relationship to writer's area: Different discipline, specialty, and general concentration
      2) Level of experience:

	Extensive	*Moderate*	*Little*
Theory			X
Design			X
Construction			X
Operation			X
Maintenance			X
Management			

c. Reader attributes affecting the document: These workers are on my level in the company hierarchy. Therefore, I must not appear to be talking down to them in my report.

A checklist would clearly be an unwieldy method if you were addressing many primary and secondary audiences with varying levels of expertise and interests. In such cases, an egocentric-audience-analysis chart or a cluster-of-audience-interest chart would be more effective means of analysis.

Audience Analysis   **19**

Egocentric-Audience-Analysis Chart[2]

The egocentric-audience-analysis chart is useful when you feel your readers' experience with your topic will be the major influence on your writing. This chart centers around you as writer and places your readers close to you or farther away on the basis of their levels of expertise. In Figure 1.3 we present the chart with the information Steve Lubahn gathered for his report to upper-level people in management, promotion and sales, and product development of Dennison Motors — a report that might be distributed to sales personnel as well.

Notice that this chart also includes all the other elements of an audience analysis: identification, classification, and characterization. Such charts ensure a thorough consideration of the technical experience of all your readers and allow you to visualize a complex audience situation — readers with varying levels of expertise — at a glance. You will then be able to select an appropriate content, arrangement, and style to satisfy the needs of all readers. For example, two of Steve's primary readers, Samuel Sage and Theresa Goodall, are placed farthest from Steve on the chart. This placement indicates to Steve that they will have little experience with the digital dashboard. He must remain general and nontechnical in the body of his report.

Cluster-of-Audience-Interest Chart

A second analytical technique Steve might have used is the cluster-of-audience-interest chart. Unlike the egocentric-audience-analysis chart, which is based on variations in readers' technical experience, the cluster-of-audience-interest chart is based on the similarity of readers' interests in the document. Therefore, this chart is most useful when you wish to group readers according to their interests rather than distinguish them by expertise.

In Figure 1.4, we show a cluster-of-audience-interest chart that Steve would have constructed for his report to Dennison Motors. Notice that such charts also include all the elements of an audience analysis. In addition, these charts alert you to the interests your readers have in common so you may make effective content and arrangement decisions. For example, two of Steve's primary readers, Samuel Sage and Theresa Goodall, require a general description of the digital dashboard, while a third primary reader, Carl Handsprecker, requires a technical discussion. However, since Samuel Sage and Theresa Goodall will decide for or against purchasing the design and must understand it to make this decision, Steve will place his general description in the body of his report. He will then meet Carl Handsprecker's information needs by appending a full technical discussion.

We have introduced you to one unwritten and three written means of audience analysis. Perhaps you will develop other techniques. However, remember to include

---

[2]The concept in this chart was originated by Dr. Dwight Stevenson and Dr. John Mathes in *Designing Technical Reports* © 1976, Figure 2-3, "Egocentric Organization Chart," page 15, permission granted by the publisher, Bobbs-Merrill Educational Publishing, a subsidiary of ITT.

FIGURE 1.3
Egocentric-audience-analysis chart

LEVEL OF EXPERIENCE

Little

Moderate

Extensive

WRITER

Steve Lubahn
Electronic engineer, Product Development
(Lab Group I)
Does laboratory research
M.S. electrical engineering
Worked on developing digital dash

ˇ Carl Handsprecker
Director, Product Development
Dennison Motors
Develops and produces products
Will work with us to develop
and produce the dash
Ph.D., mechanical
engineering
Different area of study
and specialty,
same general
concentration
Will understand some
theory behind dash,
its design, operation,
and construction,
but as yet, has
not had experience
producing it

ˇ Salespersons
Dennison Motors
Sellers of cars
Will sell cars with the dash
All educational levels
Different area of study, specialty,
and general concentration
No experience with
digital dash

ˇ Tom Consamius
Supervisor, Product Development
Acme Systems
Oversees all lab groups
Eventual superior
Ph.D. mechanical engineering
Different area of study and specialty,
same general concentration
Hands-on experience with digital dash

ˇ George Brown
Supervisor, Promotion and Sales
Acme Systems
Publicizes products
Has been working with me
on promoting the dash
MBA, advertising
Different area of study,
specialty, and general
concentration
Has knowledge of the
dash's distinctive
features but little
theoretical
background

˙ Samuel Sage
President
Dennison Motors
Directs company
Will decide on dash
MBA
Different area of study, specialty
and general concentration
No experience with digital dash

˙ Theresa Goodall
Director, Promotion and Sales
Dennison Motors
Publicizes products
Will promote the digital dash
MBA, advertising
Different area of study, specialty,
and general concentration
No experience with
digital dash

˙ Primary audiences
ˇ Secondary audiences

FIGURE 1.4
Cluster-of-audience-interest chart

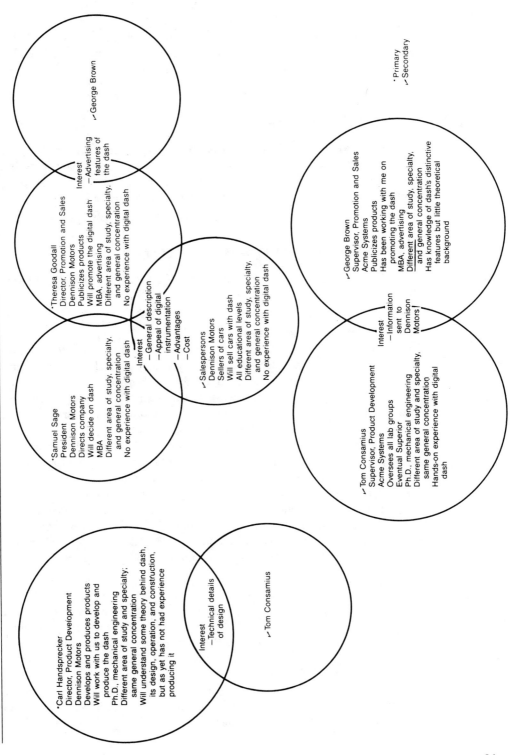

the three parts of a successful analysis — identification, classification, and characterization — because these parts tell you who your primary and secondary readers are and what they will be looking for in your document.

PURPOSE-AND-USE ANALYSIS

In addition to defining their audiences, experienced writers also consider their purposes in writing and the likely uses audiences will make of their documents. In the following sections, we discuss the elements of a purpose-and-use analysis, then present both unwritten and written means for accomplishing this task.

Elements of a Purpose-and-Use Analysis

Your purposes in writing and the likely audience uses of your document can be both general and specific.

General Purposes and Uses

Your purposes are your intentions in a document: to entertain, inform, instruct, or persuade. These purposes depend on the way you wish to affect your reader. When you entertain, you want to provide amusement. When you inform, you wish to convey information. When you instruct, you want to give the steps necessary for carrying out an action. When you persuade, you wish to motivate your reader to make a decision. Readers then use documents according to these intentions.

You can have more than one purpose for any document. However, one (your primary purpose) will usually predominate over others (your secondary purposes) and define your ultimate goal for the document. For example, let's say you wish to persuade readers to institute a computer billing system. This primary purpose is your ultimate goal. Your secondary purpose might then be to inform readers about the system. Notice that secondary purposes are important because your primary purpose can depend on them: your readers would not be able to decide for or against instituting the billing system without certain information.

Specific Purposes and Uses

In addition to general purposes and reader uses, you also have more specific purposes and uses for your documents: precisely what you want your reader to understand, do, or decide. For example, perhaps the reader of the computer-billing report must understand the physical components of the system. Providing this exact information would be a more specific purpose.

Knowing your more specific purposes and reader uses is important, because this knowledge helps you decide the precise content you will include. The methods for analyzing purpose and use in the following section illustrate this concept.

Purpose-and-Use Analysis   **23**

Methods for Purpose-and-Use Analysis

Unwritten Analyses

Experienced writers often conduct unwritten analyses of their purposes and uses when their reporting situations are not complex or when they are under the pressure of time. Cathy Hamlett's purpose-and-use analysis is an unwritten operation in which she reflects on her writing purposes and the uses Doris Wilson will make of Cathy's letter. Cathy's reasoning, if articulated, might read something like this:

General purpose and use ——— My primary purpose is to inform Ms. Wilson of my questions about their microcomputers: cost, ease of use, range ◄——— *Specific purpose and use*
of function, and maintenance. If I express myself completely, she will have the knowledge she requires to provide
General purpose and use ——— me with exactly the facts I require. My secondary purpose
is to persuade her to reply to all my questions. If I make
her reply easy by writing a well-structured and detailed ◄——— *Specific purpose and use*
inquiry, perhaps with a list of my questions, she should
decide to answer me.

Notice that Cathy's analysis has suggested content she will include.

Written Analyses

Grids

Grids are helpful for analyzing purpose and use when the communication situation is more complex. In addition, a grid will help you articulate your purposes and the audience uses of your document so you acquire skill in discerning these. In Figure 1.5, we present a purpose-and-use grid with the information Steve gathered for his report to Dennison Motors.

FIGURE 1.5
Purpose-and-use grid

Audiences	Primary purpose/use		Secondary purpose(s)/use(s)	
Primary				
1. Samuel Sage (President, Dennison Motors)	Purpose:	Persuade Mr. Sage of the soundness and sales potential of the digital dash	Purpose:	Inform Mr. Sage of the components of the design, its advantages, and its probable impact on the market
	Use:	Decide to purchase the design for use in his cars	Use:	Learn background information as a basis for decision making

24 1 | Analyzing the Communication Context

FIGURE 1.5 *continued*

Audiences	Primary purpose/use	Secondary purpose(s)/use(s)
2. Theresa Goodall (Director, Promotion and Sales, Dennison Motors)	Purpose: Persuade Ms. Goodall of the dash's marketability Use: Decide that the dash is marketable. Decide to recommend that Sage purchase the design because of its sales appeal	Purpose: Inform Ms. Goodall on the physical characteristics of the dash and its public appeal Use: Gain knowledge of the dash's physical characteristics and its public appeal
3. Carl Handsprecker (Director, Product Development, Dennison Motors)	Purpose: Persuade Mr. Handsprecker of the technical soundness of my design Use: Decide that the design is sound and ought to be used in Dennison's cars. Decide to recommend that Sage purchase the design	Purpose: Inform Mr. Handsprecker on the technical aspects of the design Use: Gain a technical knowledge of the dash's design

Secondary

1. Tom Consamius (Supervisor, Product Development, Acme Systems)	Purpose: Inform Mr. Consamius of the technical details I have sent Mr. Handsprecker Use: Gain knowledge of these details, so Mr. Consamius can work with Mr. Handsprecker on producing the dash	Purpose: Persuade Mr. Consamius the report is ready to be sent to Dennison Motors Use: Decide to send the report, on the basis of its quality
2. George Brown (Supervisor, Promotion and Sales, Acme Systems)	Purpose: Inform Mr. Brown of the details I have given Ms. Goodall Use: Gain knowledge of these details, so Mr. Brown can effectively advise Ms. Goodall	Purpose: Continue to convince Mr. Brown of the design's worth Use: Decide to support the dash in his work with Ms. Goodall
3. Salespersons (Dennison Motors)	Purpose: Inform the salespersons of the dash's design, its unique features, and its customer appeal Use: Gain information as background for promoting the dash to customers	Purpose: Persuade salespersons of the dash's selling appeal Use: Decide to promote the dash in Dennison's new cars

Notice that Steve's purpose-and-use grid has helped him make decisions on content: He must describe the design and discuss its marketability in terms of its advantages and sales appeal.

To construct a purpose-and-use grid, list both your primary and secondary audiences in hierarchical order (primary decision maker first, etc.). Then give both your general purpose and your readers' general uses of your document, and your specific purposes and reader uses. You will then have devised a list of content for your document.

CHECKLIST FOR ANALYZING THE COMMUNICATION CONTEXT

Audience Analysis

1. Have I identified my audience?
   A. Name
   B. Title or job role
   C. Affiliation
   D. Job description
   E. Relationship to writer
2. Have I classified my audience?
   A. Primary
   B. Secondary
3. Have I characterized my audience?
   A. Educational background
   B. Technical experience
   C. Reader attributes affecting the document

Purpose-and-Use Analysis

1. Purposes
   A. What is my general primary purpose with each reader — to entertain, to inform, to instruct, or to persuade? What are my specific primary purposes?
   B. What is/are my general secondary purpose(s) with each reader? What is/are my specific secondary purpose(s)?
2. Reader Uses
   A. What are the primary reader uses, general and specific, of my document?
   B. What are the secondary reader uses, general and specific, of my document?

26   1 | Analyzing the Communication Context

EXERCISES

Topics for Discussion

1. Select an article from a scientific journal or magazine. Identify, classify, and characterize the audience or audiences of the article, using one of the written methods we have provided. Indicate the basis of your judgment for each item of information you give. Then analyze the purposes and uses of the article, using the grid we have provided. In class, discuss the information you have gathered about audience, purpose, and use. Indicate the basis of your judgment for each statement you make by referring to the article.

2. Select an article from a popular magazine on the same topic as in question 1. Analyze the audiences, purposes, and uses of the article. Do the audiences, purposes, and uses differ from those in question 1? How and why?

Topics for Further Practice

1. Construct a checklist analyzing a colleague in your field as the audience of a technical document you have written or may write. Then construct a purpose-and-use grid for the document. Remember to consider both primary and secondary purposes and uses.

2. Create a job role for yourself in a firm of your choice (e.g., project engineer, Crosser Construction Company) or in an institution with which you are familiar (e.g., library aide, State College). Construct an egocentric-audience-analysis chart, indicating all the audiences to whom you might write documents. Be sure to identify, classify (if possible), and characterize your audiences. (If you do not have sufficient information about these audiences, interview a professional who holds the position you have chosen in order to gather the necessary data.)

3. Select a technical paper you have written for another class. Using the topic of the paper, create a real-world situation (e.g., you are investigating the efficiency of a local store's practices, using querying theory) and possible audiences (e.g., management of the store, fellow workers). Analyze these audiences, using both the egocentric-audience-analysis chart and the cluster-of-audience-interest chart we have provided. Then analyze the purposes and uses of the paper for these audiences and indicate the writing decisions these analyses have helped you make.

2

Gathering Information:
I. Writer-Directed Techniques

Because writers require material for their documents, gathering information is an important strategy. Structured methods for gathering this material, such as techniques for laboratory or library research, are helpful because they can be applied to many communication situations. For example, you do not have to relearn how to conduct a laboratory experiment or how to use the library each time you must perform these actions.

In this book, we show you two types of information-gathering techniques: writer-directed and topic-directed. Writer-directed techniques are ways of examining your experience to discover what you already know about a subject and where you need to research further. The writer-directed techniques we describe include brainstorming; free writing; and using relating devices, checklists, and grids.

Topic-directed techniques provide ways of exploring your environment. Examples of topic-directed techniques are laboratory research; on-site observation; requests for information by questionnaire, telephone, interview, or letter; and library research. However, because some of these techniques are highly specialized, your major professors and supervisors on the job are best equipped to provide information on them. We will cover only letter inquiries and library research.

Information-gathering techniques do not follow a necessary order. For example, while doing library research, you may discover that brainstorming your subject would clarify and direct your thinking or, after on-site observation, that a checklist would help you organize and determine what you have discovered. Therefore, employ these techniques in any order you choose.

In this chapter, we use the example of Tina Malone, Brad Garcia, and Rick Shurr, fourth-year civil engineering students at a state university. Tina, Brad, and

27

Rick are enrolled in an independent study course for which they must select a design problem with real-world applications. They choose to consider a highway bypass of Mason, Iowa, a town experiencing traffic congestion on Main Street because this street is also a major through road. They must study the problem as an engineering project, select an optimal route for the bypass, and design that route. Finally, they must report their findings to the city council of Mason as a practical solution to its traffic congestion.

The three students have already gathered some information. They have studied Mason's location and geography, both on site and from topographical maps. They have selected three possible routes, then narrowed this choice to one route south of Mason, based on such factors as environmental impact, extent and nature of the construction involved, and cost. They have also made preliminary designs of the optimal route. Tina, Brad, and Rick feel confident of their engineering skill. Now, however, they must tackle the job of presenting their findings to the city council.

In preparation for gathering more information, they reflect on their readers, their purposes, and the likely uses of their report. Their primary purpose is to persuade the city council that Mason needs a bypass so that the council can make a decision about construction. This primary purpose depends on informing the council about the bypass route and its impact on Mason, as background for deliberation. This knowledge of their audience, their purposes, and the uses of their report now influences the group's search for additional material by indicating areas they must explore.

Tina, Brad, and Rick realize they must already be familiar with some of the data they need. Therefore, they decide to apply writer-directed techniques to their topic to discover what they already know and where they must research further. In the remainder of this chapter, we illustrate the group's use of several techniques. We then cover topic-directed techniques in Chapter 3, Library Research, and Chapter 10, Technical Letters and Memorandums.

BRAINSTORMING

Brainstorming is a quick method for tapping knowledge you may not realize you have and for discovering many facets of a subject. You can brainstorm individually or in a group, in which case you pool your ideas.

To brainstorm, introduce a topic and quickly record every thought that occurs about that topic. If you are brainstorming in a group, you may find using a blackboard helpful. You can then note each person's contribution as it is given and can easily view the resulting list. The only stipulation about brainstorming is that you must not prejudge and thus discard any ideas. You are looking for a multiplicity of viewpoints, not evaluating their worth.

In Figure 2.1, we list the results of Tina, Brad, and Rick's brainstorming session on possible negative impacts of the bypass. Notice that brainstorming generates many directions for research, which you must then evaluate. First, classify related topics into groups and provide a heading for each group, identifying the common characteristic. We give Tina, Brad, and Rick's classification in Figure 2.2.

Brainstorming   **29**

FIGURE 2.1
Results of brainstorming session

Possible negative impacts of the bypass

Dislocation of farms, industries, etc.

Will incur additional maintenance costs

Possibly ineffective in removing enough traffic from Mason center

Budget may not be sufficient to cover initial construction cost

May disrupt shopping patterns

Loss of business

Noise level may disturb residents

May or may not affect number of accidents

May disturb environment, e.g., wildlife

May disrupt important facilities (graveyards, churches, hospitals)

What about swamp south of Mason? May flood bypass in rainy weather, making road impassible. Would reduce road's practicality as a solution.

FIGURE 2.2
Classification of brainstorming results

Possible negative impacts of the bypass

Effectiveness

Won't remove traffic from center of town

Won't reduce accidents

Won't be passable in rain because of swamp

Effect

Dislocation or disruption of farms, industries, facilities

Noise may disturb residents

Shopping patterns may be disrupted

Loss of business

Environment may be disturbed

Cost

Initial construction cost may be too high for budget

Additional maintenance costs

Next, decide whether or not each topic seems relevant to your information search by considering material your readers may require. You may then use other information-gathering techniques to explore the relevant topics further. Tina, Brad, and Rick decide to explore all three topics, because they all represent questions the city council may have.

30   2 | Gathering Information I. Writer-Directed Techniques

FREE WRITING[1]

Free writing is a method of using writing as exploration in order to discover new ideas. When you free write, you allow your mind to play with a subject and follow any train of thought, even if you seem to stray from the topic with which you began. Straying is the point of free writing, since you are looking for new information and directions rather than recording data.

In order to free write, place your topic at the head of your paper and write down any ideas that occur as you think about this topic, without evaluation or prejudgment. You may free write for a specific time period (15 to 20 minutes is usually sufficient) or you may free write until you run out of material. We illustrate this technique in Figure 2.3, Rick's free writing on traffic removal and accident reduction.

FIGURE 2.3
Free writing

Traffic removal and accident reduction

Once tried to pull out of a parking space on Main Street and had to wait 10 minutes for traffic to clear. Rush hour in Mason, plus through traffic on the highway. (Survey of drivers downtown at rush hour might yield useful comments.) Had to wait out three lights turning into the mall as well. (Checkpoint for traffic pattern analysis?) Could hand out questionnaire here. Frustrated drivers may favor a bypass as an alternative to current traffic routing, which is dangerous as well as time-consuming. I saw two accidents at the mall lights in one week — wonder what the statistics are on accidents downtown? Could check at the police station. Should also get figures on traffic volume through Mason: peak hour traffic, normal load, through traffic. Do a breakdown maybe. Should do traffic-pattern analysis to estimate how much traffic a bypass would remove from downtown. Perhaps a comparison of traffic removed with accident statistics would give estimates on accident reduction.

FIGURE 2.4
Evaluation of free writing

---

1. Heavy congestion
   - □ Main Street and mall are particular problems.
   - □ Survey drivers for support.

2. Dangerous situation
   - □ Get statistics on accidents downtown.
   - □ Get figures on traffic volume. Do a breakdown: peak vs. normal load; local vs. through traffic.

3. Traffic removed by bypass
   - □ Do traffic-pattern analysis.
   - □ Compare estimated traffic removed with accident statistics.

---

[1]For a more complete discussion of free writing, see Peter Elbow, *Writing With Power* (New York: Oxford University Press, 1981).

Notice that free writing raises issues and questions rather than providing answers, because of its exploratory nature. Moreover, free writing may suggest other information-gathering techniques. Here Rick mentions a questionnaire survey of local drivers and on-site research on accident statistics, traffic volume, and traffic patterns as techniques he may employ.

When you finish free writing, evaluate your results by identifying key points your free writing has generated and then listing subpoints under these keys. In Figure 2.4, you see Rick's evaluation. You can then discard ideas that do not seem relevant and further explore those that do, either by free writing or by using other information-gathering techniques.

RELATING DEVICES

Relating devices are information-gathering techniques that allow you to indicate relationships among your ideas while gathering them. Three major types of relating devices are hierarchical, sequencing, and clustering devices.

Hierarchical Devices

Hierarchical devices, which classify and arrange information in levels, include idea trees and outlines. These differ in that outlines proceed vertically down a page, while idea trees branch out. Both devices are useful. However, the branching structure of the idea tree provides more flexibility than the vertical structure of the traditional outline by allowing you to visualize, reflect on, and alter the arrangement of your information more easily than may be possible with an outline. In addition, the branching structure of the idea tree facilitates adding details to any category of the tree as you proceed, whereas the vertical structure of an outline tends to confine you to the section you are presently developing. We illustrate an idea tree in Figure 2.5, Tina, Brad, and Rick's idea tree on negative impacts of the bypass on Mason.

FIGURE 2.5
Idea tree

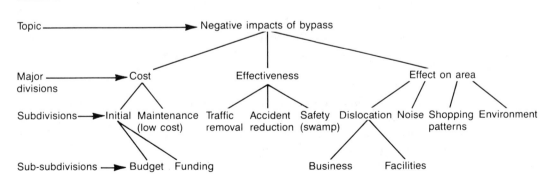

To gather ideas by using an idea tree, first place your topic at the top of a piece of paper. Then consider the major divisions of that topic and place these divisions at the ends of branches. Proceeding down the page, add more branches as necessary for subdivisions and sub-subdivisions.

An outline can be used in a similar way, by placing major divisions of your topic beside Roman numerals, subdivisions beside capital letters, sub-subdivisions beside Arabic numbers, and so on. When using an outline to gather ideas, however, be careful to leave sufficient space under each number or letter so you can return to a topic and write down more ideas as they occur to you. This space is important, because you are not organizing the data you have on hand but discovering more. Insufficient space may restrict your exploration.

Sequencing Devices

Sequencing devices, which indicate linear or positional relationships among ideas, include flow charts, diagrams, line graphs, and continuums.

Flow Charts

If you are describing a process, you might construct a flow chart as a means of discovering information. This schematic representation of the steps in the process, with arrows drawn between each step, will indicate visually the details you might include. We show an example of a flow chart in Figure 2.6: Tina, Brad, and Rick's flow chart for the process they should use when designing Mason's bypass.

To construct a flow chart for generating ideas, place the first major step in your process at the point where your flow begins. If your process proceeds from left to right, this point is at the left margin; if from right to left, this point is at the right margin; if hierarchically, this point is at the top. Then draw arrows to boxes containing succeeding steps in the direction or directions of your process. Notice that major steps may also generate substeps, which you can indicate by arrows and boxes from the appropriate major step.

Diagrams

If you are describing an item, a diagram of its parts will help you discover information. In Figure 2.7 on page 34, we show the diagram Tina, Brad, and Rick drew of the bypass's route. Because drawing such a diagram forces you to represent your item visually, you may find that the item is more complex than you had originally assumed. Therefore, you will have gathered additional facts.

Line Graphs

Line graphs will assist you if you are working with data that vary over time. Figure 2.8 on page 35 shows a line graph plotting traffic removal for three possible bypass routes over a four-year period. After you have constructed your line graph, you should assess the trends and relationships it reveals in order to gather information.

**FIGURE 2.6
Flow chart**

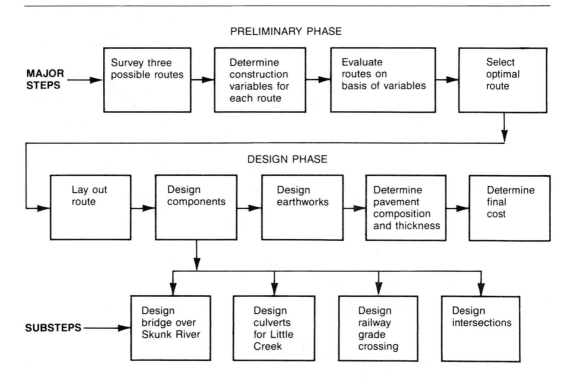

For example, Tina, Brad, and Rick note that the southern bypass will remove the most traffic by year 4 while the half bypass will remove the least. The northern bypass remains stable at figures lower than those of the southern bypass.

Continuums

Continuums may be used to visualize information that proceeds from one point to another. One familiar continuum is the time line, where data are plotted on a line according to their time of occurrence. However, other characteristics may form the poles of a continuum. Figure 2.9 on page 35 illustrates the continuum Tina, Brad, and Rick used to plot the desirability of various solutions to Mason's traffic problem.

Clustering Devices

Clustering devices indicate overlap or areas of similarity and dissimilarity among ideas. (In Figure 1.4, page 21, we illustrated a clustering device used to gather information about audiences.) Figure 2.10 on page 36 shows the clustering device Tina, Brad, and Rick constructed to compare and contrast their three bypass routes.

FIGURE 2.7
Diagram

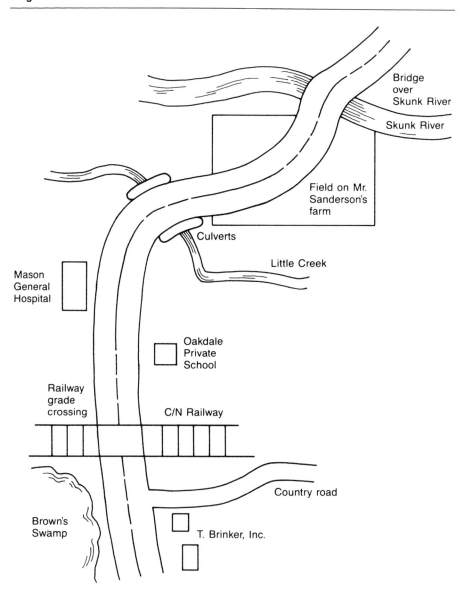

Relating Devices

FIGURE 2.8
Line graph

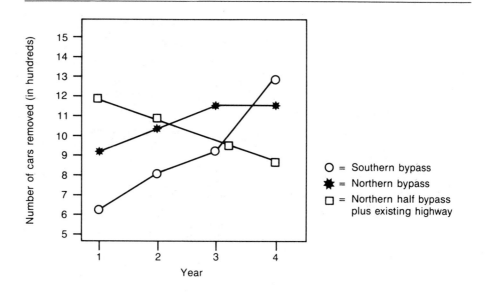

FIGURE 2.9
Continuum. To construct a continuum, place the poles you have selected at each end, then use spaces of varying lengths to separate each component of the continuum. The length of the spaces indicates each entry's relationship to the preceding and succeeding entries.

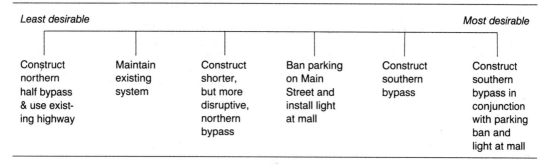

FIGURE 2.10
Clustering device

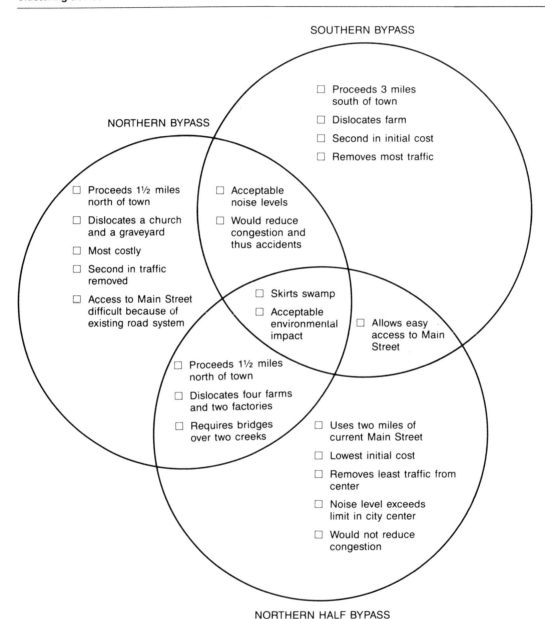

To construct a clustering device, first draw overlapping circles. Then place the details your topics have in common in the intersections of circles and the details that distinguish each topic outside these intersections. You may then use these data to compare and contrast your topics.

CHECKLISTS

Checklists are lists of items to be considered in an inquiry. These lists provide systematic means for discovering facts by structuring your inquiry procedure and ensuring that you consider many facets of a subject. In this book, we employ many such lists. Here we illustrate the reporter's formula, or the 5 W's + H.

The reporter's formula (who, what, when, where, why, how) is a simple, quick means of answering essential questions about a topic. You may then explore these answers further by using other information-gathering techniques. In Figure 2.11 on page 38, we reproduce Brad's application of the reporter's formula to the issue of funding, with the questions he generated about city funding for a bypass and the answers he has obtained thus far in his research. Notice that Brad will use other information-gathering techniques to amplify the reporter's formula: he mentions an interview with the city treasurer and a letter of inquiry to agencies in charge of funding programs.

GRIDS

Grids are devices that create patterns of intersection among ideas or variables and that generate information on the basis of the intersection. Grids are useful because the intersection of two ideas may lead you to view your data in a new way. Moreover, a grid organizes your data as you generate them, thus helping you visualize what you have discovered. (In Figure 1.5, we presented a grid for gathering information about purpose and use.) To construct a grid, place one set of variables at the heads of columns and a second set at the left of rows. Then plot the resulting information in the grid's squares. In Figure 2.12 on page 39, we reproduce Tina's grid on the effects of the bypass. Notice that a grid's headings may contain both major headings and subheadings if your topic is complex.

In this chapter, we have illustrated several writer-directed techniques for gathering information. Perhaps you will discover others to add to the ones we have presented, as you gather information.

CHECKLIST FOR GATHERING INFORMATION: WRITER-DIRECTED TECHNIQUES

1. Do I need to tap knowledge I do not realize I have or to discover many facets of my subject by brainstorming or free writing?

FIGURE 2.11
Application of the reporter's formula

Who
Who can obtain funding?

Cities that qualify for certain federal programs.

Whom does the city contact?

Don't know. Perhaps city treasurer will.

What
What are the guidelines for funding?

Depends on the program. Probably city treasurer will know programs. Write to her. Can then contact agencies in charge.

What are the limits on funding?

Depends on city's budget and resources.

Variable with program — obtain figures.

What other sources are available besides federal support?

Perhaps floating revenue bonds. Explore question with city treasurer.

When
When would federal support be available?

Depends on program, probably.

When could construction begin?

Get funding deadlines. Then calculate.

Where
Where can funding be obtained?

Get a list of federal programs.

Consider possibility of state aid.

Must look into revenue bonds.

Why
Why is funding necessary?

Give figures from city budget on road allocation.

Won't cover initial cost.

How
How does the city apply for funding?

Get guidelines for each program.

How are bonds floated?

Ask treasurer.

FIGURE 2.12
Grid

Effects of the bypass

	Human environment				Natural environment
	Businesses/Facilities	*Shopping Patterns*	*Noise*		
Present effects	▢ Businesses/ Facilities in route: T. Brinker, Inc.: heavy equipment manufacturer Farm Mason General Hospital Oakdale Private School ▢ Bypass skirts Brinker, hospital, and school ▢ One farm cut in half	▢ Businesses on Main Street Must determine the amount of trade they derive from traffic traveling through Mason on current highway	▢ Noise level will not exceed governmental limits		▢ Some farm land is in corridor for the bypass but is the least fertile land ▢ Brown's Swamp is south of Mason, near route ▢ The Skunk River and Little Creek cross the corridor
Future effects	▢ Since bypass skirts important facilities, they will not be disrupted ▢ Bypass may facilitate travel to outlying industries. Brad's traffic pattern analysis will answer this question	▢ Businesses downtown may experience increased local trade Must see how many shoppers avoid downtown because of congestion	▢ Noise level should not exceed governmental limits in the future, despite increased use of bypass		▢ Bypass skirts swamp, so no wildlife will be disturbed ▢ Bridges over the river and creek will ensure that the flow of water will not be affected

2. Does my topic involve the relationships among ideas? Would a relating device help me gather information?

   a. Would an idea tree or outline help me generate information by arranging ideas in hierarchical form?

   b. Would a flow chart, diagram, line graph, or continuum help me generate information by sequencing ideas?

   c. Would a clustering device help me generate information by indicating areas of similarity and dissimilarity among ideas?

3. Do I need to examine my topic systematically by using a checklist such as the reporter's formula?
4. Does my topic concern several variables? Would a grid help me gather information on the basis of the intersection of these variables?

EXERCISES

Topics for Further Practice

1. As an employee of Beals and Company, in a position of your choice, you have been asked to make a presentation to the board of directors on your latest research project.
   A. Brainstorm or free write on the topic. Then evaluate your major points and indicate directions you will want to explore further.
   B. Construct an outline or idea tree of the points you have made in your brainstorming or free writing. Be sure to classify your information adequately.
   C. Apply any other suitable relating devices to your topic. Indicate additional information you have generated.
   D. Apply the reporter's formula to your topic. Indicate additional information you have generated.

2. Draw a flow chart to gather information on a common process in your field.

3. Draw a diagram and label the parts of an item commonly used in your field, to gather information for a description of the item to a general audience.

4. Construct a line graph plotting the following data and indicate the information you have gathered about the data as a result.

   AZCO Tractor Sales

	1980	*1981*	*1982*
1st quarter	1 million	2 million	0.5 million
2nd quarter	1.5 million	2.5 million	0.75 million
3rd quarter	2 million	2.5 million	0.75 million
4th quarter	1.75 million	1.5 million	0.5 million

   The fiscal year for AZCO begins in January and ends in December. In addition, 1982 was a recession year.

5. Construct a continuum of the activities you follow to research and write a paper or report. Your continuum will be arranged by time:

   Indicate the length of your activities by allotting each a space on the continuum.

6. Construct a clustering device to gather information about three careers you have been considering.

7. You have been asked to compare three products on the basis of certain characteristics (e.g., three brands of tomato soup for taste and nutritional value). Complete the fol-

Exercises

lowing grid to gather information on the comparison. (You may select the products and characteristics.)

	Product #1	Product #2	Product #3
Characteristic #1			
Characteristic #2			
Characteristic #3 etc.			

3

Gathering Information:
II. Library Research

As a writer, you will frequently use library research to gather information. Perhaps you will have to survey previous work on a topic for a review of the literature in a report. Perhaps you will require background facts in order to conduct an effective inquiry. Because library research can be important in any information search, we devote this chapter to the resources libraries provide and to how these resources can best be used.

LIBRARY RESOURCES

Important library resources include the card catalog, the reference area, government documents, and computerized information-retrieval services.

The Card Catalog

The card catalog lists the books, periodicals, and reference works the library owns. Information about these works is put on $3'' \times 5''$ cards and filed in the catalog under three categories: author, title, and subject. (In many libraries, this information is now computerized rather than filed by card.) You can find books, periodicals, and reference works by looking them up under any of these three categories.

Books
Figures 3.1a, 3.1b, and 3.1c show an author card and portions of title and subject cards for a book. (We have annotated the information contained on the author card.) Notice that these three cards are identical except for the top entry: the title (on title cards) and the subject (on subject cards) are placed above the author's name.

Library Resources

FIGURE 3.1a
Author card

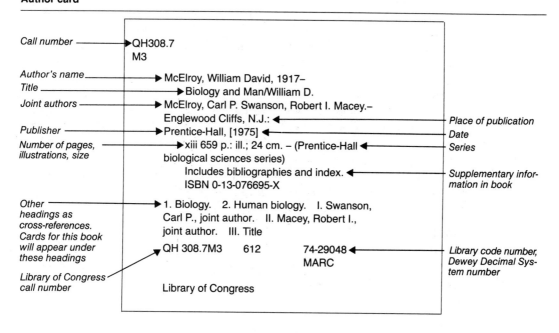

FIGURE 3.1b
Portion of a title card

FIGURE 3.1c
Portion of a subject card

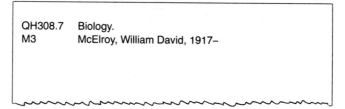

Cards in the catalog are arranged alphabetically, so the book in Figures 3.1a-c would be filed under B for Biology and M for McElroy. Remember that alphabetical listings are arranged by first *substantive* words: ignore *a, an,* and *the* when consulting the card catalog.

In addition to being cataloged by author and title, books are filed under one or more subject headings. Books are first classified according to standard subject headings from the Library of Congress classification system. For example, the book in Figures 3.1a-c has been classified under the subject heading Biology.

These standard subject headings have then been subdivided into narrower subject areas. For example, Biology contains such subdivisions as History, Philosophy, and Social Aspects. Books are then filed under these narrower divisions, which serve to pinpoint your information search.

The major heading and subheadings under which a book has been filed are given near the bottom of the catalog card. You will find cards for the book under each of these headings.

Periodicals

Many card catalogs also contain listings of periodicals and journals to which the library subscribes. These periodicals and journals are usually called serial holdings.

Cards for these holdings are entered in the catalog in alphabetical order by author (i.e., sponsoring agency or institution), title, and subject. Some libraries, however, list their periodical and journal holdings separately in books called serials catalogs. These catalogs are located at a central point, such as the main desk of the library or in the Periodicals Department. Figure 3.2 illustrates part of a page, with annotations, from a serials catalog.

Notice that the information given for each entry is the same as the information for a title card in the card catalog.

Reference Works

Card catalogs also contain entries for reference works, again under author, title, and subject headings. Figure 3.3 illustrates a title card for a reference book.

The Reference Area

Reference works are usually shelved in a special reference section of the library. Here you will find works that present facts, such as encyclopedias, dictionaries, handbooks, almanacs, biographical sources, and yearbooks; guides to literature, that is, guides to available resources; guides to specific sources, that is, bibliographies, indexes, abstracting journals; and frequently, atlases. Sometimes, however, atlases will be shelved in a separate map room.

Works Presenting Facts

Works presenting facts are either general or specialized. General works attempt to cover all knowledge, while specialized works pertain to a field of study. For example, the *Encyclopaedia Britannica* is a general reference work, whereas the *Encyclopedia of Chemical Technology* is narrower in focus and more specialized.

FIGURE 3.2
Portion of a page from a serials catalog

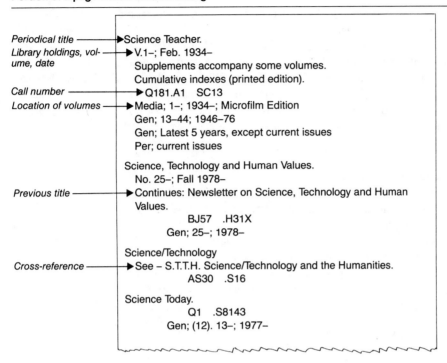

Periodical title → Science Teacher.
Library holdings, volume, date → V.1–; Feb. 1934–
  Supplements accompany some volumes.
  Cumulative indexes (printed edition).
Call number → Q181.A1 SC13
Location of volumes → Media; 1–; 1934–; Microfilm Edition
  Gen; 13–44; 1946–76
  Gen; Latest 5 years, except current issues
  Per; current issues

Science, Technology and Human Values.
  No. 25–; Fall 1978–
Previous title → Continues: Newsletter on Science, Technology and Human Values.
    BJ57   .H31X
  Gen; 25–; 1978–

Science/Technology
Cross-reference → See – S.T.T.H. Science/Technology and the Humanities.
    AS30   .S16

Science Today.
    Q1   .S8143
  Gen; (12). 13–; 1977–

FIGURE 3.3
Title card for a reference book

HF5681   Almanac of business and industrial financial ratios.
R25   Troy, Leo.
T68
1976   Almanac of business and industrial financial ratios. 1976 ed. Englewood
          Cliffs, N.J., Prentice-Hall (1975)
            XIX, 171 p. ; 28 cm.

          1. Ratio analysis   2. United States — Industries.
            I. Title
          HF5681.R25T68   1974   338.5   0973   74-161806
          ISBN 0-13-022673-4   MARC
          Library of Congress   74 (4)

45

Dictionaries, handbooks, almanacs, biographical sources, and yearbooks may also be general or specialized. Dictionaries such as the *Dictionary of Ecology* provide definitions of terms. Handbooks such as the *Business Writer's Handbook* provide useful information about the topics their titles announce. Almanacs, such as the *World Almanac and Book of Facts*, are compendiums of facts. Biographical sources include such books as the *Dictionary of Scientific Biography* or *Who's Who in Economics* and give details on the lives of well-known figures. Yearbooks provide a digest or synopsis of important events. *Facts on File Yearbook*, issued weekly, is the most complete of these yearbooks.

Guides to Literature

Guides to literature are reference books that list available resources (books, journals, indexes, and abstracting journals). For example, one well-known guide, Sheehy's *Guide to Reference Books*, describes various kinds of reference materials available in general research and in individual disciplines. Figure 3.4 shows a portion of the entries in Sheehy's for the field of geology.

Two other useful guides to literature are more specific in their coverage. Robert Malinowsky and Jeanne Richardson's *Science and Engineering Literature* lists reference materials available in various scientific fields (e.g., astronomy, physics, geoscience), while Carl White's *Sources of Information in the Social Sciences* lists reference materials for the social sciences and business.

Guides to Specific Sources

Bibliographies

Bibliographies are lists of books and journal articles concerned with certain topics. These bibliographies are useful because they provide a ready-made survey of the literature on a subject. The topics of these bibliographies can be very broad (e.g., *Bibliography of Meteorology*) or more narrow (e.g., *Bibliography of Electric Arc Welding*). The list of books and articles may also be comprehensive or more select. Therefore, you must determine the range of a bibliography when considering its entries.

Bibliographies are arranged by year, although brief bibliographies may be arranged alphabetically. In addition, bibliographies are frequently annotated. That is, they contain brief descriptions of the contents of each book or article. These annotations are useful in helping you decide whether or not to consult the work.

You can find bibliographies in the backs of books or at the ends of journal articles. You can also find bibliographies listed in the card catalog, if they are complete books.

Indexes

The indexes in the library, like those at the backs of books, are lists of items with indications on where to find each item. Library indexes will help you locate books, newspaper stories, and periodical articles.

FIGURE 3.4

Portion of the entries for geology in Sheehy's *Guide to Reference Books*

Pure and Applied Sciences | Earth Sciences | Geology   EE61

Ward, Dederick C. Bibliography of theses in geology, 1964. (*In* Geoscience abstracts, v.7, no.12, pt.1, Dec. 1965, p.101–37)   **EE49**

Continues the Chronic bibliography (above), listing an additional 682 theses.

———— and **O'Callaghan, T. C.** Bibliography of theses in geology, 1965–66. [Wash.], Amer. Geological Inst., [1969]. 255p.   **EE50**

"Published by the American Geological Institute in cooperation with the Geoscience Information Society."—*title page.*

Ward, Dederick C. Bibliography of theses in geology, 1967–1970. [Boulder, Colo., Geological Soc. of Amer., 1973]. 160p., 274p. (Geological Society of America. Special paper, 143)   **EE51**

A classed subject arrangement with subject, author, and geologic name indexes, and a directory of colleges and universities.

The continuations for 1965–66 and 1967–70 were prepared with automatic data-processing equipment, and the contents are stored on magnetic tape as part of the larger GEO-REF permanent data file.

Citations to master's and doctoral theses in geology for 1971– appear in the monthly issues of the GSA *Bibliography and index of geology,* and no further separate bibliographies of geology theses are planned.   Z6034.U49W32

Periodicals

Lomsky, Josef. Soupis periodik geologických věd: periodica geologica, palaeontologica et mineralogica. Příruční seznam s citačními zkratkami názvovými. Praha, Nakl. Československé Akademie Věd, 1959. 499p.   **EE52**

An alphabetical listing by first word of title of the geological periodicals of the world, past and present. Classified index. Gives title, publishing body, place, dates, and abbreviation.   Z6032.L6

Abstract journals

Geologisches Zentralblatt. Anzeiger für Geologie, Petrographie, Palaeontologie und verwandte Wissenschaften . . . Leipzig, Borntraeger, 1901–42. Bd.1–70. Semimonthly.   **EE53**

Frequency varies.

Ceased publication.

1932–42 in two parts: Abt. A, Geologie (no more published). Abt. B, Palaeontologisches Zentralblatt (continued in *Neues Jahrbuch für Mineralogie, Geologie und Paläontologie,* EE55).

Signed abstracts of book and periodical material in various languages in a subject arrangement, with author, geographical, and subject indexes in each volume. Cumulative indexes to Bd.1–15, 16–30, 31–50.   QE1.G494

GeoScience abstracts. v.1–8. Wash., Amer. Geological Inst., 1959–66. Monthly.   **EE54**

Ceased publication.

Superseded *Geological abstracts* (1953–58).

A classed abstract journal of publications on the geology, solid-earth geophysics, and related areas of science published in North America, or, if published elsewhere, dealing with North America Also includes abstracts of Soviet literature which has been translated and published in North America. Subject and author index published annually.

Superseded by *Abstracts of North American geology,* which was published 1966–71.

Neues Jahrbuch für Mineralogie, Geologie und Paläontologie, 1830–1949. Stuttgart, Schweizerbart'sche Verlagsbuchhandlung, 1830–1949. 192v.(?)   **EE55**

Title varies.

This bibliographical periodical has had a long and complicated history with varying coverage, but has usually included "Neues Literatur" and, from 1925 to 1942, "Referate." Abstracts generally presented in a subject arrangement, with yearly indexes. Some cumulated indexes were also published.

"Referate" became *Zentralblatt für Mineralogie, Geologie und Paläontologie,* 1943–49; superseded by EE57 and EE135.

U.S. Geological Survey. Geophysical abstracts. 1929–71. Wash., 1929–73. Monthly.   **EE56**

Frequency varies. Ceased publication with issue for Dec. 1971.

Abstracts 1–86 were issued in mimeographed form by the Bureau of Mines. On July 1, 1936, the geophysical section was transferred to the Geological Survey, which issued abstracts 87–111. By Departmental Order of Oct. 5, 1942, the geophysical section was again placed with the Bureau of Mines, and abstracts 112–27 were issued by that bureau. Beginning July 1, 1947, it was transferred again to the Geological Survey.

Offers worldwide coverage of literature pertaining to the physics of the solid earth and to geophysical exploration. Abstracts in English. Annual author and subject indexes. (Index for 1971 published 1973.)   QE500.U5

Zentralblatt für Geologie und Paläontologie. Stuttgart, Schweizerbart'sche Verlagsbuchhandlung, 1950– . v.1– .   **EE57**

In two sections (titles vary): T.1, Allgemeine, angewandte, regionale und historische Geologie; T.2, Paläontologie.

International in scope. Topical arrangement of signed abstracts, mainly in German. Annual author, subject, and topographic indexes. Supersedes in part the *Zentralblatt für Mineralogie, Geologie und Paläontologie,* 1943–49 (*see* EE55).   QE1.Z45

Encyclopedias and handbooks

American Geological Institute, Conference, Duluth, 1959. Geology and earth sciences sourcebook for elementary and secondary schools. Robert L. Heller, ed. Prep. under the guidance of the American Geological Institute, National Academy of Sciences–National Research Council. N.Y., Holt, [1962]. 496p. il.   **EE58**

A textbook and practical handbook presenting various areas of the earth sciences, with introductions, suggestions for methods and activities, problems, teaching aids, and references. Geologic topics for biology, chemistry, and physics courses are discussed. The appendix lists sources of information in geology and earth science as well as suppliers of teaching aids. Indexed.   QE41.A55

Lexique stratigraphique international. Paris, Centre national de la Recherche Scientifique, 1956– . v.1– . (In progress)   **EE59**

Contents: v.1, Europe; v.2, U.R.S.S.; v.3, Asie; v.4, Afrique; v.5, Amérique Latine; v.6, Océanie; v.7, Amérique du Nord; v.8, Termes stratigraphiques majeurs.

A lexicon of stratigraphic nomenclature in all continents and countries of the world, which is published in various parts, each covering a particular country. Gives a description of the formation, the type of locality, age, and reference wherein described.

National Research Council. Handbook of physical constants. Ed. by Sydney P. Clark, Jr. Rev. ed. [N.Y.], Geological Soc. of America, 1966. 587p. il. (Geological Society of America. Memoir 97)   **EE60**

Contains an impressive amount of physical data needed for geological and geophysical calculations. The compilations of data are grouped by topic. Index to properties. This edition revised and greatly expanded from the 1942 edition.   Q199.N25

Dictionaries

Beringer, Carl Christoph. Geologisches Wörterbuch, begründet von Carl Christian Beringer. 5. erg. und erw. Aufl. bearb. von Hans Murawski. Stuttgart, Enke, 1963. 243p. il.   **EE61**

47

48   3|Gathering Information II. Library Research

Book Indexes.   Book indexes are lists of published books, with complete bibliographic data. The following are the major book indexes:

American Book Publishing Record: monthly announcements of books published; cumulated annually

Books in Print: listing of all printed books available in a given year; issued annually

Cumulative Book Index: author, title, and subject listing of all books published in English; issued monthly

International Catalogue of Scientific Literature: listing of books in anthropology, astronomy, biology, botany, chemistry, mathematics, meteorology, paleontology, physics, and zoology

Library of Congress Catalog: subject bibliography of books cataloged by the Library of Congress; issued quarterly

National Union Catalog: author listing of books cataloged by the Library of Congress; issued monthly

Paperbound Books in Print: listing of all paperback books published; issued annually

Publishers Trade List Annual: listing by publishers of their printed books

Technical Book Review Index: review of new technical books; issued annually

If your library does not own a book you would like to consult, you can often order it through interlibrary loan. Remember, however, that books ordered may take three weeks or more to arrive, so plan your research well in advance. Also, remember that books cannot contain the latest findings in your field, because books are often a year in production. Therefore, you must check other information sources (e.g., journal articles) for the most recent research findings.

Newspaper Indexes.   Newspaper indexes provide information on newspaper articles and features. The major index for news stories is that covering *The New York Times.* This index, arranged by subject, refers you to back issues of this newspaper, which most libraries keep on microfilm. In addition, once you know the date of a particular story, you can locate more information on that topic in the back files of other newspapers.

Figure 3.5 shows an entry from the *New York Times Index,* with annotations.

FIGURE 3.5
Entry from the *New York Times Index*

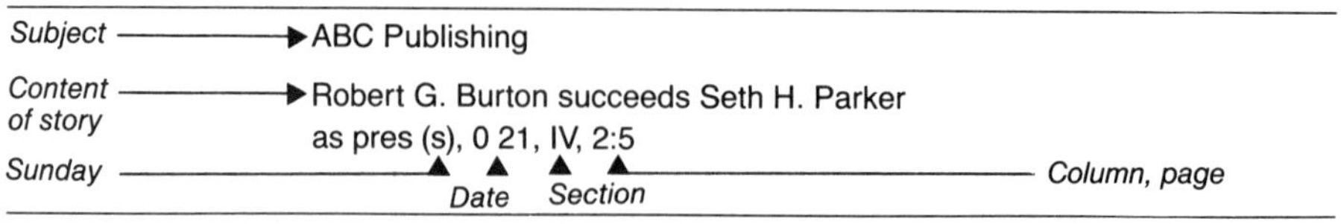

Since the indexing system used is quite different from that used in other such books, you should refer to the detailed instructions at the beginning of each yearly volume for further guidance.

Periodical Indexes. Periodical indexes provide information on articles appearing in periodicals and journals. These sources are very important because articles contain the latest research findings, often unavailable in book form. In fact, so many journals and periodicals exist that a special guide to literature, Ulrich's *International Periodicals Directory*, is available as a list of possible sources to consult.

You may already be familiar with the most common general periodical index: *Readers' Guide to Periodical Literature*, issued twice a month. *Readers' Guide* covers over 100 publications, the names of which are given in the front of each guide; it lists entries alphabetically by subject, author, and title. These publications and entries are usually nontechnical, giving general background rather than in-depth research. Figure 3.6 shows a portion of a page from *Readers' Guide*, with annotations. As with *The New York Times Index*, instructions for using the *Readers' Guide* are given in the front of each yearly volume. (The same is true for most periodical indexes and abstracting journals.)

FIGURE 3.6
Portion of a page from *Readers' Guide to Periodical Literature*

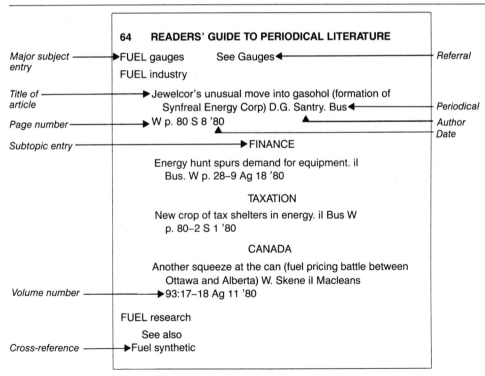

FIGURE 3.7
Portion of a page from *Applied Science & Technology Index*

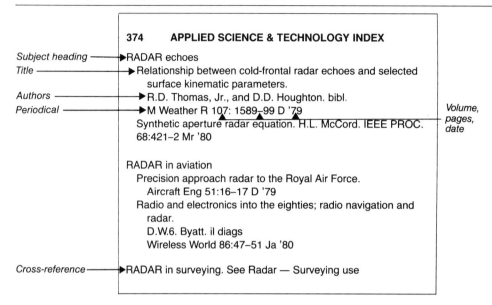

More specialized periodical indexes list journal articles in particular disciplines. *Applied Science & Technology Index*, the most general of these indexes for the sciences, lists articles in over 100 professional journals, mentioned in the front of the book, in many fields. The articles are entered under subject headings, listed alphabetically. Figure 3.7 shows a portion of a page from this index, with annotations.

The *Science Citation Index* is a second useful reference for scientific literature. This index is made up of several volumes and is arranged by authors, corporations, or citations. The citations volume gives authors and their publications as well as other writers who have referred to these authors. You may find that reading these references will give you additional material. The source index volume then provides full bibliographic listings for all references. Figure 3.8 shows a sample entry from this volume, with annotations.

FIGURE 3.8
Sample entry from the *Science Citation Index* source volume

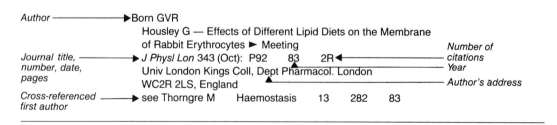

Library Resources

For the field of business, the *Business Periodicals Index* provides yearly volumes, while the *Business Index* (available only on microreader) provides a three-year accumulation of citations. For the social sciences, the *Social Sciences Index* references periodicals in the English language, while the *Social Sciences Citation Index* references foreign journals as well. This index is arranged identically to the *Science Citation Index*.

Other specialized indexes are also available, as you can see from the partial list of titles below:

Bibliography and Index of Geology	*Energy Index*
Biological and Agricultural Index	*Index Medicus*
Current Physics Index	*Index Veterinarius*

You should consult your library's holdings for other indexes in your field. In addition, *Magazines for Libraries* by Bill and Linda Katz can be used to discover the basic indexes for a variety of disciplines. This book also contains index descriptions, given alphabetically in the front of the volume.

Abstracting Journals

Abstracting journals, like indexes, reference periodical articles but provide an abstract or brief summary of the article as well. Because you can use this abstract to decide if you wish to consult that particular article, abstracting journals are more useful than indexes in helping you select articles you will read and those you will not.

Abstracting journals are discipline-specific, as you see from the following partial list:

Aerospace Engineering Index	*Forestry Abstracts*
Agrindex	*Genetics Abstracts*
Applied Mechanics Reviews	*Geological Abstracts*
Aquatic Sciences and Fisheries Abstracts	*Horticultural Abstracts*
Astronomy and Astrophysics	*International Aerospace Abstracts*
Biological Abstracts	*International Petroleum Abstracts*
Botanical Abstracts	*Metals Abstracts*
Chemical Abstracts	*Microbiology Abstracts*
Computer Abstracts	*Mineralogical Abstracts*
Dissertation Abstracts International	*Nuclear Science Abstracts*
Electrical and Electronics Abstracts	*Physics Abstracts*
The Engineering Index	*Transportation Research Abstracts*
Excerpta Medica	*Water Pollution Abstracts*
Food Science and Technology Abstracts	

Figure 3.9 shows a sample entry from *The Engineering Index*, with annotations.

FIGURE 3.9
An entry in *The Engineering Index*

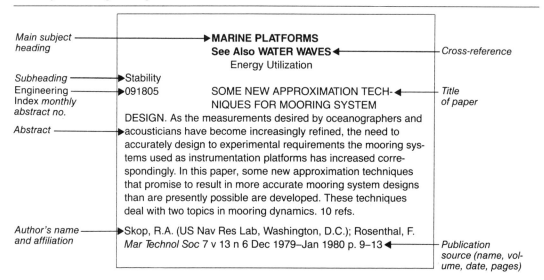

Government Documents

Most university libraries have large collections of government publications, shelved separately from other holdings. These government publications include pamphlets, periodicals, reports, and books on a wide variety of topics. Two books that may introduce you to these documents are *Government Publications and Their Use* by Laurence Schmeckebier and *Subject Guide to Major United States Government Publications* by Ellen Jackson. (The *Subject Guide* contains a section on the guides, catalogs, and indexes that list bibliographic information on government publications.)

Other guides to government publications are the following. (We have starred those you will find most helpful.)

1. *American Statistic Index* (ASI) A monthly index, with abstracts, of federal publications dealing with statistics.

*2. *CIS Annual* A listing, with abstracts, of House of Representatives and Senate hearings and reports, including an index of public laws under consideration. The *CIS Annual* is the best guide to documents originating in congressional committees.

3. *GPO Sales Publications Reference File* A listing of government publications offered for sale, only available on microfiche.

*4. *Government Reports Announcements & Index* A biweekly listing of reports on research commissioned by the federal government. Twenty-two subject areas are indexed; abstracts are provided.

*5. *Monthly Catalog of U.S. Government Publications* A listing by governmental agency of most federal publications issued each month. (Subject, author, and

Library Resources 53

title indexes are also provided.) The December catalog gives information on ordering these publications.

6. *Monthly Checklist of State Publications* A listing of major state publications.

*7. *U.S. Government Books* A subject listing, often with abstracts, of government publications of general interest; published quarterly.

8. *United Nations Documents Index* A listing of United Nations publications. *The Yearbook of the United Nations* surveys the major research activities of UN agencies each year. You may then look up publications in the *Index*.

If you wish to locate less recent government publications, the *Cumulative Subject Index to the Monthly Catalog of United States Government Publications, 1900–1971* (15 volumes), compiled by William W. Buchanan and Edna M. Kanely, indexes this information.

Computerized Information-Retrieval Services

Libraries now have computerized information-retrieval services. You can use these services to search your library's holdings for information, and you can also search data bases, bibliographic listings stored in a few computers throughout the country of the information found in major indexes and abstracting journals. Several commercial firms offer access to these data bases. Two of the best-known such commercial services are BRS (Bibliographic Retrieval Services) and DIALOG by Lockheed Information Retrieval Service, the oldest and largest service.

Computerized information retrieval is useful because such searches are fast and comprehensive. More than 200 different data bases containing more than 80 million citations may be accessed, as of 1984. However, since most of these bases date from the late 1960s or early 1970s, you must still use indexes and abstracting journals manually when searching for information published prior to these dates.

These data bases are also updated regularly. For example, *The Engineering Index* adds 8,500 entries monthly. As of 1984, it totaled 1.174 million entries. Other bases you might find useful include the following. (The date indicates the year the base begins.)

Life Sciences/Agriculture

Agricola (based on the *Bibliography of Agriculture*),1970 +

Field Crop Abstracts, 1973 +

Forestry Abstracts, 1973 +

Index Medicus, 1966 +

Pollution Abstracts, 1970 +

BIOSIS Previews (based on *Biological Abstracts*), 1970 +

Excerpta Medica, 1975 +

Food Sciences and Technology Abstracts (FSTA), 1969 +

Horticultural Abstracts, 1973 +

Review of Plant Pathology, 1972 +

Veterinary Bulletin, 1972 +

If you are in the life sciences, *BIOSIS Previews* will be a useful data base. It covers over 9,000 journals and has 3.7 million entries as of 1984. If you are in agriculture, *Agricola*, the data base of the National Agricultural Library, will be very useful. This data base covers journals and monographs on agriculture, published worldwide. A comprehensive data base, *CAB Abstracts*, also accesses agricultural and biological information. *CAB Abstracts* covers 26 abstracting indexes and journals, including *Field Crop Abstracts, Forestry Abstracts,* and several more listed above.

Physical Sciences/Engineering

CA Search (based on *Chemical Abstracts*), 1970 +

Compendex (based on *The Engineering Index*), 1970 +

Electrical & Electronics Abstracts, 1969 +

ISMEC (Mechanical Engineering), 1973 +

NTIS, 1964 +

Computer & Control Abstracts, 1969 +

Energyline, 1971 +

Georef (Geological Index), 1967 +

Physics Abstracts, 1969 +

Water Resources Abstracts, 1968 +

The most useful data bases in the physical sciences and engineering depend on your field, since these bases are discipline-specific. For example, if you are in chemistry, you will frequently use *CA Search*. If you are in engineering, you will frequently use *Compendex*.

Social Sciences/Humanities

ABI/Inform (Business Abstracts), 1971 +

Accountants Index, 1974 +

Agricola, 1970 +

America: History and Life, 1964 +

American Statistic Index (ASI), 1973 +

Art Bibliographies Modern, 1974 +

CIS/Index, 1970 +

Comprehensive Dissertation Index, 1861 +

ERIC, 1966 +

Exceptional Child Education Resources, 1966 +

Federal Register Abstracts, 1977 +

Historical Abstracts, 1973 +

Language and Language Behavior Abstracts (LLBA), 1973 +

Magazine Index, 1977 +

Management Contents (MARC), 1974 +

Modern Language Association (MLA) *Bibliography*, 1977 +

PAIS, 1976 +

Philosopher's Index, 1940 +

Psychological Abstracts (PsycINFO), 1967 +

Sociological Abstracts, 1963 +

United States Political Science Documents, 1975 +

ERIC (Educational Resources Information Center) is a useful data base for those in the social sciences and humanities. This data base accesses two indexes: *Resources in Education*, which lists unpublished papers and presentations; and *Current Index to Journals in Education*, which lists published journal articles.

A computerized information-retrieval search will vary in cost, depending on which data base you enter and how long the search takes. For example, searching *CA Search* costs $68 per hour on DIALOG; *Compendex* costs $98 per hour (1984). The average cost for an entire search will usually range from $10 to $20. Searches in the pure and applied sciences and in engineering are most costly, because these data bases are most expensive.

PROCEDURE FOR CONDUCTING LIBRARY RESEARCH

Jackie Brennecke, a nutrition major, must complete a senior research project as part of her degree program. She wants to investigate coagulants used in tofu (bean curd) in order to determine the most palatable, least costly method of gelling tofu. Before Jackie can perform her experiment and write her report, however, she must gather information on work that has already been done with coagulants, so she will not duplicate previous research. She can also include this material in a literature review section of her report. She uses the library for her information search.

An effective procedure for conducting library research involves the following steps:

1. Determining topic, purpose, and audience
2. Locating information
3. Constructing a working bibliography
4. Evaluating sources
5. Note-taking, reading, and recording information
6. Integrating sources with the text

Determining Topic, Purpose, and Audience

You could waste many hours if you do not determine your topic, purpose, and audience before conducting library research. Accordingly, you should write a topic/purpose/audience statement to ensure that you have these factors firmly in mind. Jackie's statement reads as follows:

Topic ——▶I will investigate research on coagulants used in tofu, in order to
Purpose ——— determine what other researchers have discovered.
Audience—▶I will present this research in the literature-review section of my report to my major professor. My report will also be filed in the library, so that future nutrition students may review my findings.

Notice that Jackie does not simply name a subject area (e.g., coagulants used in tofu). Instead, she also gives the purpose (to determine what other researchers have discovered) and the audience (her major professor and future nutrition students). These two factors narrow Jackie's topic and direct her research.

For example, her purpose states that she wishes to determine the *research* done on tofu coagulants. Therefore, she will seek sources of information on this subject and ignore other sources about tofu coagulants, such as the history of their use or their advantages. In addition, she will ignore sources containing information of general interest, such as light scientific articles explaining coagulants' effect on tofu, because of her readers' technical expertise. They will be interested in serious research results, not in information of general interest.

As you see, your topic/purpose/audience statement will help narrow and direct your information search.

Locating Information

After determining your topic, purpose, and audience, you are ready to locate information. You must first select appropriate library resources to consult, then discover specific sources you might use.

Selecting Appropriate Library Resources

You will not use all the library resources we have described in every information search. Table 3.1 lists each resource and indicates when you would or would not find it useful. Select the resources that suit your research task. If you must discover the most recent information on your topic, combining a search of the latest indexes and abstracting journals in your field and a search of the most recent issues of the appropriate *Citation Index* with a computer search of the past five years might be most helpful. For less recent information, you can use additional sources such as the card catalog or bibliographies.

Jackie will use the card catalogue, *Science Citation Index*, and two of the best abstracting journals in her field (*Food Science and Technology Abstracts* and *Nutrition Abstracts and Reviews, Series A*) for references published before 1969. She will run a computerized information search for references after 1969, the year the data base for *Food Science and Technology Abstracts* begins.

Jackie will not consult reference works presenting facts because she is conducting a literature search, not gathering general information on a subject. She will not require guides to literature because she is already familiar with the reference works in her field.

Locating Specific Sources

Once you have decided on appropriate resources to consult, you are ready to use them for locating specific information sources.

Procedure for Conducting Library Research 57

TABLE 3.1
Library Resources

Resource	Use to locate . . .	Do not use to locate . . .
1. Card catalog	Older works General works Extensive accounts of research listing additional sources Bibliographies	Recent sources Brief, in-depth reports of research
2. Reference works	General information	In-depth accounts of research
A. Works presenting facts Encyclopedias	Background useful for narrowing a topic Bibliographies of additional sources to consult	
Dictionaries, handbooks, almanacs, biographical sources, yearbooks	Specific facts	Extensive discussion
B. Guides to literature	Reference works available	Specific sources of information
C. Guides to specific sources		
Bibliographies	Selected works on a topic	All works published on a topic or in a field
Indexes	All works published in a field for a specified year	A select list of works on a topic
Abstracting journals	All works published in a field for a specified year Abstracts of these works	A select list of works on a topic
3. Government documents	Publications by the government on a wide range of topics	Nongovernment publications
4. Computerized information-retrieval services	Sources your library owns Sources listed in data bases A comprehensive list of sources	Sources before the year in which the data base begins

58   3 | Gathering Information II. Library Research

The Card Catalog

If you know the authors or titles of the works you need, you can look them up in the card catalog under either designation, then locate them by call number. However, if you do not know the works you need, you must locate them by subject.

Locating books by subject involves using the headings and subheadings of the Library of Congress classification system. The following procedure will help you with this task:

1. Examine your topic for a logical heading. (If you already know one book on your topic, the subject designations on its card will suggest headings you may use.) This heading should not be too broad (e.g., education), since overly broad headings will not produce specific enough references on your topic.

2. Look this heading up in the catalog. The heading may be further subdivided, in which case the subdivisions may also guide you to books on your topic. If the heading is not subdivided, examining the books classified under it may produce the references you require.

3. Check several headings related to your topic if you do not find your first subject heading listed in the catalog. If your first subject heading is broad (e.g., communicable diseases), try consulting narrower subject headings (e.g., the names of specific diseases). If your first subject heading is narrow (e.g., drone bees), try consulting progressively more general headings (e.g., drone — bees — insects — entomology).

Jackie chooses Tofu as a logical heading. She feels that a second possibility, Coagulants, would be too broad, since coagulants have many applications beyond their use in tofu.

When she looks up Tofu in the card catalog, she discovers a cross-reference to Bean Curd. She now knows that books on her topic are classified under this heading, which she will check for specific sources.

You may find that the references listed under the headings and subheadings you have located are not the ones you require or do not give sufficient information. However, noting four pieces of information from cards for these references will produce additional places to search:

1. Other headings under which these works are classified
   Note the cross-references at the bottom of the cards. Works cataloged under these cross-references may produce usable material.

2. Related headings
   Note the related headings under which works on your topic have been filed. These related headings are usually listed on cards placed before the first card for a subject.

3. Call numbers
   Note the call numbers of these works. Additional references may be filed under other major divisions of the Library of Congress classification system, as desig-

Procedure for Conducting Library Research **59**

nated by call number. You can then check these other major divisions for useful references.

4. Bibliographies

   Note works that contain bibliographies. These bibliographies may suggest additional sources to consult. The bibliographies may also suggest prominent researchers on your topic. You can then consider documents they have written.

Jackie notes that related material can be found under the subject heading Cookery. However, she does not check the card catalog under this subject heading, because she does not feel it will produce serious research works. Jackie also notes that one source she has found, *The Book of Tofu*, contains a bibliography. She will examine this bibliography for additional references.

Reference Works

If you do not know the major reference works in your field, you should consult a guide to literature such as Sheehy's to gather this information. You can then use these reference works to discover specific information.

The arrangement of most reference works is quite simple. For example, encyclopedias and dictionaries are alphabetically arranged; biographical sources are usually alphabetical within the year of the person's birth. However, indexes and abstracting journals are more complex. They may be arranged in two ways: by subject heading or by citation number. You must use a different method for locating information with each arrangement pattern.

Arrangement by subject heading is alphabetical. Usually several levels of headings will be used, just as in the card catalog. For example, in *Index Medicus*, the main subject heading Frostbite is divided into the three subheadings Chemically Induced, Pathology, and Therapy. These levels of headings identify narrower areas to check for articles on your topic.

With arrangement by citation number, you must locate references to articles in a list at the back of the index. This list may be alphabetical according to subject headings or author. (Sometimes both types of listings will be available.)

To locate references by subject headings, be sure to check several terms applicable to your topic rather than simply your topic's name, just as you did in the card catalog. Once you have found a reference to an article, you then turn to its citation number in the index for complete bibliographic information.

When Jackie consults her indexes, she sees that they are arranged by subject. She turns to the subject heading she used in the card catalog (Bean Curd) but does not find it listed. Therefore, she tries Tofu, where she discovers the cross-references Soy Products and Soya Beans. She discovers references on her topic when she turns to these subject headings.

When Jackie consults the *Science Citation Index* source volume, she decides to use soybean as a possible subject heading. She finds that the term is subdivided, with coagulating as one of the subdivisions. She then checks this subdivision for usable references.

Government Documents

You can consult the directions in the front of the volumes we have listed or ask a reference librarian for assistance in locating government material on your topic, because searching government documents often requires special knowledge.

Jackie checks the indexes in *CIS Annual*, *Government Reports Announcements & Index*, and the *Monthly Catalog of U.S. Government Publications* under Soybeans. However, she finds that the reports issued by the government and the research it has funded have dealt with economic considerations or agricultural issues rather than with processing. Therefore, she does not explore government sources further.

Computerized Information-Retrieval Services

You can search data bases using authors' names and titles of articles if you know this information and only require bibliographic data. However, you will usually search data bases by means of key words — logical combinations of words and phrases describing your topic. The search can be as broad or as specific as you wish. For example, you may want a comprehensive search on preventatives of the common cold, or you may wish to search for relatively few articles concerning vitamin C as one preventative measure.

In order to plan a computer search, first write down your topic, then list words relevant to that topic, including synonyms, plurals, and appropriate phrases. These key words should be those that would appear in the titles of articles. For example, Jackie lists the words you see in Figure 3.10a. Next, arrange your words in groups according to subject. You see Jackie's groups in Figure 3.10b.

FIGURE 3.10a
Jackie's list of words

Coagulants in tofu
Soybean
Soya bean
Soy protein
Tofu
Bean curd
Coagulants
Coagulation
Gelling

FIGURE 3.10b
Jackie's groups of words

Coagulants in tofu	
Soybean	Bean curd
Soya bean	Coagulants
Soy products	Coagulation
Tofu	Gelling

FIGURE 3.11
Overlap of key word groups

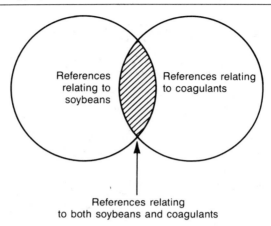

These groups of logically related words provide you with the key words for your search. The computer will discover all references where the groups of key words overlap, as we diagram in Figure 3.11.

After you have decided on your key words, determine the years your search will cover and the number of citations you require. (Jackie will search from the beginning of the data base on and will retrieve all citations, since she is involved in a literature search.) Then seek the services of your reference librarian, who will conduct the search for you.

Since many reference librarians may not be experienced in your field, a description of your topic will assist them in conducting your search. They might then be able to suggest data bases to enter or other key words to narrow your search further. Jackie provides the following description:

> Soybeans are used to make many food products. One of these products is tofu, pressed bean curd. Coagulants are used to solidify this tofu.

Sample titles a search ought to produce can also assist the librarian in conducting yours by indicating key words that ought to be used. For example, Jackie provides the following title:

> "Coagulation of Soybean Milk by Calcium Ions and Antagonistic Behavior of Sodium Salts"

This title contains two of her key words, which tells the librarian the title will be retrieved and the search will be successful.

After the operator has searched the base or bases you have selected, you can order a printout of the references you have found. This printout will list bibliographical information. (An abstract will also appear if you have searched an abstracting journal.) You can then locate the references in your library or, if your library does not own a reference, order it through interlibrary loan.

Constructing a Working Bibliography

While you are locating specific sources of information, you will be constructing a working bibliography, a listing of references that seem to contain useful information and therefore ought to be further explored. Note, however, that this bibliography is only preliminary. You may or may not use or even closely read all the sources it contains.

Compile your working bibliography by recording each information source on a 3″ × 5″ card so that you may discard any sources you do not find helpful. Place a source number in the upper right-hand corner for your use when taking notes, then include complete bibliographic data for each source. Recording these data will ensure that you have all the necessary information to locate the source and compile the final bibliography for your report. Figure 3.12 shows a sample card from Jackie's working bibliography.

FIGURE 3.12
Sample working-bibliography card

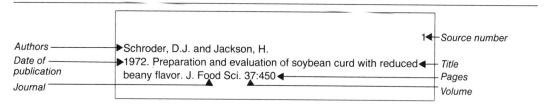

Evaluating Sources

Once you have constructed a working bibliography, you must evaluate your sources to select those you will read and those you will not, and check the quality of those you read in order to determine their reliability.

Selecting Sources to Read

Several procedures can help you determine sources you will read:

1. Examine the title.
   The title of a work can indicate its relevance to your topic. For example, Jackie's manual search of her indexes produced the following citation:
   ☐ Barker, H. L. 1966. Edibles from Soybeans.
      Food Technol. 20:788.
   She decides to discard this reference on the basis of its title: The work would probably contain only general information on tofu, not research results on coagulants.

2. Scan the table of contents or headings used.

Scanning the table of contents of a book or the headings of an article will provide an overview of the information covered. When Jackie scans the table of contents of *The Book of Tofu*, she discovers a chapter on "Ingredients." She decides to place this book in her working bibliography, because this chapter may discuss past research on coagulants — one ingredient in tofu.

3. Skim the introduction and topic sentences.

Skimming the introduction and topic sentences at the beginnings of sections (in a book) or paragraphs (in an article) will tell you the writer's major concerns. You can then decide whether or not to retain the work in your bibliography.

Determining Reliability

In addition to selecting sources for close reading, you must be certain your sources are reliable. Both external and internal clues will help you judge this reliability.

External Clues

External clues to reliability include the writer's reputation in your field and the frequency with which other researchers cite this work. If the writer has an established reputation, you can usually rely on the quality of the source. In addition, citation by other researchers indicates reliability as well as the relative importance of the source.

For example, Jackie discovers that the pamphlet "Recent Advances in Soybean Milk Processing Technology" has been cited twice in other works. She places this pamphlet on her working bibliography, because these citations may indicate the importance and quality of the source.

Internal Clues

Internal clues to reliability are based on the information the source presents. Answering the following questions will help you evaluate sources by internal clues:

1. Are the research methods reliable?
2. Do the data/results appear to be valid, based on the discussion given?
3. Are the conclusions supported by the data?
4. Are the recommendations sound, based on the data presented?

You can then discard sources that do not meet these criteria.

Note-Taking, Reading, and Recording Information

After you have decided which sources you will read, you are ready to survey them for information. To perform this task effectively, you must develop procedures for taking notes, for reading, and for recording information.

Taking Notes

Effective note-taking allows easy access to information when you are compiling it for inclusion in a document. The following procedure will help you with this task:

1. Use large note cards (5" × 8") rather than 3" × 5" cards or sheets of paper. Large cards allow you to record more information than you can on 3" × 5" cards. Note cards also assist with categorizing data as recording notes continuously on sheets of paper does not, because you place notes for one topic on each card. This categorization makes accessing and sorting your information easier.
2. Number the bibliographic card for each source you have decided to use.
3. Place the author's last name and appropriate source number in the upper left-hand margin of each note card, to identify the source. In this way, you avoid having to repeat bibliographic information on each card.
4. Write a topic heading, one per card, in the upper right-hand margin. This heading will identify the subject of the card. Since you will be reusing topic headings as you discover additional information in other sources, this procedure will allow you to correlate the cards with the outline you develop for your document.
5. Note the information pertaining to that heading from your source and the page number on which you found the information. Recording the page number will ensure that you have these data if you wish to document the information in your report.

Figure 3.13 shows one of Jackie's note cards, with annotations.

FIGURE 3.13
Sample note card

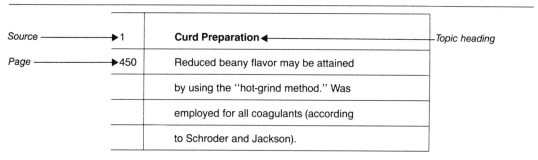

Reading Information

You should survey information sources in two ways: by skimming and by close reading.

Skimming

Skimming is a quick method of surveying sources to determine their contents. You may skim a source to decide if you wish to examine it more closely. You may also skim a source to obtain an overview of its contents before reading pertinent parts in detail or to quickly note relevant information.

In order to skim, first look for the main ideas an author presents. You can find these ideas by checking several locations:

1. Table of contents
2. Introduction
3. Headings and subheadings
4. First sentences of sections and subsections
5. Topic sentences of paragraphs

Once you have located major ideas, quickly scan the supporting information given under each. You can then decide to discard the source, to read the information in more detail, or to record it immediately if closer reading is not required.

Close Reading

You will read your important sources closely, after skimming to obtain an overview of their contents. However, you may not give equal attention to all parts of a source. Examine the abstract (if present), the introduction, and the conclusion in detail, because these sections provide important background information and major results, conclusions, and recommendations. Then skim the discussion sections, reading more closely when you discover relevant points.

While you are reading your sources, you will also be recording your information.

Recording Information

You can record information in three ways: by quoting directly, paraphrasing, or summarizing your sources.

Quoting Directly

You should quote a source directly *only* when you want to reproduce the author's exact words in the text of your report because this wording is important. When taking notes by direct quotation, however, you must carefully avoid the danger of plagiarism.

Plagiarism occurs when you reproduce an author's words without giving credit through documentation. Plagiarism also occurs when you use an author's ideas without giving credit, unless the ideas are common knowledge or have appeared in many sources (e.g., Einstein's theory of relativity). Put quotation marks around all words you have placed in your notes directly from your sources, as you see in Jackie's note card (Figure 3.13). This technique will help you avoid plagiarism by alerting you to material you must cite.

66 3 | Gathering Information II. Library Research

FIGURE 3.14
Source with paraphrase

Source

After extraction, the protein-containing extract is separated from the insoluble flake residue by appropriate mechanical devices. The major globulins are then precipitated by the addition of food-grade acid, usually hydrochloric acid. The pH of the clarified extract is lowered to about pH 4.5, at which pH the solubility of the globulins is near a minimum. This curd is then separated to remove the soluble constituents present in the whey. Because of the presence of soluble oligosaccharides, proteins, peptides and amino acids, salts and minor constituents, washing the curd increases the purity of the separated protein. The washed curd can be concentrated to a heavy slurry or cake containing from 15 to 30 percent solids.[1]

A synonym is used for "cake."

Sentence order is reversed and synonyms are used.

Order of sentence is altered and words are changed.

Jackie's Paraphrase

The order of the source is not followed.

Tofu can be solidified to 15–30%. Mechanized separation divides protein from flakes. Acids (hydrochloric) extract curd and pH is lowered (appr. 4.5). Solid curd in whey is removed by separation. Curd is washed to make it purer, removing oligosaccharides, proteins, peptides and amino acids, salts, and minor impurities. Then the curd is formed into solid block.

A synonym for "globulins" is used.

Paraphrasing

Paraphrasing is using your own words to express an author's ideas. You should paraphrase a source when you want to use the substance of the information presented, but wish to emphasize different points or follow a different order.

To paraphrase, you should record in your notes all the information the author presents. However, you can use sentence structures, phrases, and words different from those in your source and can record the information in a different order.

Figure 3.14 shows a portion of one of Jackie's sources, with the paraphrase she devised. (We have indicated the paraphrasing techniques she has used.)

Summarizing

Summarizing involves recording only your source's most important ideas. You should summarize a source when you are using it to obtain general information rather than an in-depth or a detailed discussion.

In order to summarize, you must first discover the writer's major ideas, using the locations we have discussed: the table of contents, the introduction, headings in the text, the first sentences of sections, and the topic sentences of paragraphs. Record these major ideas, then briefly note important supporting information under each. Be

---

[1]R. A. Lowrie, ed., *Proteins as Human Food* (Westport, CT: Avi Publishing Co., 1970), p. 348.

Procedure for Conducting Library Research **67**

FIGURE 3.15
Summary of the source paragraph presented in Figure 3.14

Major idea ———▶**Process for Making Tofu**

Supporting ———▶After being extracted from soybeans, the protein is separated, the
information   globulins are precipitated, the pH is lowered, and the curd is separated, washed, and concentrated.

sure to include significant results, conclusions, and recommendations, since this
information always represents support for major ideas.

Figure 3.15 shows a summary of the source paragraph presented in Figure 3.14.

Remember, even though you have paraphrased or summarized information from
your sources, you must still give credit through documentation for an author's original ideas, unless these ideas have become general knowledge. Therefore, carefully
note the page numbers where you obtained your information to facilitate later citation in your report.

Integrating Sources with Your Text

When you have gathered and recorded your information, you are ready to integrate
your sources with your text. Your first step is to sort your note cards by topic, corre-
lating these topics with the outline you have created for your document. As you
write your text, the sources you have summarized will give you background information, the sources you have paraphrased will provide in-depth knowledge, and
those you have directly quoted will provide specific support for the points you wish
to make. However, you should avoid overuse of quotations as a substitute for making
your own points. In addition, do not use quotations when the author's precise words
are not important. In this way, you avoid creating a document pieced together from
other researchers' words.

Figure 3.16 illustrates ineffective integration of direct quotations with the text.

FIGURE 3.16
Ineffective integration of direct quotations and text

Using Quotations as Substitutes for Points

"The washed curd can be concentrated to a heavy slurry or cake
containing from 15 to 30 percent solids" (Lowrie, 1970, p. 348). This
concentration is accomplished by means of coagulants.

Quoting When the Author's Words Are Not Important

Coagulation takes place after washing. The tofu is formed into "a
heavy slurry or cake containing from 15 to 30 percent solids"
(Lowrie, 1970, p. 348).

Figure 3.17 shows effective integration; since the source's exact words are not important, a direct quotation is not used.

FIGURE 3.17
Effective integration of source with text

Coagulation forms the curd into a block of tofu with 15 to 30 percent solid matter (Lowrie, 1970, p. 348).

Notice in Figure 3.17 that the material is still documented to give credit to the original source.

CHECKLIST FOR GATHERING INFORMATION: LIBRARY RESEARCH

Library Resources

1. Have I checked under major headings and subheadings in the card catalog for books, reference works, and periodicals on my topic?
2. Have I used the reference area of the library?
   A. Have I located works presenting facts (encyclopedias, dictionaries, handbooks, almanacs, bibliographical sources, and yearbooks) that might assist me with my topic?
   B. Have I used guides to literature to locate the major resources on my topic?
   C. Have I used guides to specific sources (bibliographies, indexes, abstracting journals) to locate books, newspaper stories, and periodical articles on my topic?
3. Have I used the references in government documents to locate government publications on my topic?
4. Have I conducted a computer search of the library's holdings and the data bases in my field to locate sources on my topic?

Procedure for Conducting Library Research

1. Have I directed my information search by writing a topic/purpose/audience statement?
2. Have I located information effectively?
   A. Have I selected appropriate library resources?
   B. Have I used these resources to locate specific information sources?

Exercises   69

3. Have I constructed a working bibliography by placing each source I wish to consult on a separate note card?

4. Have I evaluated my sources?

   A. Have I effectively selected the sources I will read?

   B. Have I used external and internal clues to judge my sources' reliability?

5. Have I read and recorded my information effectively?

   A. Have I followed a note-taking procedure that will allow easy access to my information?

   B. Have I skimmed or read closely, choosing the technique that suits my source?

   C. Have I integrated my sources by using paraphrases and quotations to support my points but not as substitutes for the points themselves?

   D. Have I carefully documented direct quotations and any original ideas I have paraphrased or summarized, unless those ideas have become common knowledge?

EXERCISES

1. Locate a book in the card catalog by author or title. Using the cross-references at the bottom of the card, locate three other books you might consult if you were searching for information on this topic. Give complete bibliographic references for the books.

2. Locate the three cards (author, title, subject) for a periodical or journal used in your field. Copy the information given on the author card and annotate that information. (If your library uses serials catalogs, locate the periodical or journal in that catalog.)

3. Using Sheehy's or another guide to literature, list the important reference works in your field.

4. Locate a card for a specialized encyclopedia in your field. Copy its call number and complete bibliographic data. Now find that encyclopedia in the reference section of your library.

5. List the important indexes and abstracting journals in your field. (A guide to literature and your reference librarian can help you with the task.) For each index or abstracting journal, give the method of listing entries: by subject or by citation number. Now locate a reference to a journal article on a topic of your choice and copy down the complete bibliographic information. Using this information, find the article in your library and photocopy the first page. If your library does not own the journal, fill out but do not send an interlibrary loan card.

6. List five or six of the major research journals in your field. (The journals listed in the *Science Citation Index* and *Social Science Citation Index* publish 90% of the literature in these fields. Checking these indexes or obtaining the services of a reference librarian will help you with this question.)

7. Locate a government report on a topic of your choice in *Government Reports Announcements & Index*. Copy the information given about the report.

8. Obtain information from your reference librarian on the computerized information-retrieval services available at your library. (Bring this information to class.) Then plan a search on a topic of your choice, using the steps we have provided:

   A. List key words relating to your topic.

   B. Arrange these words in groups on the basis of similarity.

   C. Determine the years of your search.

   D. Write a description of your topic.

   E. Provide sample titles.

9. List the important data bases to enter for information on the topic in Question 8. (Your reference librarian can help you with this task.)

10. Using the procedure we have discussed, gather information from the library on a topic of your choice. Be sure to follow all six steps:

    A. Determining topic, purpose, and audience.

    B. Locating information.

    C. Constructing a working bibliography.

    D. Evaluating sources.

    E. Note-taking, reading, and recording information.

    F. Integrating sources with the text.

4

Selecting and Arranging Content

In most technical investigations, you will gather much more material than you will eventually include in your document. Therefore, selecting the precise information your readers require and arranging that information for best effect are important.

Arrangement refers to the order of the content in your document (structure) and the layout of your writing on the page (format). Both aspects of arrangement are important, because they affect the way your audience processes material. If your ideas do not proceed in logical order or if your writing is not presented effectively, you will fail to communicate, even though you have selected appropriate content.

In this chapter, we discuss principles for selecting and structuring information. (We discuss format in Chapter 16.) We use the example of Duane Stall, an independent forestry management consultant who has been retained to study poorly conditioned black-walnut trees at Eagle Ridge Environmental Area and to recommend a management plan.

Content selection and arrangement depend on your analyses of audience, purpose, and use, because you choose information for specific readers and structure it for audience effect. Duane has three readers. George Hamilton, director of the county conservation board, is his primary audience, because Mr. Hamilton will decide whether or not to institute Duane's plan. Sally Beatty and Ken Hart, foresters at Eagle Ridge, are Duane's secondary readers. They will carry out the plan, if instituted.

Mr. Hamilton, who oversees the running of all county environmental areas, holds an M.A. degree in ecology. He is interested in managing black walnut for maximum timber potential. Sally Beatty has had three years of college, majoring in forestry; Ken Hart has a B.S. degree in forestry. Although they have not studied black

71

walnut in particular, they are thoroughly familiar with forest maintenance and with Eagle Ridge in particular.

Duane's primary purpose with George Hamilton is to persuade him to adopt the plan. Duane's primary purpose with the two foresters is to instruct them in the steps necessary for managing black walnut, so that they can carry out his plan. To accomplish both primary purposes, he must inform his readers about the plan.

SELECTION

In order to select appropriate information, you should first consider the amount and kind of detail your readers require, then list this content.

Amount and Kind of Detail

Your readers' educational background and technical expertise, and your purposes in addressing them, influence your decisions on amount and kind of detail. At times, you may want to include more information when writing to an audience that is outside your field or is unfamiliar with your topic. At other times, you may want to include less. You will also select different kinds of details, depending on your readers' uses of your document.

For example, George Hamilton requires more discussion of the specific problem with black walnut at Eagle Ridge because he does not work at the area and is unfamiliar with the situation. On the other hand, Sally Beatty and Ken Hart do not require these details because they are quite familiar with the problem.

However, George Hamilton does not require a detailed discussion of the management plan, in part because his area of specialization (ecology) is different from Duane's. Therefore, Mr. Hamilton might not understand these details. Moreover, since he will not be carrying out Duane's plan, a brief discussion will provide him with the background for decision making. On the other hand, Sally Beatty and Ken Hart need theoretical background and a full technical discussion in order to understand and carry out the plan.

Lists of Content

After you have considered the amount and kind of detail each reader requires, you should list the specific content you will include. (Your purpose-and-use grid, which we discussed in Chapter 1, Analyzing the Communication Context, will assist you with this task.) Figure 4.1 shows Duane's list. You can then use this list when structuring your document.

ARRANGEMENT

You should proceed from larger to smaller sections of your document when deciding how to structure it. First, segment your document into major divisions and order these divisions. Then order the details within divisions.

Arrangement **73**

FIGURE 4.1
Duane's list of content

George Hamilton

1. Description of the state of black walnut, with recommendation
2. Single-tree management
3. Plantation management
4. Cost
5. Maximum timber potential
6. Profit

Sally Beatty/Ken Hart

1. Theoretical basis of the plan
2. Step-by-step procedures for implementing the plan

Major Divisions

Standard reporting forms and your list of content will help you decide on the major divisions of your document and their order.

Standard Reporting Forms

Standard reporting forms are traditional structures for technical documents, e.g., the structure of a justification, a comparison, or a laboratory report. (We discuss these standard forms in detail in Chapter 14, Final Reports.) Because audiences in technical fields have read many of these documents, they expect a certain content and order. Therefore, the standard forms provide you with guidelines for structuring your document.

However, many times, no standard reporting form will precisely suit your communication situation. At other times, you may have to fulfill the differing needs of a variety of readers. When these situations occur, your list of content is the basis for structuring your document.

List of Content

Classifying Details

When using your list of content to structure your document, you should classify the details on your list into the largest possible groups, on the basis of similarity. These groups will then form the major divisions of your document. For example, the topics of single-tree and plantation management on Duane's list constitute a discussion of his management plan. Therefore, these details should be classified as one major division in his report. In addition, since both maximum timber potential and profit are effects of the plan, these details should be classified to form another division. Figure 4.2 shows Duane's classified list of major divisions.

FIGURE 4.2
Duane's classified list of major divisions

George Hamilton

1. Description of the state of black walnut, with recommendation
2. Discussion of the plan
3. Cost
4. Effects of the plan

Sally Beatty/Ken Hart

1. Theoretical basis of the plan
2. Step-by-step procedures for implementing the plan

Prioritizing Reader Needs

If you have only one reader or several readers with the same information needs, you can use your classified list of major divisions as the basis for structuring the body of your document. However, if you have several readers with different information needs, you must prioritize these needs so that you can decide where to place the different details readers require. Ranking primary readers in order of importance, as you did on your purpose-and-use grid (e.g., primary decision maker first), will assist with this task. Your primary decision maker's requirements then become the basis for structuring the body of the document and dictate its level of technicality.

The needs of Duane's primary reader, George Hamilton, are the basis for structuring the body of his report, which will be nontechnical in nature because of Mr. Hamilton's level of expertise.

Segmenting the Document

The information requirements of secondary decision makers and secondary audiences should also be met. You can accomplish this task by segmenting the document and directing each segment to different readers: the introduction, conclusion, and appendices can be written differently to provide these readers with information they require. For example, in a document written for both technically and nontechnically versed audiences, the introduction and conclusion should always be written to satisfy the requirements of the nontechnically versed reader. The appendices can be used to provide other readers with the technical details they require.

Duane will satisfy his secondary readers' needs by appending a theoretical discussion of the plan and step-by-step procedures for carrying it out.

Arranging Major Divisions

After you have decided where you will place your information, arrange the major divisions from your list of content in a logical order by using either a deductive or an inductive structure and sequencing the remaining divisions.

Using Deductive or Inductive Structure. With a deductive structure, the main ideas (your conclusions and recommendations) are placed first in the document, either

Arrangement   **75**

in a section by themselves or in the introductory section. With an inductive structure, the main ideas are placed last. You should use a deductive structure when readers will not resist your conclusions or recommendations. You should use an inductive
structure when readers may not agree with your conclusions or recommendations or
when readers require explanation in order to understand these points.

Since Duane will discuss very favorable profit and cost data, he decides to structure his report deductively with his conclusions and recommendations first.

Sequencing Divisions. We show the sequence of divisions in Duane's report in
Figure 4.3, then discuss his reasons for this arrangement.

FIGURE 4.3
Sequence of divisions in Duane's report

George Hamilton	Sally Beatty/Ken Hart
1. Description of the state of black walnut, with recommendation	1. Theoretical basis of the plan
2. Discussion of the plan	2. Step-by-step procedures for implementing the plan
3. Effects of the plan	
4. Cost	

For George Hamilton, Duane's general description of the state of black walnut
will provide a context and background information for the discussion of the plan.
Therefore, Duane decides to put this description first in his report, preceding his discussion. Once George Hamilton understands the state of black walnut and the plan
itself, he will be prepared for understanding the plan's effects. If he is convinced by
these effects to institute the plan, he will be ready to consider cost.

For Sally Beatty and Ken Hart, Duane's technical discussion of the plan's theoretical basis will provide the necessary background and motivation for carrying out
the steps.

Once you have decided on the major divisions of your document and their order,
the next step is to order details within these divisions.

Details Within Divisions

Traditional patterns of development and principles of emphasis help you to arrange
details within divisions: subsections of your document, paragraphs, sentences, and
series of words.

Traditional Patterns of Development

Traditional patterns of development are sequences appropriate to certain kinds of
material. These patterns help make your writing coherent, because readers are famil-

iar with them and thus expect your thoughts to proceed in a certain order. The patterns can be used to structure both paragraphs and longer sections of documents.

Ordering Paragraphs

A paragraph groups information on one topic or idea. Usually, this topic is announced or predicted by one sentence — the topic sentence — which controls the information in the paragraph. This topic sentence may also suggest an order for the paragraph, based on a traditional pattern of development. Therefore, knowing the traditional patterns your topic sentences suggest will help you decide how to write your paragraphs.

We illustrate this concept by presenting the traditional patterns of development with sample topic sentences for each pattern. (The words that suggest the pattern are italicized.) We then discuss how these patterns assist in composing paragraphs.

1. Chronological patterns (progression over time)

   □ Narration: sequencing events or occurrences
   In the *past ten years,* a *series of events* has led to increased interest in managing black walnut for maximum timber potential.

   □ Process description: sequencing steps in a procedure
   Plantation management consists of *the following steps.*

   □ Cause/effect or effect/cause: proceeding forward to outcomes or backward to causes
   Because Eagle Ridge has not been under a management plan, production of black walnut has declined. (Cause/effect)

   Investigation determined the following *reasons* for this decline. (Effect/cause)

2. Spatial patterns (physical relationship)

   □ Description: sequencing parts of objects in space
   Eagle Ridge is *made up of* two areas.

3. Relating patterns (similarities or differences)

   □ Comparison: giving similarities among subjects
   The management techniques for all single trees are *identical.*

   □ Contrast: giving differences among subjects
   Managing black walnut will *vary with region.*

   □ Definition: separating a subject from others
   Pruning is a method of caring for trees that removes useless or dead branches.

   □ Evaluation: weighing subjects against each other
   Under some circumstances, black walnut will produce marketable timber, *but under others*, it will not.

4. Analytical patterns (division into aspects)

   □ Enumeration: listing aspects of a subject, then discussing each in the order given
   Releasing is superior to other management techniques *in two ways.*

- Exemplification: providing examples of aspects of a subject
  Further rejuvenation of black walnut can also be accomplished, as the following *examples* indicate.
- Classification: dividing a subject, then grouping the divisions on the basis of common characteristics
  The trees at Eagle Ridge *can be classified* into two *groups:* single-tree and plantation.
- Problem/solution: identifying a problem, then providing solutions
  Solutions to the *problem* of ineffective management can be found.
- Summary: presenting the major points of a subject
  The results of the management plan can be *summarized* as follows.

5. Progressive patterns (a predetermined developing order)
   - Deduction/induction: proceeding from general to specific or specific to general. Placing the main idea first or last
   - Most to least important, or least to most important: proceeding from the most vital information to the least, or vice versa
   - Order of preference: using an order the writer determines to be effective with specific readers, depending on their needs

In paragraphs organized by progressive patterns, the topic sentences do not suggest the pattern. Instead, progressive patterns are used along with other patterns. For example, in deductive paragraphs the topic sentence occurs first, while in inductive paragraphs it occurs last, as you see in the following examples:

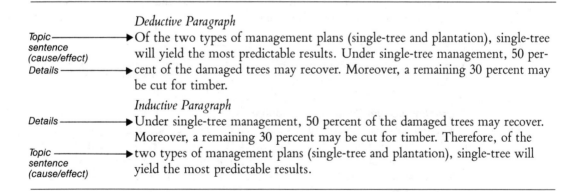

In paragraphs organized by the pattern of most to least important or least to most important, the information proceeds according to the weight it will have with your reader. For example, in the enumerative paragraph (under 4 in previous list), "Releasing is superior to other management techniques in two ways," the most important advantage could be discussed first or last. In paragraphs organized by order of preference, the writer determines the most effective sequence. For example, in the paragraph arranged by contrast (under 3 in previous list), "Managing black walnut will vary with region," Duane would have to decide in which order to discuss the regions (e.g., north to south, large to small).

These traditional patterns of development assist you in writing paragraphs by indicating details you ought to include and the order these details ought to follow. For example, in the narrative paragraph (under 1 in previous list), Duane could discuss in chronological order the events that have increased interest in black-walnut management. In the process description (also under 1), he could discuss chronologically the steps in plantation management. In the descriptive paragraph (under 2), he could discuss the two areas at Eagle Ridge according to a physical relationship he determines. (He decides to proceed from the center of the area to the outlying plantations.) All the details in these paragraphs must relate to his topic sentences. The details must fit the patterns these sentences predict, or the paragraphs will not be coherent. The following examples illustrate this idea.

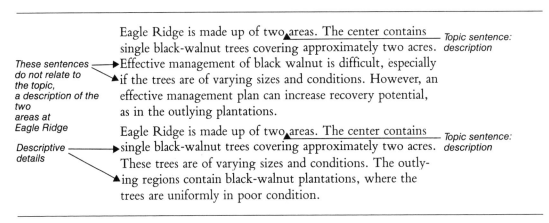

Notice that the second paragraph is coherent: All the details relate to the topic sentence and follow the descriptive pattern it suggests.

Ordering Longer Sections of Documents

These traditional patterns of development can also be used to arrange longer sections of documents. For example, Duane plans to include an entire section of process description — the procedure for managing single black-walnut trees — as a major division of his document. He will also include a section of description — black-walnut growth at Eagle Ridge.

When ordering sections of documents by traditional patterns, follow the same rules as for paragraphs. All the details in the section should relate to the pattern you have used and follow the order you have chosen.

Principles of Emphasis

Principles of emphasis are a second aid for arranging details in your document. Three criteria affect the emphasis given to details: position, amount of space, and repetition.

Position

The beginning position in a section of a document or a whole document receives most stress. The ending position receives second. The middle position receives third and can be used to deemphasize details. If you feel your readers will not resist the most important information you are conveying, or if they do not require explanation to understand your details, you should place this information in the beginning position. However, if your readers will not agree with or understand your most important information, you should place it last, after you have prepared them for it. If you must convey negative information, you should place it in the middle position between more favorable details.

Duane uses these principles to arrange the major divisions of his report. Because George Hamilton is his primary audience, Duane places the information this reader requires first in his report, so that the information will receive appropriate stress. Moreover, he places the most important section in terms of his purposes and George Hamilton's needs — the description of the state of black walnut — first in his discussion for the same reason. He also places his recommendation at the beginning of the report, because he feels George Hamilton will not resist instituting a financially favorable plan.

Duane also uses these principles of position to order details within sections of his report. He decides to arrange his discussion of the effects of his plan (profit, maximum timber potential) from most to least important, so that the favorable profit figures he has gathered will receive maximum emphasis.

Position affects emphasis in paragraphs as well as in documents and sections of documents. You should arrange your paragraphs deductively (topic sentence first) when you feel your reader will not resist your main idea or when you wish to stress it, and inductively (topic sentence last) when your reader requires preparation to accept your point or when you wish to build up to your point. For example, Duane chooses a deductive arrangement for his paragraph on single-tree management (page 77), because he wishes to stress the predictable results of this management.

The same principles can be used to construct sentences and arrange series of words. Consider the following sentences:

Under single-tree management, 50 percent of the damaged trees may recover.

Fifty percent of the damaged trees may recover under single-tree management.

Sentence 1 stresses single-tree management, while sentence 2 stresses 50-percent recovery. Because Duane wishes to stress the particular aspect of his plan that will yield a 50-percent recovery rate, he chooses sentence 1. He arranges the following list of words according to the principles of position as well:

Increased profit and maximum timber potential will result from adopting this management plan.

He places increased profit, his most important effect, first, so that it will receive maximum stress.

Amount of Space

The amount of space you devote to material also affects the emphasis it receives. Usually, the more discussion you include, the more emphasis a detail will receive. However, extreme brevity can also be used to achieve emphasis, as in one-sentence paragraphs. These paragraphs receive greater stress than surrounding material, because they are much shorter.

Duane uses the principle of amount of space when writing his document. He includes only a brief, nontechnical discussion of his plan in the text addressed to George Hamilton, so that this section will not receive greater emphasis than his plan's effects or cost. He supports this deemphasis by placing the nontechnical discussion in the middle of his report.

Repetition

Repetition of words or ideas also creates emphasis, as the following sentences show:

> The cost of managing black walnut can be reduced by effective management techniques. This cost can also be offset by increased profit from timber sales.

Repetition of the word *cost* and the idea of reduction adds to the stress these details receive. Because offsetting of costs will interest George Hamilton, this stress helps Duane achieve his persuasive purpose.

Ordering Devices

After you have decided on the sequence of your details, constructing an ordering device for your material will help you visualize the relationship of your ideas. Ordering devices differ in type (e.g., lists, outlines, idea trees, flow charts, diagrams) and also vary in complexity from simple ''scratch outlines'' to full-sentence outlines, or from simple flow charts to complex ones. The particular type of ordering device you choose and its complexity depend on your preferences as a writer and your communication situation.

The *diagram* is useful for ordering descriptions of items, while the *flow chart* is useful for processes. *Lists* and *outlines* can be used to order other types of documents. When these documents are brief, you may find a simple list adequate for ordering your ideas. For longer, more complex documents, you may prefer a more extensive outline.

Because Duane's report is fairly long, he uses a more extensive outline as his ordering device. Figure 4.4 shows a portion of that outline. Notice that even this more extensive outline is not complete. Duane has noted points he will want to include, but he has not listed the details because he is quite familiar with them.

The ordering device you use will serve as a guide to writing your first draft. Although you may not begin writing with the first point in your ordering device (many writers begin with the easiest or the most essential parts of documents), this

FIGURE 4.4
A portion of Duane's outline

Management Plan for Black Walnut at Eagle Ridge

I. Single-Tree Management
  A. Three techniques used
    1. Releasing
    2. Pruning   (Describe each process for
    3. Harvesting   George Hamilton's benefit)
  B. Potential for recovery as the result of management

II. Plantation Management
  A. Adverse conditions cause poor potential
    1. Poor soil
    2. Lack of management for timber potential
  B. Two techniques used
    1. Leave plantation 1 to mature as is
      a. Is in worst condition
      b. Some merchantable timber will result
    2. Clip the trees off at ground level in plantation 2
      a. May sprout back and start over
      b. Survival cannot be guaranteed

device will help you recall the points you wish to include and the sequence they should follow.

CHECKLIST FOR SELECTING AND ARRANGING CONTENT

1. Have I considered the amount and kind of detail my readers require?

2. Have I listed the content each reader needs?

3. Would a standard reporting form help me structure my document?

4. If no standard reporting form suits my material, have I structured that material effectively, proceeding from larger to smaller sections of my document?

   A. Have I classified my list of content into major divisions?

   B. Have I prioritized my readers' needs?

   C. Have I satisfied my most important primary reader's needs in the body of my document?

   D. Have I segmented my document to satisfy other primary and secondary readers?

   E. Have I arranged the major divisions of my document in logical order by using a deductive or inductive structure, then sequencing the remaining divisions?

82   4 | Selecting and Arranging Content

F.   Have I used the following traditional patterns of development to arrange details within sections of my document, and in my paragraphs?

1) Chronological patterns

Narration
Process description
Cause/effect or effect/cause

2) Spatial patterns

Description

3) Relating patterns

Comparison
Contrast
Definition
Evaluation

4) Analytical patterns

Enumeration
Exemplification
Classification
Problem/solution
Summary

5) Progressive patterns

Deduction/induction
Most to least important/least to most important
Order of preference

G.   Have I used principles of emphasis (position, amount of space, repetition) to arrange details in sections of my document, and in paragraphs, sentences, and series of words?

5.   Have I constructed an appropriate ordering device for my document?

EXERCISES

Topics for Discussion

1.   The following topic sentences predict traditional patterns of development. In class, discuss the traditional patterns each topic sentence predicts and indicate the predicting words. Then discuss the details you would select if you were writing a paragraph for each topic sentence.

A.   The process of modeling usually entails starting with a black box.

B.   While working for Farmers' Home Administration, I performed several loan officer's functions.

C.   Parvo virus, contracted from infected animals, may have disastrous effects on puppies and older dogs.

Exercises | **83**

D. The three recreational areas in the Skunk River Greenbelt are identical in facilities offered.
E. The Pioneer Venus probes have sent back descriptive details on Venus' cloud formations.
F. A number of factors influence choice of material to be used in constructing the chassis.
G. Cue Agency's needs involve both loading and spatial considerations.
H. The following steps have been taken to limit maintenance costs.
I. A series of events led the Carver City Planning Commission to consider redesigning Group Home II.
J. Geothermal energy is caused by heat sources within the earth.
K. The tractor models differ greatly in ripping ability.
L. The four basic modes of failure in carbide tools are deformation, diffusion, attrition, and thermal fatigue.

2. As a basis for class discussion, decide how you would sequence the details you have selected for the above paragraphs. Use the traditional patterns of development and principles of emphasis to help you.

Topics for Further Practice

1. Select a report you have written. Analyze the audience, purposes, and uses of the report. Then perform the following operations, as a basis for understanding selection and arrangement:
   A. Outline the major divisions of the report. Consider your audience and your purpose-and-use analyses, then reflect on your reasons for including each division. Why do your readers require this material? Do they require any additional information or a different kind of information?
   B. Consider the outline you have made. Are the sections classified? If not, how would you change them?
   C. Prioritize your readers' needs, then discuss how you have segmented the report to meet these needs.
   D. Consider the arrangement of major divisions in the report. Is it arranged deductively or inductively, and why? Do the sections follow a logical order? If not, how would you change the order?
   E. Outline the major details within each section. Do they follow a logical order? If not, how would you change the section?
   F. Choose a paragraph in the report and locate the topic sentence. Is the paragraph deductive or inductive, and why? Does the topic sentence suggest another traditional pattern of development? Which one?
   G. Choose a sentence and a series of words from the report. Have you used principles of emphasis to arrange the sentence and words? Discuss how.

2. Locate a report or journal article in your field, and repeat the activities in the previous exercise.

5

Planning
Style

Style refers to the way you express your thoughts in language. Thus, style involves such considerations as how to construct sentences or which words to choose. You will make many stylistic decisions while writing and rewriting documents. However, you must also set guidelines for style before you compose, in order to direct writing and revising.

We illustrate stylistic plans with the example of Virginia Quay, a regional consultant with the Nursing Home Association of America. Virginia's job is to survey problems that member nursing homes may be experiencing and recommend solutions. This particular consultation for Independence Nursing Home in Carroll, Missouri, involved a survey of supportive environments for the aged in the home and recommendations for improvement.

Your first step in planning style is to consider your audience and your purpose-and-use analyses, since this information guides your stylistic choices. Virginia has one primary and three secondary readers. Susan Root, administrator of the home, will decide whether or not to implement Virginia's recommendations. Susan must be persuaded that these recommendations are effective. Nancy Loos, director of nursing; Richard Ziebner, physical therapist; and Lois Colby, activities director, will read the report to discover the impact of Virginia's recommendations on their programs. They must be adequately informed about these facts. Of these four readers, only Susan Root has an overall knowledge of administering the home or a theoretical background on supportive environments.

This information about audience, purpose, and use assists in making stylistic plans.

Stylistic Plans   **85**

STYLISTIC PLANS

Stylistic plans involve setting guidelines for sentence construction, word choice, voice, and tone.

Sentence Construction

Planning sentence construction means considering the length, structure, and complexity of sentences.

Sentence Length

Sentence length can sometimes be correlated with ease of reading: Shorter sentences generally require less sustained attention; longer sentences generally require more. The following list illustrates this correlation:[1]

Very easy	8 words or fewer
Easy	11
Fairly easy	14
Average	17
Fairly difficult	21
Difficult	25
Very difficult	29 words or more

However, difficulty of the material presented also affects reading ease: A shorter sentence containing technical material will be harder for nonexpert audiences or those with a less sophisticated reading ability to comprehend than a longer sentence with less technical material. The following sentences illustrate this idea:

Supportive environments in nursing homes depend on ethos, constituency, personnel, and facilities.

Supportive environments in nursing homes depend on four factors: the character of the home, the types of clients, the people who work at the home, and the building space and equipment available.

Although sentence 2 is longer than sentence 1, sentence 2 is easier to understand, because it is expressed in a less technical way.

In general, then, plan to use shorter sentences when writing for audiences that include nonexperts, or express your material in a less technical way.

Sentence Structure and Complexity

Structure is the way you arrange the parts of your sentences; complexity is the number of ideas included in each sentence.

---

[1] See Dr. Rudolph Flesch, *The Art of Plain Talk* (New York: Harper, 1946), p. 38.

The simplest sentence you can write in English consists of at least a subject (S) and a verb (V), as in the example:

S   V

The client reads.

Common variations of the SV sentence are Subject-Verb-Object (SVO) and Subject-Verb-Complement (SVC). In the SVO sentence, the subject performs an action affecting an object:

S   V   O

The client reads the newspaper.

Two types of SVC sentences exist. The subject may perform an action in a particular way:

S   V   .C

The client reads quickly.

Or the subject may be described in a particular way:

S   V   C

The reader is quick.

Documents are never written entirely in simple sentences because, although these sentences are not necessarily short, they are only useful for stating single, unqualified observations. Simple sentences do not allow you to indicate relationships of ideas other than the linear relationship of one fact following another (e.g., ''The client reads the newspaper. He enjoys the news.''). Therefore, you should combine sentences to express complexities of ideas. For example, you might combine the previous two sentences in the following way:

The client reads the newspaper because he enjoys the news.

Joining these sentences with the word *because* indicates that your ideas are related by cause and effect.

We discuss combining sentences in Chapter 6, Revising the Draft, because you will consider different combinations of ideas when you are reviewing your document. However, when planning your document, remember that the general complexity of your sentence structures, like sentence length, should be suited to the audience you address. An audience with less technical knowledge or a less sophisticated reading ability might have difficulty comprehending a style where many ideas are combined in a few sentences. On the other hand, an audience technically versed in your subject or an audience of more able readers could comprehend a more complex style. The following examples illustrate this idea:

Although the desirability of supportive environments in nursing homes has been recognized for a number of years, nursing home personnel have just begun to explore the effects of these environments. A growing body of research indicates that they will slow the onset of senility as well as contribute to the emotional well-being of the clientele and create a pleasing atmosphere for the home.

Stylistic Plans   **87**

The desirability of supportive environments in nursing homes has been recognized for a number of years. However, nursing home personnel have just begun to explore the effects of these environments. A growing body of research indicates that they have three effects:

1. Slowing the onset of senility
2. Contributing to the well-being of clients
3. Creating a pleasing atmosphere for the home

Example 1 contains only two sentences. Example 2 conveys the same information in three sentences and a listing device. If the audience were not technically versed in the subject of supportive environments, example 2 would be easier for these readers to comprehend. Example 1, on the other hand, could be written to technically proficient readers.

When setting guidelines for sentence construction, then, note the educational backgrounds of your readers and their levels of technical experience. For example, Virginia notes that all her readers are college educated, which means she could write longer, more complex sentences. However, her three secondary readers are outside her own area of specialization (nursing home administration) and have no background in theoretical material. Therefore, Virginia feels she should write shorter, less complex sentences with fewer ideas included in each whenever she discusses the technical aspects of her program.

Word Choice

Planning word choice involves setting guidelines for the difficulty of your language and selecting nondiscriminatory expressions.

Difficulty of Language

Your readers' educational backgrounds and technical expertise affect decisions on difficulty: If you feel they may have trouble comprehending difficult language, or if they do not have your technical experience, you should use words they will understand.

The following list gives types of words that will automatically increase the difficulty of your language:

1. Abbreviations
   Rh factor
   bp
2. Hyphenated or compound words
   Health-related symptoms
   Relationship patterning
3. -tion words created from verbs
   Computation
   Deterioration

4. Words of three syllables or more, if uncommon or technical
Senility
Transluminary

5. Abstract words or concepts
Ethos

6. Any word or acronym (an abbreviation made from the first letters of the parts of a compound word) used only in a given field; often called jargon
Subcutaneous
CPR

When writing for audiences with a less sophisticated reading ability, for those outside your field, or for audiences with varying levels of expertise, you should use synonyms for, or define, all difficult words. For example, you might say "physical and mental decay with age" instead of *senility* or define *subcutaneous* as "under the skin." In this way, you will ensure that you communicate with all your readers.

You should also carefully consider point 6 in the previous list, even when writing to audiences in your field. Discipline-specific words can aid communication with these readers, because technical terms are often more precise and shorter than synonyms or definitions. However, some jargon is meaningless because of abstraction, as the following examples indicate:

We must *implement alterations in functionality.*
= We must *make practical changes.*

Our goal is to *optimize a cost minimization.*
= We want to *reduce costs as much as possible.*

Since abstract jargon hinders communication, avoid using it at all times.

Virginia considers these guidelines when planning the difficulty of her language. Her primary reader, Susan Root, is in Virginia's area of specialization and would understand a more technical vocabulary. However, her three secondary readers are outside her area of specialization and have no background on supportive environments for nursing homes. Therefore, she decides to define or use synonyms for all her technical words in order to communicate with her entire audience.

Nondiscriminatory Expressions

A discriminatory expression is language that stereotypes people on the basis of sex, occupation, race or ethnic origin, or religion. For example, consider the following sentence:

The receptionist will receive clients. *She* will then refer them to the appropriate doctor. *He* will recommend treatment.

This sentence stereotypes on the basis of sex and occupation, because the sentence presupposes that all receptionists are female and all doctors are male. When consider-

Stylistic Plans   **89**

ing discriminatory expressions, you should be especially careful of words containing *man*, singular pronouns, and names and titles.

Words Containing Man

In the past, the word *man* was used to refer to humanity in general. However, many people now feel that alternative expressions ought to be found. The following list gives some words containing *man* and possible alternatives:

Word Containing Man	*Alternative*
Chairman	Chairperson/chair
	Head
	Person presiding
Congressman	Member of Congress
Mailman	Mail carrier
Manhood	Adulthood
Mankind	Human race

Notice that synonyms can also be used, as with ''head'' or ''person presiding'' for ''chairman'' or ''human race'' for ''mankind.''

Singular Pronouns

Unlike other languages, English does not contain a singular pronoun designating *either* a man *or* a woman. In the past, *he* was used to refer to a human being in general. However, applying one of the following techniques will allow you to avoid the bias implicit in using the word *he* to stand for either a male or a female:

1. Omit the pronoun.
   The doctor will visit ~~his~~ patients at 10 A.M.

2. Use the plural.
   Doctors will visit *their* patients.

3. Use the passive.
   Patients *will be visited* by their doctors at regular intervals.

4. Use both pronouns.
   Each doctor will visit *his or her* patients at 10 A.M. (This construction can become awkward with overuse.)

5. Repeat the noun.
   Each doctor will visit patients at 10 A.M. *The doctor* will then recommend treatment.

Names and Titles

Names. Men and women should be treated equally with regard to names. For example, a man should not be designated by his last name and a woman by her first.

Instead, both should be addressed in parallel fashion, by their first names, their full names, their names and courtesy titles, or (less often) their last names:

Smith and Linda	(incorrect)
Bob and Linda	(better)
Bob Smith and Linda Carley	(better)
Mr. Smith and Ms. Carley	(better)
Smith and Carley	(better, but less accepted)

Titles. A man can be designated by the courtesy title ''Mr.'' and a woman by the courtesy titles ''Miss'' or ''Mrs.,'' if her marital status is known. However, the generic title ''Ms.'' is appropriate for all women regardless of marital status and is preferred by many over the other two.

The title ''Sir,'' traditionally used in the salutation of business letters, causes special problems of biased language. If you are unsure of the sex of the person you are addressing, the following replacements for ''sir'' may be useful:

1. A name
   Dear Dr. Rouse:
2. A job role
   Dear Personnel Manager:
3. Dual titles
   Dear Ladies and Gentlemen:

Virginia keeps these guidelines in mind. She will be especially careful to avoid using discriminatory expressions, a problem which may arise because she is addressing a mixed audience, since these expressions may anger her readers and interfere with her persuasive purpose.

Voice

A sentence can be written in either the active or passive voice, as the following examples illustrate:

Active voice: The doctor prescribed the medicine.

Passive voice: The medicine *was prescribed* by the doctor. (form of *to be* + past participle)

Notice several points about these sentences:

1. In an active sentence, the subject (doctor) performs an action (prescribed) on the object (medicine).
2. In a passive sentence, the subject does not perform the action: The medicine is not prescribing. In addition, three operations have occurred:
   □ The object has moved to the subject position.

Stylistic Plans **91**

□ The subject has moved to the object position, as object of the preposition *by*.

□ The verb has become a form of the verb *to be* plus the past participle of the main verb.

The passive voice stresses the action rather than the actor, because the actor is placed later in the sentence and is subordinate to the object acted on. In fact, sometimes the actor drops out of the sentence entirely:

Active voice: I completed the tests.

Passive voice: The tests were completed [by me.]

The idea ''by me'' completes sentence 2, but that idea does not need to be expressed because it is understood.

Since the passive voice stresses the action rather than the actor, the passive is less direct:

Active voice: We understand we delivered the wrong wheelchairs to you.

Passive voice: We understand the wrong wheelchairs were delivered to you.

As you can imagine, the writer of sentence 2 has consciously chosen to be indirect. He or she does not want to assume responsibility for delivering incorrect goods and so employs the passive voice.

The passive voice also creates a more impersonal, objective tone, because the actor does not intrude:

Active voice: I will complete your chart before I discharge you.

Passive voice: Your chart will be completed before you are discharged.

The passive voice is a useful stylistic device. However, we must make four points about its use:

1. The passive voice may be wordier than the active voice, since you must use a longer verb form and sometimes the prepositional phrase *by* + *actor*. For instance,

The doctor prescribed the medicine. (5 words)

The medicine was prescribed by the doctor. (7 words)

In the interest of being concise, you may want to use the active voice or eliminate the actor from the sentence when the actor is understood.

2. The passive voice may create dangling modifiers. A modifier is a word ending in -ed or -ing that looks like a verb but is really an adjective. As such, a modifier must describe a noun, usually the subject:

On *entering* the lab, she saw the damaged chromatograph.

The modifier *entering* refers to the subject of the sentence — *she*; *she* is doing the entering. Now write this sentence in the passive voice:

On *entering* the lab, the damaged chromatograph was seen [by her].

The chromatograph *is not* performing the action of entering. Therefore, the modifier dangles. Here is a second example:

After *having surveyed* the damage, she reported it.

After *having surveyed* the damage, it was reported [by her].

In sentence 2, "it" is not surveying the damage. Thus, this modifier dangles as well. Sentence 1, on the other hand, is correct.

Although the problem of dangling modifiers does not occur with simple sentences in the passive voice, you should be careful of this error whenever you combine ideas and write in the passive voice.

3. The passive voice with *it* can involve a problem with pronoun reference:

It was decided that tests be conducted in St. Louis to better simulate the conditions of your nursing home.

The initial pronoun *it* does not refer to a specific noun. Therefore, this use of *it* could be confusing. The following sentence is clearer and more direct:

Tests were conducted in St. Louis to better simulate the conditions of your nursing home.

You could also replace the passive with the active voice:

I decided to conduct tests in St. Louis to better simulate the conditions of your nursing home.

4. The passive voice may result in awkwardly separating your subject and verb, as in the following example:

Procedures for assisting clients with emotional difficulties, recommending therapy, and prescribing medication *were reviewed*.

Because the subject and verb are separated by a lengthy list, the reader may forget the subject by the end of the sentence. The sentence could be revised as follows:

Procedures were reviewed for. . .

            or

We reviewed procedures for. . .

Therefore, carefully consider your use of the passive voice. It is helpful in the following cases:

1. When you want to deemphasize the actor or when the actor is not important

2. When you wish to be less direct and more impersonal, or to create a more objective tone

Virginia sets guidelines for voice in her report. She feels she will write largely in the active voice, because she will be describing a situation at Independence Nursing Home and her solution for it. Therefore, she will not need to deemphasize the actor (the situation) throughout most of her report. However, she decides to use the passive voice whenever she refers to work *she* has done, because her role as actor is not important (e.g., "The following procedure was developed. . ."). She also decides to

Stylistic Plans

use the passive voice in her recommendations section (e.g., "Three recommendations have been formulated. . ."), because she wishes to deemphasize her role as the person submitting the recommendations and to focus on the specific requests.

Tone

Tone refers to the way you approach your reader — the particular stance you take or attitude you convey. Consider the following sentences:

John, please send the specifications for building use to Dick as soon as possible.

Specifications for building use should be sent immediately to Richard Beale, facilities scheduler.

The information these sentences convey is the same, but they differ greatly in tone. The first sentence is informal and relaxed, while the second sentence is formal, distant, and businesslike.

You create tone by means of sentence construction, word choice, and voice. Usually short, sharp sentences will convey a sense of urgency, curtness, or abruptness, as the following example shows:

Send building-use specifications to the facilities scheduler immediately.

On the other hand, longer sentences may seem less abrupt. They may also seem more formal, if they contain many ideas.

Word choice is a second influence on tone. Notice how the word *please* in the first sentence softens its tone. In addition, words may have different emotional overtones or connotations. For example, "as soon as possible" seems less demanding than "immediately" or "at once," even though all three terms convey the urgency of the writer's request.

Voice is a third influence on tone: The passive voice contributes a sense of distance to a sentence.

The following list gives possible tones, with general techniques for creating them.

Informal/Personal/Colloquial	*Formal/Distant/Businesslike*
Longer sentences	Shorter sentences
Contractions	Spelling out contractions
Colloquial words (e.g., "The choice *to go with* this design. . .")	Formal words (e.g., "The decision *to implement* this design. . .")
Active voice	Passive voice

Audience and purpose should determine your tone. The sentence "John, please send specifications for building use to Dick as soon as possible" might be written to a colleague who is also a friend or someone of equal rank with the writer. The sentence "Specifications for building use should be sent immediately to Richard Beale, facilities scheduler'' might be written to a person of lower rank in the company than the writer or to a group of people as general information. The sentence ''Send building-use specifications to the facilities scheduler immediately'' might be written when a previous instruction had been ignored. You should thus consider audience and purpose when thinking about tone.

Virginia does consider audience and purpose and decides to maintain a rather formal tone. She feels this tone will suit her role as consultant. Moreover, she does not want to appear condescending toward her readers, whose levels of expertise do not equal hers. She is afraid a more informal tone might create this appearance, thus increasing reader resistance to her recommendations.

Her use of the passive voice will help create this formal tone, since the passive voice is more distant. In word choice, she will avoid contractions (e.g., ''Supportive environments *aren't* necessarily costly to implement'') and colloquial expressions. Thus, she hopes to achieve an appropriate approach to her readers as an aid in fulfilling her persuasive purpose.

CHECKLIST FOR PLANNING STYLE

1. Have I set guidelines for sentence construction?
   A. Should I use shorter sentences for communicating technical material to the nonexpert or the less sophisticated reader, or longer sentences? Should I express difficult material in a less technical way?
   B. Should I write less complex sentences with fewer ideas in each to communicate technical material to the nonexpert or less sophisticated reader, or more complex sentences?
2. Have I planned word choice?
   A. Have I considered the level of difficulty of language?
      1. Will I need to use synonyms for, or define, words that will be too difficult for my readers?
      2. Can I use meaningful jargon appropriate to those in my field?
   B. Have I considered how to avoid discriminatory expressions?
3. Have I set guidelines for voice?
   A. Have I planned to use the passive voice when I wish to deemphasize the actor or create a formal, objective tone?
   B. Will I remember to avoid dangling modifiers, pronoun-reference problems, and awkward separation of subject and verb, and to eliminate the actor from the sentence where possible?
4. Have I planned my tone?
   A. Have I decided which tone to use: informal/personal/colloquial or formal/distant/businesslike?
   B. Have I considered how sentence construction, word choice, and voice will create this tone?

Exercises **95**

EXERCISES

Topics for Discussion

1. The following sentences are long and contain technical material. Simplify them for a lay audience by writing the information in several sentences.
   A. In the event of an accident or occurrence, written notice of particulars sufficient to identify the insured, and also reasonably obtainable information with respect to time, place, and circumstances, and names and addresses of the injured and of available witnesses must be submitted.
   B. Boolean algebra is a mathematical system used to operate on bivalued variables, generally represented by two mutually exclusive values denoted as 1 and 0, or to express the logical nature of a system by the designations *true* and *false* or *high* and *low*.
   C. Neutrophils are phagocytes, or "eating cells," capable of engulfing bacteria and destroying them or destroying other cells that are infected with viruses or have become cancerous by various mechanisms and thus protecting the body from damage that might occur at the sites of cuts or burns, or from infection.

2. The following groups of sentences contain technical material expressed in complex sentence structures. Rewrite the sentences for a lay audience so that the sentences are less complex in structure.
   A. The properties of the neutron explain the results of the beta-decay process and are necessary to balance the energy-conservation equations of this process. These properties include the fact that the neutron must have no charge, because its charges are balanced by the proton and electron, resulting in neutral electricity. The neutrons must also be practically massless, which is discovered by measurements of the particles involved, and must have a quantum number 1/2 spin, in order to conserve spin. The neutron also travels at the speed of light.
   B. Four basic types of ratios that can be used in analyzing a firm's financial position are liquidity ratios, which indicate a company's ability to convert current assets into cash to meet short-run debt obligations; leverage ratios, which indicate a company's ability to meet short-term and long-term debt obligations; activity ratios, which indicate how efficiently a company is using its assets; and profitability ratios, which indicate how effectively a firm generates profits. Different numbers can be used to calculate each type of ratio, depending on the information one desires, and several subtypes exist for some ratios. For example, two main liquidity ratios are the current ratio, or current assets divided by current liabilities, and the quick ratio, which equals current assets that can quickly be converted into cash divided by current liabilities.

3. The following words would be difficult for the nonexpert to understand. In class, discuss why each word is difficult. Then supply a synonym or express each in a simpler way.

analyzation	thermal convention
endeavor	digit
mass	capacity
F.O.B. job site	pressure
evaporation	NASA

4. The following paragraph contains abstract jargon. Express the paragraph in words a lay audience could understand. In addition, simplify the sentence construction.

   The subscribers at the Home Insurance Association, herein called the Company, have severally reached an agreement with the insured, named in the declarations made a part hereof, in consideration of the payment of the premium and in reliance upon the statements in the declarations and subject to the limits named herein and other terms of this policy and in further consideration of the agreement heretofore made and executed by the insured named herein.

5. Identify the voice of each of the following sentences (active/passive). In class, discuss the effects of that voice. Then rewrite the sentence in the opposite voice and discuss the effects of the change.
   A. Yield can be increased by continued application of our product.
   B. The veterinarian will inoculate your dog for rabies.
   C. Extensive tests will be conducted to determine the cause of the pipe's failure.
   D. Storm damage to telephone lines will be repaired in order of priority.
   E. Scientists can now control Dutch elm disease by injections into the bark of infested trees.

6. The following sentences in the passive voice have mechanical problems. Correct the sentences.
   A. On leaving the building, the cause of the fire was sighted.
   B. It has been determined that Atlantic Hotel would be an excellent relocation site.
   C. Means for debiting accounts, cashing $1000 and over checks, transferring money from savings to share drafts, and depositing funds have now been established.
   D. After assessing the fire's effects, the amount of reimbursement was set.
   E. When considering the impact of environment on road construction, a statement must be submitted to the federal government.

7. Identify the tone of the following sentences and the means by which it is created. In class, discuss a situation when this tone would be appropriate. Then rewrite the sentences, conveying a different tone for a different situation you devise, and identify the means you have used to change the tone.
   A. We would appreciate the use of proper procedures in compiling material for your audit.
   B. Keep off the heavy equipment.
   C. The results will be sent to you when the study is completed.
   D. I've checked on the prices of new air conditioners and found we can't afford them.
   E. All personnel should contribute to the company's goals.

8. Read the following passages. The first is written for an audience of experts, the second for an audience of nonexperts. Analyze specific differences in sentence construction, word choice, voice, and tone. Give reasons for the differences you cite, based on the author's probable stylistic decisions.
   A. Power loss can be limited by choosing highly transparent polymers and also by constructing the wave guide so that a minimal amount of the signal power propagates in the cladding. If the core material is pure fused silica, then two broad categories of polymers satisfy the refractive index requirement — fluorocarbons and silicones.

Exercises **97**

B. Some of these huge molecules are able to "crosslink" or chemically join together into a tough, resilient network. A thin layer of a polymer protects the surface of a glass fiber in the same way a coat of acrylic paint protects the body of an automobile from rust. Although some water vapor can penetrate the polymer coating, the vapor's concentration is low enough that danger of corrosion is minimized.

Topics for Further Practice

1. Choose a report you have written. Analyze the audiences, purposes, and uses of the report. Then list your guidelines for sentence construction, word choice, voice, and tone. Indicate your reasons for those guidelines, based on your audiences and purposes.

2. Change the audience of the report to readers with less technical expertise than yours. Now reconsider your guidelines for sentence construction, word choice, voice, and tone. Would you change your guidelines? Why or why not? Indicate the changes you would make.

3. Locate an article in a scientific journal. Analyze the audiences, purposes, and uses of the article. Then list the author's general guidelines for sentence construction, word choice, voice, and tone. Are these guidelines effective? In each case, state why or why not, based on audience and purpose.

4. Now locate an article in a popular scientific magazine. Analyze the audiences, purposes, and uses of the article. Then list the author's general guidelines for sentence construction, word choice, voice, and tone. Are the guidelines effective? State why or why not. Compare the author's decisions with those of the author in Question 3. State specific differences in sentence construction, word choice, voice, and tone and indicate the reasons for those differences, based on audience and purpose.

6

Revising
the Draft

A DEFINITION OF REVISING

When you revise, you reflect on and alter your document. Therefore, revision involves seeing alternatives for your writing and selecting the most effective ones.

Revising can take place at any point in the writing process. In fact, from the moment you begin your first draft you will find yourself altering choices you have made. You can perform this operation mentally. For example, you can write one word after having considered several alternatives, or a sentence after you have reformulated it in your head. You can also revise after you have actually written the word or sentence by erasing or crossing out the old choice and substituting the new. In either case, the ability to reflect on your writing and see alternatives is important to successful composing.

This reflection involves your prewriting analyses, where you set the preliminary guidelines for composing. While writers are working on drafts of their documents, they recall facts about their audience, their purposes, and the likely reader uses of their documents. They also remember their preliminary decisions on content, arrangement, and style. They then alter their documents, if what they have written does not fulfill their intentions.

We must stress the fact, however, that these prewriting analyses and decisions are only *preliminary* guides. Because writing is dynamic and creative, you may discover more effective alternatives to your prewriting guidelines while you are composing and revising your document. You then incorporate the new choices you have made.

98

In the remainder of this chapter, we introduce you to techniques for revision using the example of Don Sperry, a mechanical engineer with Engineering Design, Inc., of Chicago. Don has been working on the design of a solar-heated swimming pool. He has been asked to submit a report on his design to Mr. Edgar Hampton, president of Swimmers, Inc., a St. Louis-based firm that might market such a pool. Don is now revising his report.

TECHNIQUES FOR REVISING

Writers consider the following criteria when reflecting on the documents they compose:

1. Amount and kind of detail
2. Appropriate emphasis of material
3. Logical progression
4. Stylistic appropriateness
5. Mechanical accuracy

In addition, they often proceed from larger to smaller units of their documents, solving major problems before revising for minor ones. In this way, they avoid making changes in sections, paragraphs, or sentences that they may later omit.

In the following sections, we discuss the first four criteria for revision. Our handbook in Part IV will then assist you with mechanical accuracy.

Amount and Kind of Detail

Amount and kind of detail refers to the quantity and type of content you include. You may fail to communicate with your readers because of two faults: including too little or too much detail, or including inappropriate details. Role playing (putting yourself in the place of) your readers will assist with revising for amount and kind of detail.

For example, when Don reviews the major divisions of his report, he notices he has included a discussion section with all the technical details of his work:

1. Basic design data
2. Operating principle of the solar unit
3. Methods for constructing the design (e.g., sizing the solar modules, building the collector plates)
4. Process of operation (e.g., flow configuration, control technique)
5. Results of using the design (e.g., heating and cooling criteria, quantification of gains)

Don role plays his audience: Edgar Hampton is a business person with no formal training in mechanical engineering. Don now realizes that this lengthy technical discussion is not appropriate for Edgar Hampton's needs. He substitutes a brief, nontechnical description instead.

Don then reviews smaller parts of his report. He is particularly troubled by the amount and kind of detail he has included in sentences. He reads the following sentence with Edgar Hampton's needs in mind:

> The flow configuration of the solar-heated swimming pool can be seen in Figure 3.

Don now realizes Edgar Hampton may not be familiar with the term *flow configuration*, so Don substitutes more familiar words:

> The flow of heated water throughout the system of the pool is seen in Figure 3.

Don continues to role play his audience until he is satisfied with the amount and kind of detail he has included.

Appropriate Emphasis of Material

Appropriate emphasis refers to including and arranging material for psychological effect. You will use the principles of emphasis we discussed in Chapter 4, Selecting and Arranging Content (position, length, repetition), to revise your document for appropriate stress. Don considers these principles when reviewing the major sections of his report:

Justification for the Solar-Heated Swimming Pool

I. The problem with conventional heating methods

II. Description of the design

III. Standards of justification

IV. Cost

He contemplates reversing sections III and IV, because the solar-heated pool has a high initial cost and Don does not wish to place unfavorable details in the position of second emphasis, at the end of his report. However, he decides to retain his original order, feeling that the fuel savings he will discuss as one of his standards will offset the information about cost.

Don also uses length to revise for appropriate stress of major sections. He alters his description of the design so that section II is the briefest in the report. He feels that the remaining three sections are more important in persuading Edgar Hampton to purchase the design.

Techniques for Revising

Don then uses these principles of emphasis to revise shorter units in his report. For example, he had ordered the standards of justification in the following way:

1. Increased fuel savings
2. Decreased maintenance
3. Environmental impact

He now feels that increased fuel savings should be placed last in his discussion, in order to build a case for his solar-heated pool and counter the high initial cost. Therefore, Don rearranges this section as follows:

1. Environmental impact
2. Decreased maintenance
3. Increased fuel savings

He considers the emphasis of his sentences:

> Although the solar-heated swimming pool has a high initial cost, this cost is more than offset by increased savings in fuel.

On rereading, Don realizes that the idea of high initial cost receives greatest stress in this sentence. He rearranges the sentence, placing this detail at the end to deemphasize negative information:

> The initial cost of the solar-heated swimming pool is more than offset by increased savings in fuel, even though the initial cost is high.

Don also considers emphasis of individual words:

> The solar-heated swimming pool conserves fuel and maintenance costs.

Don feels his audience will be more interested in conserving fuel, so he rearranges his sentence as follows, placing the *fuel* costs at the end:

> The solar-heated swimming pool conserves maintenance and fuel costs.

With these revisions, he hopes to achieve the psychological effect he desires.

Logical Progression

Logical progression refers to sequencing material in order to provide connections between ideas. If readers cannot follow your train of thought, they cannot make effective use of your document. Constructing skeletal outlines will help you when revising for logical progression.

Skeletal Outlines

A skeletal, or brief, outline of your document will reveal whether or not the sections follow a logical order. To construct this outline, list the major topics or divisions in your document. Then examine your list to see if it develops logically.

Skeletal outlines will also help you revise the development of sections of documents and paragraphs. To construct a skeletal outline of sections of your document, list the topics developed within those sections. For example, Don lists the topics he included in his discussion of the problems with conventional heating methods for swimming pools:

I. Problems with conventional heating methods
    A. Problems
        1. Rising fuel costs
        2. Increased demand for fuel
        3. Decreased supply of fuel
    B. Types of conventional heat

This skeletal outline shows him that this section does not develop logically. Edgar Hampton must know the types of conventional heat before he can relate to the problems they entail. Don revises this section to place that information first.

To construct a skeletal outline of individual paragraphs, find the predicting or topic sentence for each paragraph, then list supporting details underneath. Don constructs the following skeletal outline for his paragraph on increasing fuel demands:

I. As the number of heated swimming pools increases, the demand for fuel also increases.
    A. Electricity is in least demand — most costly
    B. Natural gas is in most demand — has been cheapest
    C. Fuel oil is in the middle — cost is approaching electricity

This outline shows Don that his paragraph does not develop logically. A more logical order would be to proceed from least to most costly, or vice versa. Don decides to use the order of most to least costly, because the increased demand for natural gas has led to a decreased supply, the topic of the next paragraph in Don's report.

These skeletal outlines will indicate the structure of your information and show whether or not it follows a logical progression. However, these outlines will not reveal a second problem you may have: apparent (not real) illogical progression. Your material will not appear to develop logically if you do not signal the connections within your document.

Signals of Connections

Signals of connections indicate the relationships among your ideas and lead the reader from one idea to the next. Thus, these signals make your writing coherent. Phrases or entire sentences can be used as signals. The following examples show a paragraph without a signal and the same paragraph with a phrase signal:

Paragraph without a Signal

    Maintenance expenses may run to $400 per year for conventionally heated swimming pools. The units are complex, making repairs costly and time-

Techniques for Revising   **103**

consuming. Parts are sometimes difficult to obtain locally. Overhaul must be done yearly.

Paragraph with a Signal

Phrase signal ——————— Maintenance expenses may run to $400 per year, *for the following reasons.* The units are complex, making repairs costly and time-consuming. Parts are sometimes difficult to obtain locally. Overhaul must be done yearly.

Paragraph 2 could also be written with a sentence signal:

Sentence signal ——————— Maintenance expenses may run to $400 per year. *The following reasons account for this cost. . .*

Repeated words can also be used to provide coherence, as you see from the underlined words:

Conventional heating systems are <u>expensive</u> to operate, difficult to <u>maintain</u>, and inefficient in heat production. Heating <u>expenses</u> may run to $200 per month. <u>Maintenance</u> problems persist. The pipes are inaccessible, making repairs costly and time-consuming. Parts are difficult to obtain. The systems require a complicated overhaul every year to assure proper operation. The temperature of the pool ranges from 60° on cold winter days to 80° on warm ones: Effective controls do not exist.

However, the information in this paragraph is still difficult to follow because other signaling words called transitions have been omitted. Consider the following rewritten paragraph: We have underlined transitions Don has added and annotated the relationship these transitions suggest.

Conventional heating systems are expensive to operate, difficult to maintain, and inefficient in heat production. <u>First,</u> ◀ *Suggests enumeration of items*

Suggests examples will follow: exemplification ——————— heating expenses may run to $200 per month. <u>In addition,</u> ◀ *Signals second item: addition*

maintenance problems persist. <u>For example,</u> the pipes are inaccessible, making repairs costly and time-consuming.

Signals second example: addition ——————— Parts are <u>also</u> difficult to obtain, <u>and</u> the systems require a complicated overhaul every year to assure proper operation. *Coordinating conjunction signals last example: addition*

Signals last item ——————— <u>Lastly,</u> the temperature of the pool ranges from 60° on cold winter days to 80° on warm ones: Effective controls do not exist.

Notice that these transitional words may occur at the beginning of or within sentences.

In the following list, we give a number of these transitional words and indicate the logical relationships they suggest.

6 | Revising the Draft

Transitional Words

1. Addition

again	as well as	moreover
also	furthermore	too
and	in addition	

2. Causation

   | because | since |
   | consequently | therefore |

3. Chronology

after	in turn	since
at length	later	subsequently
before	next	then
earlier	now	while
first	prior	
in the meantime	second	

4. Comparison

   | in comparison | in the same way | similarly |
   | in the same manner | likewise | |

5. Conclusion

accordingly	hence	thereupon
after all	in closing	thus
as a consequence	in conclusion	to conclude
as a result	lastly	to sum up
at last	so	
finally	therefore	

6. Contrast

but	nevertheless	still
conversely	on the other hand	yet
however	to the contrary	
in contrast	otherwise	

7. Enumeration

   | first | second, etc. |

8. Exemplification

   | for example | namely | to illustrate |
   | for instance | specifically | |

9. Explanation

   | in fact | put another way | simply stated |
   | in other words | | |

Techniques for Revising

10. Spatial relationship

adjacent to	next to	under
alongside	on	where
beyond	on top of	
inside	opposite	
nearby	over	

11. Summary

all in all	in summary	on the whole
in brief	in retrospect	to summarize

Stylistic Appropriateness

Stylistic appropriateness involves suiting the way you write to the audiences and purposes of your document and includes considering sentence construction, word choice, voice, and tone.

Sentence Construction

When revising your sentence constructions, you may want to combine sentences or break them apart.

Combining Sentences

Combining sentences allows you to express them using fewer words. However, combining sentences also allows you to indicate relationships among your ideas and to stress certain ideas over others. Therefore, in order to revise by combining sentences, you must know the relationship you want to express and the information you want to emphasize.

Relationship. If your ideas are of equal importance, you can revise by coordinating your sentences:

Solar-heated swimming pools are energy efficient.

Solar-heated swimming pools are also cost effective.

= Solar-heated swimming pools are *energy efficient and cost effective.* ◄——— *Coordinated elements and coordinating conjunction*

Notice, however, that the idea of cost effectiveness receives less emphasis in the combined sentence, because the idea no longer stands alone.

You can revise so your coordinated ideas are expressed in single words or phrases, as in the examples. You can also retain the two complete sentences, when such coordination does not involve unnecessary repetition:

Solar-heated swimming pools may
be used to heat living spaces.

Such pools will not be efficient if
insulation is inadequate.

=

Solar-heated swimming pools may
be used to heat living spaces, *but* ◄— *Coordinated*
such pools will not be efficient if   *elements and*
insulation is inadequate.   *coordinating*
                                          conjunction

The coordinating conjunction between complete sentences can also be replaced by a semicolon:

Solar-heated swimming pools may be used to heat living spaces; these pools will *Semicolon*
not be efficient if insulation is inadequate.   *replacing*
                                            ▲————————   *coordinating*
                                                                   conjunction

If your ideas are not of equal importance, you can revise by subordinating one idea to another. You must then consider the relationship between your ideas in order to choose the proper subordinating conjunction. For example, the ideas in the following sentences are related by cause/effect, which the subordinating conjunction must signal:

Customers must choose an option.
Otherwise, they will be unable to
use their swimming pools during
the winter months.

=

If customers do not choose an option, ◄— *Subordinated*
they will be unable to use their   *sentence and*
swimming pools during the winter   *subordinating*
months.   *conjunction*

or

Unless customers choose an option, they
will be unable to use their swim-   *Subordinated*
ming pools during the winter   *sentence and*
months.   *subordinating*
                                          conjunction

When subordinating, you must also consider placement of your ideas, to assure they have received appropriate emphasis.

Emphasis. When revising sentences you achieve appropriate emphasis in two ways: by placing the most important idea in the independent clause (the part of the sentence that can stand alone) and by changing the position of the subordinated idea. You should place your most important information in the independent clause, because this placement will increase the stress the information receives. Therefore, in

Techniques for Revising **107**

the revised sentence on choosing an option (see page 106), Don placed inability to use the pools in the independent clause as his most important idea. If he wished to place more stress on choosing an option, he could have written the sentence this way:

Independent —— *Customers must choose an option* if they are to be able to use their swimming
clause   pools during winter months.

Position of the ideas in a sentence also affects the emphasis they receive. Use the principles of emphasis we discussed in Chapter 4, Selecting and Arranging Content, to position your ideas appropriately. An independent clause in the beginning position will receive most stress; a subordinate idea in the ending or in the middle position will receive less stress. For example, if Don wished to deemphasize choosing an option as completely as possible, he could have subordinated this idea and placed it at the end rather than at the beginning of the sentence:

Independent   Customers will be unable to use their swimming pools during the winter
clause ——   months if they do not choose an option.
Subordinated ——
idea at the
end of the
sentence

Separating Ideas

When you separate ideas, you divide a sentence containing several ideas into several sentences. You can separate ideas in many ways, depending on the emphasis you desire:

Combined Sentence
When installing solar-heating devices in their swimming pools, customers usually consider overall cost, as offset by fuel savings and maintenance expenditures.

Separated Sentences
1. When installing solar-heating devices in their swimming pools, customers usually consider overall cost. This cost is offset by fuel savings and maintenance expenditures.

    or

2. Customers may install solar-heating devices in their swimming pools. When doing so, they usually consider overall cost, as offset by fuel savings and maintenance expenditures.

In example 1, offsetting overall cost is stressed, because this idea is placed in a separate sentence, introduced by the repeated word *cost*. In example 2, the idea that customers may install these devices is stressed, because this idea is expressed in a separate sentence.

Word Choice

When revising for word choice, you must consider several factors: denotation/connotation, precision, level of generality, and economy.

Denotation/Connotation

Denotation refers to the dictionary meanings of words, connotation to their emotional or psychological impact on readers. Consider a revision Don has made in the following sentence:

> This design of a solar-heated swimming pool makes conventional heating
>
> *obsolete*
> old-~~fa~~shioned.

The words *old-fashioned* and *obsolete* have the same meaning, but their psychological impact is very different. *Old-fashioned* has pleasant overtones of being slightly out-of-date but valued. *Obsolete* gives a sense of being totally inadequate because of new developments, the connotation Don desires.

Precision

Precision refers to choosing the exact word or words that will convey your meaning, rather than vague or inappropriate terms. Consider this sentence from the original version of Don's report:

> The solar-heated pool showed *good* results on all tests, when compared to the conventional pool.

The word *good* is an example of vagueness, because the term is not precise. It can be applied to any object from dogs to ice cream cones. In the following, Don has tried out more precise word choices, in which he specifies exactly what is favorable about the results.

> The solar-heated pool produced *better-than-average results* on all tests, when compared to the conventional pool.
>
> or
>
> The solar-heated pool *out-performed* the conventional pool in all tests.
>
> or
>
> The solar-heated pool *showed a heat increase of 25%* over conventional pools in all tests.

Don finally chooses sentence 3 as being the most precise he can devise, because it includes exact figures.

Consider another sentence:

> The following table *expresses* this fact through a comparison of energy efficiency (in Btu's) and initial and maintenance costs.

Techniques for Revising **109**

The word *expresses* is an example of inappropriate diction: a term close in meaning to, but not precisely, the one Don needs since *expresses* is generally applied to people rather than things. Don considers using *supports* or *illustrates*.

The following table   supports this fact. . .
                        illustrates

He finally chooses *supports*, because the table bears out conclusions he has stated.

Level of Generality

Level of generality concerns concreteness and abstraction. Concrete language is highly specific; abstract language is broader and more general. The following example illustrates this concept:

The solar-heated pool   performs well.   *Most abstract*
                          is energy efficient.
                          saves 720 Btu's per year.   *Most concrete*

Notice that these terms become progressively more specific as they proceed from abstract to concrete: "saves 720 Btu's per year" tells exactly how well the solar-heated pool performs.

When revising for abstraction and concreteness, you must suit the level of generality to your purpose. For example, Don finally chooses the following words:

The solar-heated swimming pool is cost effective and energy efficient, which make it appealing to consumers.

Because he is summarizing the solar pool's appeals, he does not choose the most concrete detail. But because he wants to specify how the pool performs well, he does not choose the most abstract detail either. However, we should caution you about one fact: Too high a level of abstraction can result in meaningless prose. Since concrete details narrow and specify your ideas, they add meaning to your writing.

Economy

Economy involves using as few words as possible to express your ideas. When revising for economy, consider several factors: initial terms, nouns for verbs, -tion or -ing + *of* phrases, *which* and *that* phrases, redundancy, and internal wordiness.

Initial Terms. Initial terms creating a wordy (and indirect) style are the pronoun *it* and the word *there*. Consider these examples:

It is advisable that the solar-heated pool be installed on the south side of the dwelling. (17 words)

The solar-heated pool should be installed on the south side of the dwelling. (13 words)

There are three requirements for all solar-heated pools. . . (8 words)

Solar-heated pools require. . . (4 words, followed by the three requirements)

Using *it* and *there* has added words to each sentence without adding information. In addition, placing *it* or *there* first in a sentence has given maximum stress to a weak word.

Nouns for Verbs. Verbs are the action words of your sentences. However, verbs can be changed to nouns, as in the following examples:

Noun
Homeowners must make *a decision* between these two options. (9 words)
Verb
Homeowners *must decide* between these two options. (7 words)

In sentence 1, the verb *decide* has been changed to the noun *decision* and joined with the verb *must make.* The verb *make* is called a weak verb because it does not express the action of the sentence: Homeowners are not "making"; they are "deciding." Thus, the action of the sentence is buried in the noun *decision.* Don correctly chooses to use sentence 2 because it is stronger as well as shorter; the action is kept in the verb.

-*tion* or -*ing* + of *Phrases*. Phrases with -tion or -ing + *of* include a noun and a prepositional phrase. Consider these original and revised sentences:

The construction of a solar-heated pool is not difficult. (10 words)

The constructing of a solar-heated pool is not difficult. (10 words)

Constructing a solar-heated pool is not difficult. (8 words)

Omitting the article *the* and the preposition *of* has allowed Don to reduce sentence length in sentence 3 by two words.

That *Phrases*. *That* phrases are parts of sentences beginning with *that:*

The collector panel *that is* tilted at a 90° angle will realize most solar gain. (15 words)

The collector panel tilted at a 90° angle will realize most solar gain. (13 words)

Eliminating a *that* phrase has reduced the sentence by two words.

Redundancy. Redundancy refers to needless repetition, where several words are used to refer to the same concept. Consider these examples:

At this point in time, solar-heated swimming pools are not widely used. (13 words)

At this point, solar-heated swimming pools are not widely used. (11 words)

Since "at this point" refers to time, Don could eliminate one word or the other as unnecessary. The following are other common redundant expressions and suggested revisions:

Techniques for Revising   111

Redundant	Revised
add up	add
advance planning	planning
assemble together	assemble
because of the fact that	because
collect together	collect
connect up	connect
consensus of opinion	consensus
continue on	continue
divide into parts	divide
each and every	each, every
entirely completed	completed
fewer in number	fewer
inside of	inside
most unique	unique
might possibly	might
a range all the way from	a range from
repeat again	repeat
round in shape	round
the same identical	the identical, the same
the solar pool in question	this solar pool
throughout the whole	throughout
yellow in color	yellow

Internal Wordiness.   Internal wordiness means using several words within a sentence where fewer would convey the same information because the information is implied or because the reader already knows it. Consider these sentences from Don's unrevised letter of transmittal. (We have annotated revisions he makes later on the basis of internal wordiness.)

Obvious fact ———— Enclosed ~~with this letter~~ is the report you requested, which includes my design for a solar-heated swimming pool. The information ~~contained in this report~~ will assist you in eval- ———— *Obvious fact*
uating and deciding on this design.

Voice

When revising for voice, you should consider parallel and correct use of the active and passive voices. For parallelism, you should maintain the voice you have planned throughout the majority of the document. However, you can alter this pattern when you want to create a certain effect. For example, consider the second sentence in Don's revised letter:

The information will assist you in evaluating and deciding on the design.

Don originally wrote this sentence in the passive voice (i.e., "You will be assisted by this information...") but he changes to the active voice in order to eliminate awkwardness, to reduce internal wordiness, and to stress "information" rather than "you." He then continues his letter in the passive voice:

Tests were conducted using the design in two pools...

When you read for correct use of voice, remember the grammatical problems the passive can create — pronoun reference and dangling modifiers — which we discussed in Chapter 5, Planning Style.

Tone

You should revise for parallelism when reconsidering tone: The tone of each sentence should suit the overall tone of your document. You should not abruptly shift your tone unless you have a reason for this shift. For example, consider the following paragraphs from Don's unrevised and revised letter of transmittal:

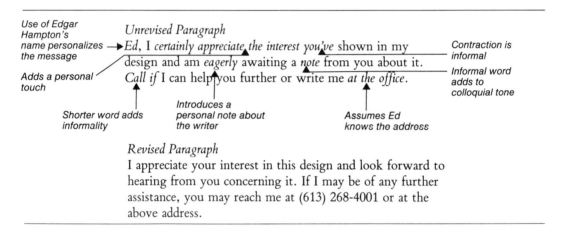

Don chooses paragraph 2 to maintain a more formal tone throughout the letter.

CHECKLIST FOR REVISING THE DRAFT

1. Have I revised for amount and kind of detail?
   A. Have I role played my audience to ensure that I have provided an appropriate amount of detail?
   B. Have I suited the kind of details I include to my reader's needs?
2. Have I revised for appropriate emphasis?
   A. Have I used principles of emphasis to ensure that I have stressed appropriate material?
   B. Does my material have the psychological effect I desire?

Exercises **113**

3. Have I revised for logical progression?

   A. Have I used skeletal outlines to ensure that major sections of my document and my paragraphs follow a logical progression?

   B. Have I provided signals of connections in my document to lead my reader from one idea to the next?

4. Have I revised for stylistic appropriateness?

   A. Have I combined sentences, where necessary, to express relationships among my ideas? Have I separated sentences, where necessary, to simplify my sentences? Have I combined or separated these sentences in ways that will stress appropriate material?

   B. Have I considered denotation and connotation, precision, level of generality, and economy when revising for word choice?

   C. Have I used the same voice throughout most of my document, altering voice when necessary for achieving an effect on my reader? Is my use of voice mechanically correct?

   D. Have I maintained parallelism of tone throughout my document?

EXERCISES

Topics for Discussion

1. The following paragraphs do not develop logically. In some cases, several unrelated ideas have been included in one paragraph. In other cases, information off the topic of the paragraph, as indicated by the topic sentence, has been incorporated, or transitions have been omitted. Construct a skeletal outline to identify the problem for each paragraph. Then revise the paragraphs to achieve logical development.

   A. Modeling techniques are becoming increasingly sophisticated and more accurate. Science will benefit from the information these models generate. Continuing research in model building is necessary if we are to learn more about the world in which we live.

   B. The Reeves 505 Melter is a water-cooled, controlled-atmosphere, direct-arc melter of metals and alloys. The 505 Melter produces the energy to melt metals and alloys from an electrical arc struck between a tungsten electrode and a metal cathode. The 505 Melter operates in either a vacuum or an inert-gas atmosphere. The straight-polarity design of the 505 Melter allows it to operate in a large melting-temperature range and maintain the direct arc to the work piece. Some of the new options of the 505 Melter are a diffusion pump, drop casting, and additional viewing ports. The 505 Melter has the improved capacity of producing up to four buttons or two fingers simultaneously. The improved capacity, versatility, and ability to eliminate critical impurities give the Reeves 505 Melter an advantage over previous Reeves Melters.

   C. Farming today has entered a new era. Today's successful farmer must still be hardworking but must also be a business manager. The farmer of the present is running an agribusiness. Financial survival in agriculture now requires good management.

D. A close working relationship between our facilities should yield productive results. The Va-Ga composites would be produced at our laboratory, then sent to State University for analysis and testing. Block Laboratory has had extensive metals-preparation experience, since the uranium purification for the New York Project of the mid-1940s. Our reputation for producing ultra-high purity materials is known countrywide.

E. As petroleum products and quality aggregates become scarce, recycling pavements is emerging as a standard for highway and street construction. The Federal Highway Administration has turned to pavement recycling as a cost-effective means of handling rising expenses for new pavements and declining budget allocations. Highway administrators realize pavement recycling must be implemented to maintain the American road system.

F. I am a former resident of Clinton and very much appreciate the rich link to the past found in places such as Riverfront Harbor and the Van Allen Building. I am soon to be a graduate of State University with a B.S. degree in community and regional planning. I have three years of experience with zoning legislation. Related projects in which I have participated include historic preservation on the State campus and the restoration of the Sherman Hills subdivision in Ottowatomie. Through association with State University, I have access to a 1.2 million-volume library. I feel well qualified to undertake Clinton's restoration project.

2. The following sentences are awkwardly choppy. Combine the ideas, indicating your reasons for joining the ideas as you did. (e.g., What relationship did you wish to indicate? What did you intend to emphasize or deemphasize?)

A. The whole-farm analysis is now complete. We submit the following report to you. It contains management plans for your crops. It also deals with dairy and swine enterprises.

B. Cloud formations on Venus indicate two scales of motion. These scales are small cells and large planetary waves. Data on the subject are still incomplete.

C. The NEP nuclear reactor is located in an underpopulated region. The area is not subject to severe weather or geological upheavals. The site is ideal.

D. A wind turbine will overspeed if the electrical load is removed. This factor results in destroying the machine. A wind turbine must have a reliable speed control system.

E. Our design was limited by time. The project had to be completed in three months. We feel another six months could be spent refining the design.

F. At the time of this report, only one preliminary test has been performed. This test produced a successful Doppler signal. The Doppler signal was produced with a Percutaneous-Transluminal-Coronary-Angioplasty Doppler catheter in the aorta of a dog.

G. High interest rates have created problems for the housing industry. The Whitney Estates are not selling. Installation of solar-energy devices would increase their desirability.

H. Windbreaks may provide many benefits for your farm. These windbreaks must be carefully designed, planted, and managed. A complete plan is necessary.

3. The following sentences are awkwardly long. Separate them into several sentences containing fewer ideas. In each case, indicate your reason for separating the sentences as you did.

A. Because of increased publicity on the topic of foreign investment in the United States, agriculture could benefit from an analysis of the situation, in order to make prompt, accurate decisions regarding any future legislative action.

B. With growing interest in this area and a drastic need for the beef producer to become more efficient in order to survive, we feel that this research is warranted, especially since our firm has the necessary capabilities and facilities to perform the experiments.

C. As a student of international studies at State University, I have noticed a severe lack of knowledge and understanding of foreign cultures, languages, and peoples on the part of the American population, even in the university setting, which is supposed to be a most informed, enlightened, and tolerant atmosphere where peoples of all nations are accepted.

D. This imbalance, which has been the main reason for falling grain prices during the last two years, has caused situations in which either the profit margin is only slightly higher than production costs or in which production costs have been higher than the returns from grain sales.

E. Engineers studying the rerouting of tracks for the Chicago-Northwestern Railroad in order to avoid Chester Center have determined that a section of abandoned track south of town could be renovated to serve both railroad and civic needs, thus providing a solution to the problem of railway congestion where tracks in Chester are open only 20 minutes in every hour.

4. The following sentences contain imprecise or vague words. Revise the sentences, using more precise words.
A. Group homes are something every community needs.
B. Merchandisers must have a high level of energy.
C. A good knowledge of Fortran is necessary for a computer programmer.
D. I would like to acknowledge the great amount of work that Steve Ungs spent in running the tests for this report.
E. Cemented-carbide cutting-tool inserts have proven to be the best alternative in the never-ending quest of the perishable-tool engineer.
F. With surmounting interest rates and reduced profits, the cattle producer cannot continue to feed cattle in an unprotected environment.
G. The foundation's worldwide expertise includes economists, agronomists, ecologists, and others, all of whom will conduct the research.
H. This Agricultural Conservation Program (ACP) is a cost-sharing venture designed to encourage landowners to construct practices which will conserve soil.

5. Provide more concrete words for the abstract terms in column 1 and more abstract words for the concrete terms in column 2. Then indicate a situation in which each word would be suitable.

Abstract Terms	Concrete Terms
Furnishing	Rose
Precipitation	80 mph
Living thing	Vitamin A
Building	Model T Ford
Body of water	Collie
Geometric shape	Maroon

6 | Revising the Draft

6. The following sentences are wordy. Revise them for economy.
   A. Because of what I learned about career opportunities with the Federal Land Bank when working for Farmers Home Administration, and from your brochures, I am applying for a position as loan officer.
   B. For this reason, it is important to ensure that a proposed group home will be similar in outward appearance to residential homes in the area.
   C. Many of the farms which Farm Counsel has analyzed were deficient in the management of their resources.
   D. There are a number of computerized systems available for your use.
   E. The completion of the wind-turbine design involves collecting together a number of parts.
   F. It is the inner catheter which contains the balloon that expands the stenosis.
   G. This is the report you asked me to prepare to go along with the presentation on cemented carbides that I will be giving to the Perishable-Tool Department at a date later this month.

7. The following paragraph is imprecise, abstract, and wordy. Revise it for precision, concreteness, and economy.

   Subject to the consent of this company and subject to the premiums, rules, and forms then in effect for this company, this policy may be continued in force by payment of the required continuation premium for each successive policy term. Such continuation premium must be paid to the company prior to the expiration of the then current policy term and if not so paid the policy shall terminate.

8. The following sentences are not parallel in voice. Is the lack of parallelism effective or ineffective and why? If it is ineffective, revise for parallelism.
   A. When considering hedging, investors must be aware of both short-term and long-term effects. If these effects are not considered, ill-advised investment may result.
   B. Stump sprouting is an easy forest-propagation technique. The forester just cuts seedlings at the optimal time. Then sprouts are allowed to grow from the remaining stump.
   C. The biomedical engineer applies engineering principles to the solution of medical problems. For example, artificial limbs and organs may be constructed for use in the human body. Biomedical engineers also produce devices for monitoring bodily functions.

9. The following paragraph contains tone shifts. Identify the tones and revise the paragraph twice for parallelism, using a different tone in each revision.

   We must inform you, Mrs. Jones, that your credit balance has exceeded your limit. We'd appreciate a prompt payment of the balance. We'd also appreciate your using the record booklet we've enclosed to keep track of your credit purchases. Use of this booklet will ensure future limitation of these purchases.

Exercises   **117**

Topics for Further Practice

1. Reread a document you have written for this class or another class. Role play the audience of the document. Then indicate specific revisions you would make at all levels of the document in amount and kind of detail, appropriate emphasis of material, logical progression, and stylistic appropriateness. In addition, edit the document for mechanical accuracy.

PART III

APPLYING
THE STRATEGIES
TO TECHNICAL FORMS

7

Definitions

THE PURPOSE OF DEFINITIONS

Definitions rarely make a complete document. However, they are often important parts of documents. Suppose that, as a nuclear engineer, you have been asked to compose a report on nuclear reactors for concerned citizens in your town who wonder about a nuclear facility located nearby. You want to explain how a nuclear reactor operates but must first define a nuclear reactor before your audience will understand its operation. Because the skill of definition is so important in writing technical documents, we devote this chapter to it.

When you are defining, you are asking a very specific question: "How is this object or concept different from all others?" The purpose of defining, then, is to separate this object or concept from similar ones. For example, if you were defining the fuel pin as one part of a nuclear reactor, you would want to distinguish this pin from the other parts of the reactor.

You can define an object or concept in several ways, depending on the amount of detail your readers require. At times, your readers will need only a brief definition. You can then insert it informally into the text by enclosing the definition in parentheses or by surrounding it with commas:

A fuel pin *(the container for the fuel of a nuclear reactor)* consists of three parts.

or

A fuel pin, *the container for the fuel of a nuclear reactor,* consists of three parts.

You can also define the object or concept more formally in a complete sentence:

A fuel pin is the container for the fuel of a nuclear reactor.

121

Notice that the parentheses or commas tend to subordinate the definition to other information, while the one-sentence definition focuses on the definition because it stands alone in the sentence. You may feel, however, that your readers require more information than an informal definition or a formal, one-sentence definition provides. In this case, you can expand your definition by giving additional facts. For example, you could expand the above definition of the fuel pin by describing each part, then explaining how the fuel pin works together with all the other parts of the nuclear reactor.

Expanded definitions also vary depending on reader use. At times your audience may simply need to be informed about the term you are defining, as in the report on the nuclear reactor. At other times, they may need to perform an action on the basis of your definition. For example, you might need to define an electron microscope to a beginning physics class before instructing them in its operation. In this case, you would write an operational expanded definition and select details pertaining to function or use.

In this chapter, our illustration concerns Bret McCleary, a highway engineer with a consulting firm whose major work is recommending highway surfaces. In particular, Bret's job entails studying clients' highway-surfacing problems and determining the optimal surface for a given set of conditions.

Bret has been asked to submit a report to the Highway Planning Commission of Slater, Oregon, on the best highway surfaces to use for constructing a new highway and for repairing and maintaining several old roads in the town. Bret has investigated the conditions and has determined that particular types of asphaltic surfaces would best fulfill Slater's needs in terms of durability and cost. Now he must write his report.

Bret realizes, however, that the planning commission's major work has been mapping highway routes for the town, not considering road surfaces. Because Bret suspects that his readers may be unfamiliar with asphaltic surfaces, he feels he should include a definition in his report. He wonders, however, about the type of definition.

The first task in defining is to choose the kind of definition to use: informal, formal one-sentence, or expanded. If you select a one-sentence or expanded definition, you must then decide on its placement. You could insert the definition in your text if you feel your primary readers require the definition at a particular point to understand the information you are presenting. However, if you feel the definition would interrupt the text, or if secondary readers require the definition but primary readers do not, you could place it in a footnote at the bottom of the page or in a glossary at the beginning or the end of your document.

Bret feels that the Highway Planning Commission of Slater will require more information than an informal or a one-sentence definition provides if they are to understand asphaltic surfaces sufficiently to weigh his recommendations. Moreover, they must have this knowledge before they can make a decision on the basis of the information. Therefore, he will write an expanded definition of asphaltic surfaces and include it as a section of his text rather than placing the definition in a footnote or a glossary. He now plans that definition.

BRET'S DEFINITION OF ASPHALTIC SURFACES

Prewriting

Analyzing the Communication Context

Analyzing the Audience

One primary audience for Bret's definition is the Highway Planning Commission of Slater, since they have hired his firm to perform this work. However, the mayor and the city council are also primary readers, because both must approve any major financial decisions the commission makes. Bret's colleagues in his firm and his superior will also receive copies of the report, and one copy will be placed in his company's files, but readers of these copies are secondary audiences.

Analyzing Purpose and Use

Bret's primary purpose with the Highway Planning Commission, the mayor, and the city council is to inform them about asphaltic surfaces, so they will be able to distinguish them from all other road surfaces and will gain other information about them, necessary for understanding his report. His secondary purpose is to persuade his readers that types of asphaltic surfaces are the best surfacing materials for them to use. If he informs them effectively about these types, he should accomplish his secondary aim.

Bret's primary purpose with his remaining readers is to persuade them by the effectiveness of the report, including his definition, that the document is ready for distribution. This purpose depends, in part, on the quality of the information he provides.

Gathering Information

A checklist of possible details to include in an expanded definition will help you gather information. We give such a checklist in Figure 7.1 on page 124 with sample information pertaining to asphaltic surfaces.

All of these items of information will not be useful in every communication situation. In order to discover which items you ought to select, you must consider your readers and their needs.

Selecting and Arranging Content

Selection

When you have analyzed your audience, purpose, and use, select details your readers require from the checklist. We present Bret's selection here with the reasons for his choice.

Analysis into Parts. The makeup of asphaltic surfaces distinguishes them from other road surfaces, a fact Bret's primary audience needs to know.

Cause and/or Effect. The effects of using asphaltic surfaces (e.g., greater dura-

FIGURE 7.1
Checklist of possible details to include in an expanded definition

1. Analysis into parts
   - ☐ The components of asphaltic surfaces
2. Cause and/or effect
   - ☐ How asphaltic surfaces are made
   - ☐ The effect of using asphaltic surfaces instead of other surfaces
3. Comparison/contrast
   - ☐ The similarities of asphaltic surfaces to other surfaces
     The similarities between types of asphaltic surfaces
   - ☐ The differences between asphaltic surfaces and other surfaces
     The differences between types of asphaltic surfaces
4. Etymology (word derivation)
   - ☐ Where and how the name "asphaltic surfaces" originated
5. Examples
   - ☐ Examples of the use of asphaltic surfaces
   - ☐ Examples of types of asphaltic surfaces
6. History or background
   - ☐ The invention and development of asphaltic surfaces
7. Illustrations
   - ☐ Visual aid picturing the makeup of asphaltic surfaces
8. Location
   - ☐ Where asphaltic surfaces are used
9. Negative statement
   - ☐ What asphaltic surfaces are *not* (e.g., they are not cement concrete)
10. Operating principle
   - ☐ The principle behind the formation of asphaltic surfaces
11. Physical description
   - ☐ What asphaltic surfaces look like
12. Special materials or conditions
   - ☐ Materials needed to make asphaltic surfaces
   - ☐ Conditions under which asphaltic surfaces are made

bility, lower cost) will help Bret persuade his readers of the surfaces' effectiveness.

Comparison/Contrast. Bret must compare types of asphaltic surfaces to explain his choices for various conditions in Slater.

Examples. Bret will be giving examples of the types of asphaltic surfaces in his comparison.

Location. Bret must tell his readers the locations where specific types of asphaltic surfaces should be used in Slater.

Physical Description. The analysis into parts constitutes a physical description.

Special Materials or Conditions. The analysis into parts also includes the materials used in making asphaltic surfaces. The location indicates conditions under which certain types of asphaltic surfaces are used.

Notice that Bret does not plan to include all the details from the checklist. Etymology, history or background, and negative statement might be of interest to audiences reading for general information on asphaltic surfaces. However, Bret's primary readers have more particular interests: precisely what asphaltic surfaces are, where they ought to be used, and why. They are busy readers who would find general information disruptive rather than interesting.

In addition, Bret does not plan to include the theory (operating principle) behind the formation of asphaltic surfaces. Although this fact might interest an audience reading for technical knowledge, such information is too detailed and theoretical for Bret's primary audience.

Arrangement (Structure)

The traditional pattern of development for an expanded definition is very general: introduction (overview of the term), details selected for expansion, conclusion. Therefore, this pattern will not provide much guidance in determining the sequence of your information. You must evolve an effective order yourself.

First, consider the list of details you have selected to discern a possible sequence. Place information distinguishing your term from all others and details providing a general overview in your introduction. Then arrange the remaining details on the list to follow the sequence you have devised. Your rearranged list will be an ordering device for your definition. We give Bret's rearrangement here, with the reasons for its sequence.

Analysis into Parts. The components of asphaltic surfaces are their distinguishing characteristics. Since Bret's purpose in defining is to separate these roadway surfaces from all others that exist, analysis into parts is a logical beginning. The components of asphaltic surfaces will also provide a general context for understanding particular types of asphaltic surfaces.

Comparison/Contrast. After Bret's audience understands what asphaltic surfaces are, they will be prepared for the comparison of types of asphaltic surfaces, which is also a statement of the ones he has chosen for various locations in Slater. Bret orders this statement from most to least important, beginning with the surface he has chosen for constructing the highway and ending with the surface for maintaining existing roads. Thus, his readers will have the answer to their major question (the surface to use for their highway) first and answers to their secondary questions later.

Cause and/or Effect. The effects of using the recommended types of asphaltic surfaces (increased durability, decreased cost) are a logical conclusion to Bret's definition, since they aid him in persuading his primary audiences to adopt his recommendations.

Arrangement (Format)

Your major formatting decision in an expanded definition is the use of headings to guide your readers. If your definition is lengthy, you can identify and highlight important points with appropriate heads. For example, Bret decides to use the names of the types of asphaltic surfaces as headings. They will call attention to this information, which he wishes to stress because it constitutes his recommendation.

Writing

Expanded definitions often begin with one-sentence definitions, because your readers may require a concise statement of the meaning of your term before you give further details about it. A one-sentence definition will separate your term from all others, thus fulfilling the purpose of defining.

In order to write this one-sentence definition, you first name your term or species: for example, *asphaltic surfaces*. You then add a form of the verb *to be* and place the term in a class of similar terms, or genus:

> *Species*   *Genus*
> *Asphaltic surfaces*   are   *road surfaces. . .*

This classification begins the separation process, because you have narrowed the group of items to which your term can belong. However, other items also belong to this genus. For example, other types of road surfaces exist (e.g., crushed stone, gravel), so you must further distinguish the term. You complete this separation by giving the term's distinctive characteristics, or differentia:

> *Species*   *Genus*   *Differentia*
> *Asphaltic surfaces*   are   *road surfaces*   consisting of   *asphalt (a dark brown to black, cementlike substance) and aggregate.*

You select your differentia by considering the basic, or distinguishing, property of your term — in this case composition, material of construction, or makeup. No other road surface except an asphaltic surface is made up of asphalt and aggregate. Other possibilities for basic properties include the following:

1. *Appearance:* ''A two-bitted axe is a firefighter's tool with a sharp edge at each end of the head.''

2. *Function or use:* ''The fuel pin in a nuclear reactor is the part of the apparatus containing the fuel.''

3. *Location:* ''A subway is a train that runs through tunnels underground.''

Informal definitions inserted in your text are similar to formal one-sentence definitions in that informal definitions also include the term's genus and differentia:

Genus and differentia

Asphaltic surfaces, *road surfaces consisting of asphalt (a dark brown to black, cement-like substance) and aggregate*, may be of use on Slater's roads.

When writing a one-sentence or informal definition, you must take a number of factors into account:

1. Limit the genus in order to classify your term in as small a group as possible and thus aid separation. For example, compare the following one-sentence definition with Bret's:

   Asphaltic surfaces are *substances* composed of asphalt and aggregate.

   The genus in this definition is very broad: Thousands of substances exist. Therefore, the genus does not effectively begin the process of separating asphalt from other terms.

2. Avoid circularity, which occurs when you repeat a key term from your species in either the genus or the differentia. This operation will not further your readers' knowledge: If they do not know what your original term means, they will not understand the genus or differentia. Bret avoids circularity by providing an informal definition of asphalt. If he had not defined this word, he would have repeated *asphalt* in his differentia and failed to inform his readers of the distinguishing characteristic of this group of surfaces.

3. Use words with which your audience will be familiar in the genus and differentia. Your purpose in defining is to inform your readers. If they cannot understand the words you have chosen, you will not accomplish this purpose.

4. Phrase your definition in the traditional pattern of the one-sentence definition (species = genus + differentia) or the informal definition (genus + differentia). You could write an example definition, e.g., "Asphaltic surfaces result from mixing asphalt and aggregate." However, example definitions do not by their phrasing separate the term from all others. Since Bret's audience would be unsure whether or not other products could result from this mixture, he would not have accomplished the purpose of defining. In addition, avoid the unacceptable form, "Asphaltic surfaces are when . . ." at all times.

5. Avoid opinions in your definition, which must be objective and verifiable. Consider the following example:

Species		*Genus*	*Differentia*
Asphaltic surfaces	are	*the best surfaces*	*for use on highways.*

   Although this statement uses the form of the one-sentence definition, it is not a definition because it contains an opinion rather than objective, verifiable fact.

When you have composed your one-sentence definition, expand it by discussing the details you have decided to include in the introduction, expansion, and conclusion.

128   7 | Definitions

We give Bret's expanded definition below, with annotations on his writing decisions:

INTRODUCTION

Asphaltic surfaces are road surfaces consisting of asphalt (a dark brown to black, cementlike substance) and aggregate. This aggregate is simply small rock or sand that strengthens the asphalt.

◄ *One-sentence definition separates the term from all others.*

◄ *One-sentence definition gives the purpose of aggregate, which supports asphaltic surfaces' durability.*

Asphaltic surfaces can be classified into two groups: asphalt concrete surfaces and penetration systems. Both types of asphaltic surfaces will be of use on Slater's roads.

Analysis into parts organizes the expansion for the reader and predicts its major sections. The parts are then repeated as headings, to reflect this organization.

EXPANSION

Asphalt Concrete Surfaces ◄

Asphalt concrete surfaces often come ready-to-mix as plant mixes. The recommended plant mixes are hot mix and cold mix.

Hot mix is defined informally. This definition includes hot mix's genus and differentia. ————

1. *Hot Mix*

→ Hot mix, the highest grade of asphalt concrete, is manufactured by heating the aggregate and asphalt to 300°F, mixing the two ingredients, and spreading the mixture while it is still hot. Hot mix is recommended for Slater's new heavy-duty highway.

Example definition clarifies the term hot mix. (Readers do not need to separate this term from all others.)

2. *Cold Mix*

Cold-mix asphalt concrete, a slightly lower grade than hot mix, is combined at room temperature. Cold mix is recommended for repairing Slater's existing roadways.

◄ *Cold mix is also defined informally, with its genus and differentia.*

A second example definition clarifies cold mix, for the same reason given above.

Penetration Systems

Penetration systems are asphaltic surfaces in which single applications of asphaltic material are applied to existing asphalt surfaces to seal cracks, then covered with a layer of aggregate. The aggregate is embedded by rolling. Penetration systems are recommended for maintaining all roads in Slater where cracking has allowed moisture to penetrate the surfaces.

◄ *One-sentence definition further clarifies penetration systems.*

CONCLUSION

The use of hot mix asphalt concrete for constructing Slater's highway, cold mix for repairing existing roads, and penetration systems for maintaining them will give Slater's roads durable surfaces at the least possible cost.

Rewriting

Your major criteria for revising an expanded definition should be amount and kind of detail and logical progression. If you have included too little or too much information, your readers may not understand your term. If you do not order that informa-

Checklist for Definitions **129**

tion effectively, your readers may become lost in the material you present. In either case, you will not have fulfilled the purpose of defining.

When Bret rereads paragraph 2, he decides he has not adequately explained asphalt concrete surfaces and penetration systems, technical terms unfamiliar to his primary audiences. Therefore, he adds the details that follow, as necessary background for his discussion of these systems:

> Asphalt concrete surfaces include those in which the aggregate is coated with asphalt by mechanical mixing.
>
> Penetration systems, often called seal coats, are asphaltic surfaces. . .

Bret also decides he should further distinguish hot mix from cold mix in terms of their quality, in order to support his choices. He adds the following details:

> . . . Hot mix is recommended for Slater's new heavy-duty highway *because of the mix's durability, despite its higher cost.*
>
> . . . Cold mix is recommended for repairing Slater's existing roadways *because, although less durable than a hot-mix surface, cold mix is quite adequate as a repair medium and costs less than hot mix.*

Bret finds no changes he wishes to make in logical progression. He feels his definition is now ready to be included in his report.

CHECKLIST FOR DEFINITIONS

Informal Definitions

1. Have I included the term's genus and differentia?
2. Have I inserted the informal definition appropriately in the text, using commas or parentheses?

One-Sentence Definitions

1. Have I used the appropriate form: species = genus + differentia?
2. Have I limited the genus by classifying the term in as small a group as possible?
3. Have I used the term's basic property as the differentia so my term is distinguished from all others in its class?
4. Have I avoided repeating key words from the species in either the genus or the differentia, which would create circularity?
5. Have I used words my audience will know?
6. Have I avoided example definitions, if my purpose is to distinguish my term?
7. Have I remained objective by avoiding opinions?

Expanded Definitions

1. Have I chosen appropriate details for my audience, from the following list?
   A. Analysis into parts
   B. Cause and/or effect
   C. Comparison/contrast
   D. Etymology (word derivation)
   E. Examples
   F. History or background
   G. Illustrations
   H. Location
   I. Negative statement
   J. Operating principle
   K. Physical description
   L. Special materials or conditions

2. Have I arranged these details in a logical order by considering my audience's needs?
   A. Have I begun with a one-sentence definition?
   B. Have I placed material providing an overview in the introduction?
   C. Have I arranged my remaining details logically?

SAMPLE DEFINITIONS

One-Sentence Definitions

In these sets of examples, we illustrate the effect of audience on one-sentence definitions. In each case, the first sentence of the pair was written for readers with a technical background on the subject, while the second was written for a general audience with no technical experience. We follow definitions 1–4 with annotations on specific differences arising from audience needs. Definitions 5–9 are unannotated, so that you may analyze them.

Definitions 1–4 were written by a student in the field of nursing home care and administration.

Definition 1
Osteoporosis is a disease of the elderly characterized by severe bone demineralization.

Osteoporosis is a disease of the elderly in which bones become porous because of the loss of calcium and other minerals and are easily broken.

In the first definition, the writer uses the term *demineralization*, which a technical audience could be expected to know. However, for her nontechnical readers, she defines *demineralization* and adds content (the porous, mineral-deficient bones are "easily broken"), because these readers do not know the effects of demineralization. A technical audience would be aware of that fact.

Sample Definitions **131**

Definition 2

A decubitus ulcer is an external lesion caused by inadequate circulation to the afflicted area.

A decubitus ulcer is an open skin sore, often referred to as a bedsore, caused by restriction of blood flow when patients lie or sit in one position too long.

The writer uses the term *an external lesion* as the genus in the first definition, a term a technical audience would know. In her definition for nontechnical readers, she substitutes a more common synonym ("open skin sore") and adds content ("often referred to as a bedsore"). As she said, "A layperson is likely to have heard of a bedsore." In the second definition, she also provides an illustration of how circulation, for which she uses the synonym *blood flow*, might be restricted. As she commented, "A general audience can easily understand the illustration of lying or sitting in one position too long. Indeed, they might be more interested in the fact that a patient can develop a bedsore in this way than in the exact cause, restriction of blood flow."

Because of these additions of content, the tone of the two one-sentence definitions differs considerably. The tone of the first is more formal, because of the conciseness of the technical diction, while the tone of the second is more colloquial.

Definition 3

Senile dementia is a lay term for organic brain diseases of two types: the atherosclerotic and the Alzheimer group.

Senile dementia is a medical term used to describe several organic brain diseases that show themselves in the progressive loss of various mental capacities, including memory, orientation, and judgment.

The writer omits the names of the types of organic brain diseases in her nontechnical definition because her audience would not understand these terms. Instead, she includes the characteristics or results of these diseases as differentia. Thus, she both omits technical material and adds nontechnical details. Notice that the added content is simpler and less precise than that omitted. A selection of the "various mental capacities" lost is given, not all the capacities, because a nontechnical audience only requires examples rather than an exhaustive list.

Definition 4

A skilled nursing facility is a nursing home that provides in-patient skilled nursing and restorative care services, has a transfer agreement with one or more hospitals, and meets specific regulatory certification requirements.

A skilled nursing facility is a nursing home that provides health care for patients who require the supervision of a registered nurse on a 24-hour basis.

The writer includes all the distinguishing characteristics of a skilled nursing facility for her technical audience, because they require a more precise definition of the term to suit their professional needs. For her nontechnical audience, she defines the term by giving only the basic or major distinguishing characteristic, which is sufficient for those who are not in the health field.

The next two definitions were written by a student in economics.

Definition 5
Disposable personal income is a person's gross wages less personal tax and certain other nontax payments, such as permit fees and traffic tickets.

Disposable personal income is the dollar amount of a person's income available for spending on goods and services.

Definition 6
The autarky price is the equilibrium price of a good with respect to the domestic demand and domestic supply only.

The autarky price is the amount a good would cost in the marketplace if international trade were not allowed.

The next definition was composed by a chemical engineering student.

Definition 7
Thermal convection is heat transfer caused by the macroscopic mixing of fluids because of density difference.

Thermal convection is the transfer of heat because of the mixing of a liquid or gas.

The last two definitions were written by a student in food science and nutrition.

Definition 8
Phenylketonuria is an inborn error of amino-acid metabolism characterized by insufficient activity of phenylalanine dehydroxylase and leads to the conversion of phenylalanine to phenyl pyruvate, high concentrations of which cause severe mental retardation and skin lesions.

Phenylketonuria is a genetic disease in which the amino acid phenylalanine cannot be properly used by the body and produces a poison that causes severe mental retardation and skin diseases. (Amino acids are the chief components of protein.)

Definition 9
Pepsin is a proteolytic enzyme secreted by the peptic cells of the stomach mucosa.

Pepsin is an enzyme in the stomach that breaks down protein.

Expanded Definitions

Our next examples are expanded definitions, the first of antenna power gain and the second of an integrated circuit. The definition of antenna gain was written to inform

Sample Definitions · **133**

beginning electrical engineering students about a concept important in their field. We have annotated the definition, indicating the kinds of details included and the reasons for the inclusion.

DEFINITION 1
ANTENNA POWER GAIN

One-sentence definition separates the term from all others. ▶ The power gain (or simply the gain) of an antenna is the mathematical ratio of the maximum radiation intensity of a given antenna to the maximum radiation intensity of a standard reference antenna, assuming both antennas receive the same input power. The gain value combines in a single quantity two important

Analysis into parts structures the remainder of the definition. ▶ aspects of antenna performance: the efficiency of an antenna and the ability to focus transmitted power.

One-sentence definition of the first part distinguishes the term from all others. ▶ The efficiency of an antenna is the ratio of the power available at the input terminals of the antenna to the power actually radiated by the antenna. All other factors being equal, a more efficient antenna will have a higher gain

This term is not defined because its meaning is obvious. ▶ value.

Operating principle explains focusing ability. ▶ The ability of an antenna to focus its radiated power is significant because antennas do not radiate power equally in all directions. In general, practical antennas direct (i.e., focus) a large fraction of their radiated power in a certain preferential direction. The remaining portion of the radiated power is distributed uniformly in the remaining directions and is sometimes regarded as being useless or even detrimental to the performance of the antenna. A stronger

Background information relates focusing ability to gain. ▶ focusing ability indicates that an antenna can concentrate a larger fraction of its radiated power in a certain desired direction in preference to other directions. A stronger focusing capability also confers a higher gain value.

Conclusion reviews the parts of antenna gain and indicates the importance of the term. ▶ Because the gain value combines in a single quantity two important parameters of antenna performance (the efficiency of the antenna and its ability to focus transmitted power), the gain of an antenna provides a useful description of the antenna's overall performance capability.

Expanded definition 2 defines an integrated circuit. The definition was written to familiarize electrical technicians in a small engineering firm with integrated circuit technology and appeared in a company newsletter.

Although the engineers were all college-educated, they had been out of school a number of years, working with practical applications rather than theory. Therefore, the writer's primary purpose was to inform these technicians about new developments in the field so that they might keep up with current technology. His secondary purpose was to persuade them of the benefits of these developments.

DEFINITION 2
MODERN INTEGRATED CIRCUITS

Integrated circuits are electronic semiconductor devices designed to perform specific circuit functions. Each integrated circuit contains many transistors, diodes, resistors, and other electronic components in one small package, or chip, as it is commonly called. The semiconductor junctions formed in an integrated circuit are similar to ordinary discrete (individual) transistors in theory and operation.

Size

Originally introduced to reduce the size of electronics products, integrated circuits are presently available in chips smaller than a pencil eraser. An example of this recent reduction in size is Regnetics RN 6488 integrated circuit, shown in Figure 1.

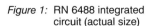

Figure 1: RN 6488 integrated circuit (actual size)

The RN 6488 contains the complete circuit for an AM/FM clock radio on a single ⅜" × ¾" plastic chip. A comparable clock-radio circuit using discrete devices would contain about 50 individual components and require a circuit area about the size of a paperback book to accomplish the same function as a single RN 6488 chip.

Manufacture

Integrated circuits are mainly composed of silicon and germanium. Integrated circuit fabrication begins with a substrate, a pure slab of the semiconductor, silicon. The substrate is usually the electrical ground of the circuit. A diagram of the circuit is etched on the substrate by means of a photonegative process similar to that used for developing photographic slides. Fine gold wires thinner than a human hair then connect the substrate to the external pins. Finally, the entire device is encased in a plastic case to protect the circuit from atmospheric variations.

Current Drain

Integrated circuits usually operate at power-supply voltages of 5–15 volts DC. The small size of integrated circuits and their low operating voltage allow operation at low current levels of about 1 mA per gate, compared to about 15 mA for discrete transistor gates. Lower current drain results in lower operating costs and greater product reliability, because of reduced heat dissipation in the circuit.

Exercises   **135**

Cost

An integrated circuit can usually accomplish the same function as a discrete
component circuit at a lower cost. The RN 6488 chip can be purchased for
about $2, while a comparable circuit using discrete electronics might easily cost
over $10. Integrated circuits also save on labor costs, since only a few chips are
used in a circuit previously requiring many manufacturing steps to assemble.

EXERCISES

Topics for Discussion

1. Discuss the basic property or distinguishing characteristic of each of the following
   terms. Consider physical makeup or composition, appearance, function or use, or
   location.

an asset	blood pressure	gravity
salt	a beaker	DNA
a personal check	a conduit	a resume
a cloud	horticulture	a tsetse fly
a knife	a catalyst	a star
electric current	cybernetics	a barometer
an iceberg	a bid	a ruler
a leaf	an erasure	hydroponics
friction	Fortran	a calorie
a pitchfork	a lathe	a backhoe

2. The following one-sentence definitions are ineffective. Discuss the reasons and the
   specific revisions you would make.
   A. A fork is a utensil used to pierce food.
   B. A motorcycle is a motorized vehicle.
   C. Water is a substance that flows.
   D. Baseball is America's favorite pastime.
   E. A potato is a tuber.
   F. A full adder is an adder that adds two bits plus a carry-in from the previous bit
      position.
   G. A two-bitted ax is what firefighters use to combat fires.
   H. Chocolate cake is everyone's favorite.
   I. An avalanche is caused by thawing conditions.
   J. Real property is real estate.
   K. Diffusion occurs when atoms migrate.
   L. Dry wall is rigid material fastened to studs to form a wall or ceiling.

Topics for Further Practice

1. Determine an appropriate genus and the basic characteristic for five terms used in
   your field. Then write a one-sentence definition of each term for a specific experienced audience you designate.

2. Select a specific lay audience (e.g., high school biology and chemistry students with an interest in the medical field as a profession). Write a one-sentence definition of each term in Question 1 for this audience. Then specify the differences between your two definitions and the reasons for those differences.

3. Write each of your definitions above informally, inserting it into a sentence by using parentheses or commas.

4. Choose one term from Question 1 for expansion.
    A. Analyze a possible experienced audience and the purposes and uses of your definition. Then select your content and arrange your definition. Label the details you have chosen from the list of possibilities (page 124), and give the reasons for your choices as well as for your arrangement of those details.
    B. Write your expanded definition.

5. Choose a lay audience for your definition in Question 4.
    A. Analyze this audience and the purposes and uses of your definition. Select your content and arrange your definition, giving the reasons for your decisions.
    B. Write the second expanded definition.
    C. Discuss specific differences between the two definitions you have written, giving reasons for the differences.

8

Descriptions
of Items

THE NATURE OF ITEM DESCRIPTION

Like a definition, the description of an item usually forms part of a longer report. For example, if you were writing a feasibility study on transfer lines (conveyor belts used in machining parts) for management in your company, you might have to describe the appearance of the lines. If you were writing instructions for beginning forestry workers on planting a tree by the bar-slit method, you might have to describe the planting bar. As you can see, composing effective description is a skill you will frequently use.

If you were asked to describe an object you would probably begin by asking several questions:

1. *What* is the item?
   What does it look like?
   What are its parts?

2. *Who* uses it?

3. *Where* is it used?

4. *When* is it used?

5. *Why* is it used?

6. *How* is it used?

The italicized words indicate that this list of questions is an application of the reporter's formula, an information-gathering technique particularly suited to item description.

137

Notice that the first question — *What* is the item? — involves definition: The name of the item (species) = the class to which the item belongs (genus) + the distinguishing characteristics of the item (differentia). For example, the answer to the question "What is a transfer line?" might be "A transfer line is a conveyor belt used to carry parts from one station to another when they are being machined." The skills you will learn in this chapter, then, depend on the skills you acquired in Chapter 7, Definitions. In fact, the entire description of an item is one type of expanded definition: First you classify your item, then further describe the item's distinguishing physical traits.

Technical description is also similar to definition in a second way. Unlike other forms of description you may have written, technical description usually focuses on the item in general rather than on an individual sample. For instance, suppose you are describing a safety walker for the aged to a group of nursing students. The particular walker you are observing has red plastic handgrips, but on other walkers these handgrips may be made of rubber and may be black. You would stress the common or generic element, handgrips. The color and material of the handgrips in this particular case would be incidental to the description, although you might decide to mention the varieties of colors and materials available as two of your details.

You would usually maintain this focus on generic elements even if you were describing a more specific subcategory within a larger category of items, for example, in specifications for a particular kind of machine or a particular manufacturer's model. Thus, in representing a plain external grinder or grinder model A24 (as distinguished from metal grinders in general), you would still concentrate on the features *all* plain external grinders or A24 grinders have in common and ignore characteristics of the one you happen to have (e.g., the headstock that broke and had to be welded). An exception would occur, however, if you had to describe a specific item — for example, the particular fungus you observed on a farmer's corn.

The remaining questions on our list involve two kinds of information: appearance and function. Although you may include other facts in your description — the history of the item or a comparison to other items — details of appearance and function will be very important facts, because of the purpose of description: to inform by producing a mental picture of the physical item. This purpose distinguishes description from other kinds of expanded definitions and means that the way the item looks and the way it moves or is used will be important details for your audience.

The kind of details you choose, however, will differ with audience uses of your description. Consider the possible examples with which we began: descriptions of a transfer line and a planting bar. Those reading the description of the transfer line — the management of a company — would not be operating a transfer line. They would require general information on the line's appearance and function, not facts related to maintenance and use. For instance, the author might omit the position of on/off switches from the description, although that detail would be necessary for the transfer-line operator. Those reading the description of the planting bar, however, would be using and maintaining the apparatus involved. To satisfy their information needs, the author would write an operational description, focusing on details of appearance as they pertain to upkeep or use. For instance, he or she might include the finish on the planting bar, because the finish must be properly renewed to avoid rust.

In our illustration, we follow Mary Foster through the process of composing an operational description of a hand auger. Mary, a fourth-year civil engineering student, is working for the summer at a dam construction site. She has been given the writing project we describe as a routine part of a young engineer's responsibilities. She wants to handle the project well, since she knows a top-quality report may increase the possibility of a job offer after graduation.

MARY'S DESCRIPTION OF A HAND AUGER

A group of freshman and sophomore engineering students work with Mary at the dam construction site, where one of their jobs is to test the soil compaction of the dam wall, using a hand auger. These students have had some field experience in construction but no classes specifically in soil testing as Mary has had. Therefore, her supervisor, James Clark, has asked her to write an operating manual on testing procedures. Her operational description of a hand auger, the testing apparatus used, will form part of that manual.

Prewriting

Analyzing the Communication Context

Analyzing the Audience

Mary has three sets of readers. The technicians working with her at the dam site and any future technicians are the primary audience of her description, since they will directly use it. James Clark, Mary's supervisor, is one secondary reader; Barry Terrance, in charge of personnel, is another. Of these secondary readers, James Clark has had extensive experience with the hand auger, but Barry Terrance has had none.

Analyzing Purpose and Use

Mary's primary purpose with the technicians is to inform them of the appearance and use of the hand auger, so that they have the background necessary for operating it. Her secondary purpose is to instruct them in details of maintenance and of assembly and disassembly, so that they can perform these actions effectively.

Mary's primary purpose with James Clark is to persuade him that she is able to handle writing assignments independently. The quality of her information should allow him to make this decision.

Mary's primary purpose with Barry Terrance is to inform him about details important to a technician's work, so that he can communicate these details to people he interviews.

140   8 | Description of Items

FIGURE 8.1
Mary's application of the reporter's formula

1. What is the hand auger?
   Soil-sampling tool used for taking disturbed (not *in situ*) samples
   What does it look like?
   Resembles a shovel with a curved head. Has a wooden handle. Rod is a hollow tube.
   Cutting head is metal. Both have finishes that must be renewed
   What are its parts?
   Handle, rod, cutting head

2. Who uses it?
   Two technicians. One can use it but not as effectively

3. Where is it used?
   Construction sites. For my audience, I'll only emphasize use at the dam

4. What is it used for?
   Tests soil compaction to 25 ft.

5. How is it used?
   Handle acts as lever. Rod connects head to handle

Gathering Information

The application of the reporter's formula we gave on page 137 provides a quick but systematic way to survey your item. Figure 8.1 shows Mary's application of this formula, with the answers she provided.

A second information-gathering technique for description is to draw a diagram of the item and label its parts. This technique, particularly useful in answering the question ''What are the item's parts?'' ensures that you will not forget any parts and helps you discover their interrelationship. We show Mary's information-gathering diagram in Figure 8.2.

Notice that Mary's labeling of parts has forced her to identify them all, whereas she had included only three parts in her reporter's formula. Observing an item, drawing a diagram, and labeling the parts may show you that the item is more complex than you had at first assumed.

A third information-gathering technique is to list the parts you have identified, then organize your list. To do so, you must consider the basis you have used to partition, or divide, your item. Basis is the principle or rationale used for partitioning an item. Four possible bases exist: spatial arrangement, function, assembly, and importance. Knowing which basis to use is important, because this knowledge will help you group your parts and will suggest an order for your description. The following list of bases illustrates this concept.

1. *Spatial arrangement — the location of parts in space.* Spatial arrangement is most useful for sequential items. For example, you might partition a pencil into the lead, the shaft, and the eraser. If you choose spatial arrangement as your basis, you must then select an appropriate sequence, e.g., left to right, right to left,

FIGURE 8.2
Mary's information-gathering diagram

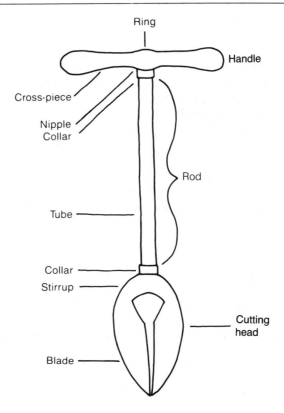

top to bottom, bottom to top, outer to inner, and so on. You should also be careful to maintain this sequence in your description. For example, if you partition the pencil from bottom to top, you should not skip from the lead to the eraser, then proceed to the shaft.

2. *Function — the use of each part.* Function or use is most helpful for nonsequential items. For example, you might partition a pipe tool into the tamper, the spoon, and the pick, because each part has a different function, and describe the parts in order of use.

3. *Assembly — the order in which parts are put together.* Assembly is most useful when your audience will be putting the item together. For example, you might partition a model airplane into the body, the wings, the propellers, and the tail because a person would assemble these parts to make the model. You could then describe the parts in the order of assembly.

4. *Importance — a basis devised by the writer when the above bases do not apply.* At times, you might have to evolve a basis. For example, you might partition a check according to the contents of the parts a person fills out: date, recipient,

amount, and signature. You could then describe the parts in the order in which a person would fill them out.

You may find that several bases apply to your item. For example, a pencil might also be partitioned on the basis of function, since the lead, the shaft, and the eraser are used in different ways. You must then select the basis that best suits your purpose or the basis that yields the most effective order. Because function would involve skipping from the shaft as the part grasped, to the lead as the part that writes, and then to the eraser as the part that deletes writing, spatial arrangement would be a more logical basis for the pencil.

Mary has found that several bases apply to the hand auger: Spatial arrangement (if she proceeds from the top to the bottom) and assembly give the same parts in the same order, since the auger is assembled from top to bottom. Function, however, might group the handle as the receiver of power, the rods as the transmitters of power, the cutting head as the point where power is applied, and the collars as connectors. Mary decides assembly is her most logical basis because the technicians will assemble and disassemble the auger. However, if she were describing the auger to readers interested in the theory behind its construction, function would be a more logical choice.

Once you have partitioned your item according to a basis, you must list your parts, then classify them into major parts and subparts. Major parts are the most comprehensive or largest divisions of an item; subparts are segments of major parts. In addition, complex items can have sub-subparts as well. You determine the divisions of your item by function: Each major part has a different use. If the use is the same (e.g., the two lenses in a hand magnifying glass), the parts can be grouped as one division (e.g., lenses). Figure 8.3 shows Mary's list of parts and classification. (We have indicated the use of each major part.)

FIGURE 8.3
Mary's list of parts and classification

handle	I. Handle ◄—————————	Used as a gripping area and to give leverage
cross-piece	A. Cross-piece	
ring	B. Ring	
nipple	C. Nipple	
collar	II. Rod ◄—————————	Used to transmit power
rod	A. Collar	
tube	B. Tube	
collar	III. Cutting head ◄—————	Used to bore into the soil
stirrup	A. Collar	
cutting head	B. Stirrup	
blades	C. Blades	

You should construct your classification in the order suggested by your basis: Since the auger is assembled from top to bottom, Mary begins with the handle and proceeds down the auger to the cutting head.

A fourth information-gathering technique is a checklist of possible details to include in an item description. We give this checklist in Figure 8.4, with annotations on the purpose of the details, then discuss the details further.

FIGURE 8.4
Checklist of possible details to include in an item description

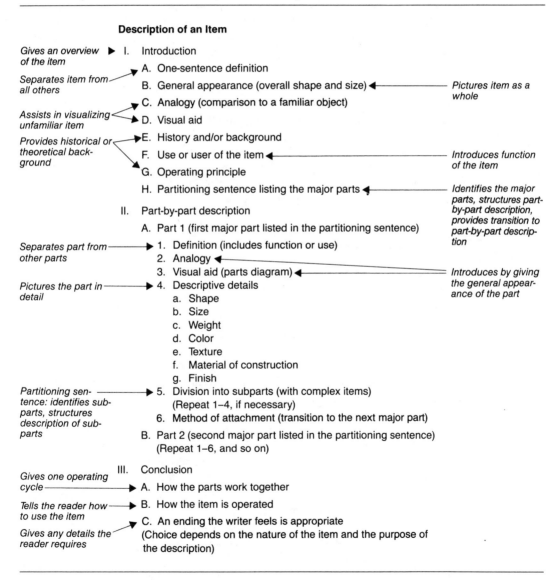

Your audience will frequently require an orientation to or overview of your item before reading a description of specific parts. Your introduction gives this overview. A definition separates your item from all others. The remaining entries on our checklist of possible details provide a general idea of the appearance of the item (overall shape and size, an analogy, a visual aid) or give background on the item's use. The last sentence in your introduction, the partitioning sentence, identifies the major parts for the reader who is unfamiliar with them, structures the part-by-part description, and forms a transition to it.

You should begin your description of the first major part with a definition when your reader is unfamiliar with the part. This definition, which identifies the part, often includes function or use as the detail that distinguishes parts. If necessary, you can then introduce the part by giving its general appearance or by using an analogy or a visual aid. This visual aid, called a parts diagram, pictures the subparts of the part and indicates their interrelationship.

The specific description begins when you have reached the part's smallest divisions. Simple parts can be described in detail after they are introduced; complex parts must be subdivided before the detailed description begins. You then end this detailed description with method of attachment, as a transition to the next major part.

Your conclusion should be suited to your purpose in describing the item. If your audience will be using the item, how it is operated is an appropriate ending. If your audience will not be using the item, how the parts work together can inform them about it or you can devise an appropriate conclusion.

Selecting and Arranging Content

Selection

You will not use all the information we have presented in every description, because your audiences will not require every detail. Therefore, listing the details to include in your particular description will help you select your content. We give Mary's list in Figure 8.5 and discuss her reasons for including or omitting information.

Since Mary's primary audience has never had a class in soil testing or much field experience, she cannot assume that they know what a hand auger is. Therefore, she will begin by defining it. The definition will also give the use and the user of the auger. She will then include a visual aid to assist readers in picturing unfamiliar equipment. This visual aid will give a clearer picture of the auger in less space than a narrative would do. She will omit an analogy (e.g., comparing the auger to a drill) because an analogy would be too general to fulfill her readers' need for precise description and because she wants to avoid the condescending tone she feels such information would impart. She will omit details on history or background and operating principle because general or theoretical information of this type is not necessary for using the auger. She will then conclude her introduction with a partitioning sentence.

In the part-by-part description, Mary will define the parts so the reader can distinguish them by function. She will give a general idea of appearance, then include parts diagrams for the complex parts to identify subparts and their interrelationship,

Mary's Description of a Hand Auger **145**

FIGURE 8.5
Mary's selection of details

I. Introduction
   A. One-sentence definition
   B. Visual aid
   C. Partitioning sentence

II. Part-by-part description
   A. Part 1
      1. Definition (function)
      2. General appearance (shape)
      3. Visual aid (parts diagram)
      4. Division into subparts
         a. Definition
         b. Shape
         c. Material of construction
         d. Finish
      5. Method of attachment
   B. Part 2
      1. Definition (function)
      2. Appearance
         a. Size
         b. Shape
         c. Material of construction
         d. Finish
      3. Method of attachment
   C. Part 3
      1. Definition (function)
      2. Visual aid (parts diagram)
      3. Division into subparts
         a. Definition
         b. Material of construction
         c. Shape
         d. Finish

III. Conclusion
   How the item is operated

but omit the diagram for the simpler parts. She will then subdivide the major parts before giving her detailed description.

After subdividing, Mary will define each subpart to separate it from others and will give its shape to provide an idea of its general appearance. She will also include information about construction and finish, since these details are important for proper maintenance. She will omit color, texture, and weight because these details are not important to the user. Method of attachment will then form her transition to the next major part.

Since Mary's readers will be operating the auger, she will conclude with details on assembly and disassembly — necessary facts for using the auger.

8 | Description of Items

Arrangement

Your list of details to include in your item description will help you with selecting and arranging material: Simply insert specific details about your item and arrange those details in a logical way. Figure 8.6 shows Mary's more specific list for her part-by-part description.

FIGURE 8.6
Mary's more specific list for the part-by-part description

I. Handle
    A. Horizontal bar acting as lever
    B. Subdivide (show parts diagram)
        1. Cross-piece
            a. Horizontal portion
            b. Wood
            c. Varnished
        2. Ring
            a. Metal
            b. Anticorrosive paint (mention renewal)
        3. Nipple
            a. Metal
            b. Screws into rod

II. Rod
    A. 3 ft. long
    B. Subdivide
        1. Collar
            a. Standard
            b. Screws into handle or rod
        2. Tube
            a. Hollow
            b. Steel
            c. Anticorrosive paint (mention renewal)
            d. Screws into cutting head

III. Cutting head
    A. Boring device
    B. Subdivide (show parts diagram)
        1. Collar
            a. Standard
            b. Welded to stirrup
        2. Stirrup
            a. Horseshoe shape
            b. T-section steel
            c. Sub-subpart (prongs)
            d. Anticorrosive paint (mention renewal)
        3. Blades
            a. Tempered steel
            b. Beveled
            c. Bare metal

When listing details in your part-by-part description, you should be careful to give the same details in the same order for each part as much as possible, because you set up expectations in readers' minds on the details you will include and the sequence of your information. For example, Mary usually mentions material of construction, then finish. However, notice that she includes details for some parts but omits these details for others (e.g., the analogy for the stirrup) when the details do not apply. She also varies her order when another seems preferable (e.g., material of construction, then shape for the blades).

This more specific list becomes an ordering device for your first draft.

Writing

Mary's first draft follows, with annotations on her writing decisions.

One-sentence definition ▶ The hand auger is a soil-sampling tool with an oval cutting ◀ *Appearance*
head, normally operated by two technicians and used for ◀ *Function*
taking disturbed samples to depths of 25 ft. A good example is the 6-in. auger manufactured by Soiltest Inc.

Partitioning sentence; Mary will now discuss these parts in the order listed ⟶ The hand auger consists of a handle, a rod (or rods), and a cutting head, as shown in Figure 1.

Visual aid. Only the major parts are labeled ⟶

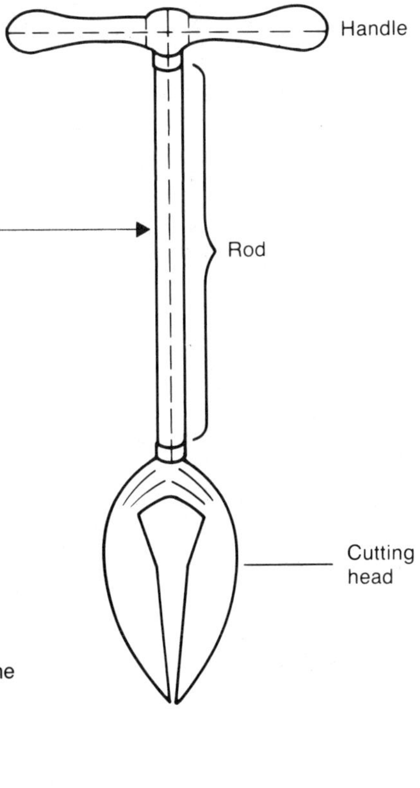

Handle

Rod

Cutting head

Figure 1: Parts of the hand auger

148 8 | Description of Items

One-sentence defini- ⟶ The handle is a horizontal bar that acts as a lever to
tion of major part 1 increase rotational force when grasped at each end. This
Partitioning sentence ▶ handle has three subdivisions: a cross-piece, a ring, and a
subdivides part 1 nipple. The handle's subparts are seen in Figure 2.

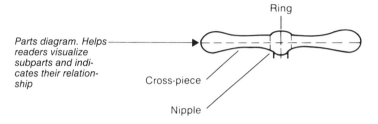

Parts diagram. Helps readers visualize subparts and indicates their relationship

Figure 2: Subparts of the handle

    The cross-piece is the horizontal portion of the handle. — *Example definition (shape) of subpart 1*
Material of construc- ▶ This piece is made of wood, sanded smooth and varnished — *Subpart 2: size,*
tion, finish to prevent damage. The broad metal ring around the mid- *material of construction, shape*
dle of the cross-piece is coated with anticorrosive paint for
the same reason. This ring is welded to a metal nipple, — *Finish*
Shape, method of ⟶ threaded to screw into the top of the rod. — *Subpart 3: material of construction*
attachment
    The rod is a 3-ft.-long hollow steel tube that connects
One-sentence defini- the handle and the head. A standard plumbing collar is fit- — *Subpart 1: This term is not defined.*
tion of major part 2: ted and welded to one end of the rod. Either the nipple of *Although Barry Terr-*
size, appearance, the handle or the end of another rod screws into this collar. *ance may not know*
material of construc- The entire rod is coated with anticorrosive paint, which *it, Mary's primary*
tion, function should be reapplied when worn. The other end of the rod *audience will, and*
Subpart 2: detail screws into the collar of the cutting head. *she wants to avoid*
emphasizes informa- *talking down to them*
tion important in
assembling and   The cutting head is the part of the auger which, when — *Finish: important for proper maintenance*
adjusting the auger rotated, digs into the soil. This head consists of a collar, — *Method of attachment*
a stirrup, and two scoop-shaped blades, as shown in — *One-sentence defini-*
Analogy pictures the Figure 3. *tion of major part 3*
shape of the blades — *Partitioning sentence: used because the major part and subparts are more complex than those for major part 2*

Collar
Stirrup

Parts diagram. Helps readers visualize subparts and indicates their relationship

Blade

Figure 3: Subparts of the cutting head

Mary's Description of a Hand Auger   **149**

Subpart 1 ——————▶ The collar is a standard plumbing collar. The threaded ◀ *Method of attach-*
ment
end of the rod screws into the collar, which is also welded
to the stirrup. The stirrup is a strip of T-section steel bent ◀ *Subpart 2: material of*
construction; second
into a horseshoe shape so the flange of the T is on the out- *analogy, for visual*
Sub-subparts ——————— side. The stirrup has two prongs, to which the blades are *effect*
Method of attach- ——▶ welded. Both the collar and the stirrup are coated with
ment
anticorrosive paint. The blades are made of tempered steel ◀ *Subpart 3: material of*
Finish: gives a detail sheet so they resist bending and breaking in dense or rocky *construction*
important for proper
maintenance soil. The leading edge of each blade is beveled to enhance ◀ *Shape: gives a detail*
important to avoid
cutting but not sharpened. The blade finish is bare metal. *maintenance error*
Finish: gives a detail The overall length of the auger can be adjusted by
important for proper
maintenance; no pro- increasing or decreasing the number of rods. If the auger
tection is necessary has been twisted into the soil to a depth which makes the
Detail important to ——▶ handle too low to be comfortable, simply remove the han- ◀ *Details on adding*
the user: when to
adjust auger dle, join an additional rod to the first, and replace the han- *rods, important to the*
user
dle. The auger can also be disassembled for ease of handling
or storage. Only two large plumber's wrenches are needed ◀ *Detail important to*
the user: materials
to disassemble or assemble the auger or to add the rods. *needed to adjust the*
auger

Rewriting

Your most important criteria when revising the description of an item are amount
and kind of detail and logical progression. If your audience is to use the description
effectively, you must include all the details they require, but no more. For example, if
Mary had not described the auger's finish, her readers would not have known what
material should be reapplied for proper maintenance. On the other hand, if she had
described the shape of the blades in great detail, her readers might have become confused by facts they could not use. (In our samples, we include a description of the
hand auger to manufacturers, who *could* use these details.) You must also arrange
these facts in a logical way.

When Mary rereads her first draft for amount and kind of detail, she decides she
has omitted some essential details. In the description of the cross-piece, she has not
stressed the fact that the finishes on the cross-piece and ring must be renewed. Therefore, her audience might not realize the importance of the mention of finishes. She
adds the italicized sentence:

> The broad metal ring around the middle of the cross-piece is coated with anti-
> corrosive paint for the same reason. *These protective coatings must be reapplied*
> *when they become worn.* This ring is welded to a metal nipple. . .

In the description of the rod, she has omitted a detail on appearance: The tube is
threaded at both ends to allow connection to the handle and the head. Since her read-

ers must know this fact to assemble and disassemble the auger, she adds the detail:

> The rod is a 3-ft.-long hollow steel tube, threaded at both ends, that connects the handle and the head.

In the description of the blades, on the other hand, she feels she has included a detail her readers do not require: the reason why the blades are made of tempered sheet steel. Since the technicians will only be maintaining the auger and not constructing it, they do not need this explanation. Mary omits it from her draft.

Logical progression — effective partitioning and classification and an orderly sequence of information — is essential to assist readers unfamiliar with the item's appearance. Otherwise they may become confused by seemingly unrelated details. Therefore, you should carefully review your basis, the resulting classification, and the sequence of the details in your part-by-part description.

Mary feels rearrangement is necessary. In the section on the cutting head, she decides to put the detail on finish of the collar and stirrup before the sub-subdivision of the stirrup, in order to form a better transition to the next subpart, the blades. We give Mary's rearrangement below:

> . . . on the outside. The stirrup has two prongs, to which the blades are welded. [Both the collar and the stirrups are coated with anticorrosive paint.] The blades are made of . . .

CHECKLIST FOR DESCRIPTIONS OF ITEMS

1. Have I gathered appropriate information for my description?

    A. Have I used the reporter's formula to quickly survey my item?

    B. Have I diagrammed my item to discover its parts?

    C. Have I partitioned my item according to the most effective basis and listed the parts? Have I classified them into major parts and subparts?

    D. Have I selected details applicable to my item and audience from the list of possible details to include in a description (page 143)?

2. Have I constructed an ordering device for my description by inserting specific details about my item in the list I selected?

3. Have I arranged those details in a logical order?

4. Does my introduction provide my reader with an overview of my item? Have I used such techniques as a definition, an analogy, or a visual aid effectively?

5. Have I identified the major parts of my item, forecasted the order of my description, and provided a transition to my part-by-part description with a partitioning sentence?

6. Does my part-by-part description develop logically?

    A. Have I taken up each major part in the order given in my partitioning sentence?

Sample Descriptions of Items   **151**

    B. Have I introduced each major part or subpart to my reader?

    C. Have I mentioned the same descriptive details in the same order as much as possible for each major part and subpart?

    D. Have I used parts diagrams to help the reader picture complex parts?

    E. Have I used method of attachment as a transition to the next part?

7. Is my conclusion appropriate for my description?

    A. How the parts work together

    B. How the item is operated

    C. A conclusion I devised

SAMPLE DESCRIPTIONS OF ITEMS

Our first example is a description of a hen's egg. The description was written for inclusion in a manual to be used in a food preparation and nutrition class. This class introduced students to the structural components and uses of common food ingredients. Thus, the description is informative and operational, since the audience would be expected to understand the nutritional value of the egg and use it effectively in various food products.

We annotate the description, giving the details included and the reasons for inclusion, based on audience needs.

DESCRIPTION 1
DESCRIPTION OF A HEN'S EGG

One-sentence definition gives the scientific meaning of the egg, important to those studying it as a subject rather than simply using it ▶ A hen's egg is the oval, white or brown reproductive structure of the common chicken. This egg can be cooked and ◀ *Background information important to users of the egg who must understand the theory behind its use* eaten in a variety of ways and performs a number of functions in the food products in which it is used. Eggs act as an emulsifier in salad dressings and a gelling agent in custards and provide a structural material in shortened cakes. When whipped to a foam, eggs are a means of incorporating air into meringues, angel food cakes, and soufflés.

Background important to understanding the egg's nutritional value ▶ As a food, eggs contribute significant amounts of protein, vitamin A, and phosphorus and are a major part of most Americans' diets. Eggs are composed of three main ◀ *—The egg is partitioned on the basis of spatial arrangement: outer to inner* parts: the shell, the white, and the yolk, as shown in Figure 1 on the next page.

The writer adds a detail on composition to the one-sentence definition. Again, the detail is important to those considering the egg scientifically ▶ **The Shell.** The shell of a hen's egg, made up mostly of calcium carbonate crystals in a carbon network, surrounds and supports the parts of the egg used for food. Some shells are white and some are brown. This coloring *Detail on appearance (color) gives readers background for judging quality of eggs*

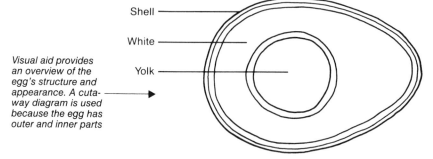

Visual aid provides an overview of the egg's structure and appearance. A cutaway diagram is used because the egg has outer and inner parts

Figure 1: Cross section of a hen's egg

depends on the breed of hen and has nothing to do with the quality of the egg.

Mentioning the shell's pores, as subparts, is important in a scientific understanding of the egg

The shell of the egg has thousands of pores too small to be seen and a few just large enough to be visible without a microscope. These pores allow the chicken fetus to breathe.

Method of attachment forms the transition to the next major part

Just inside the shell are two membranes made mostly of keratin (the main protein in hair) and mucin (another type of protein), which allow exchange of gases and protect the developing fetus from bacteria that could enter through the pores. The outer membrane clings very tightly to the shell, while the inner one surrounds the white.

— Division into subparts provides the audience with a theoretical understanding of the membranes they see when breaking an egg

A one-sentence definition includes information on composition, important in a scientific understanding of the egg

The White. The egg white, composed mainly of water (88%) and protein (11%), provides the developing chicken fetus with some nutrients. The remaining 1% is fat and minerals. About two thirds of the protein is known as ovalbumin, which coagulates when heated and gives the white its characteristic color and gum-like texture.

— Detail relates theory to the familiar appearance of the white when cooked

Theoretical background gives reasons for incorporating whipped eggs into products, important to a technical understanding of this operation

If egg whites are whipped into a foam, they make many tiny air bubbles surrounded by films of water and protein. When heated, the water turns to steam and expands the air bubbles until the protein film gets hot enough to coagulate, holding the whole foam in its inflated shape. This inflation is what happens when a soufflé or an angel food cake is baked.

The egg white is partitioned into subparts by spatial arrangement: outer to inner

The egg white is made up of three layers: the outermost thin white, the thick white, and the inner thin white. The ratio of thick white to thin depends on the hen and the storage of the egg. Grading of eggs results from the amount and firmness of the thick white: the more and firmer the thick white, the higher the grade.

— Detail on appearance gives readers theoretical background for distinguishing eggs

Sample Descriptions of Items

Partitioning into sub-subparts gives important structural information about the egg

On the innermost surface of the inner white is the vitelline membrane, or yolk sac, which is connected to two chalazae, twisted rope-like structures that extend to the outer membrane of the outer white. Figure 2 illustrates these parts.

Parts diagram provides an illustration of material difficult to present verbally

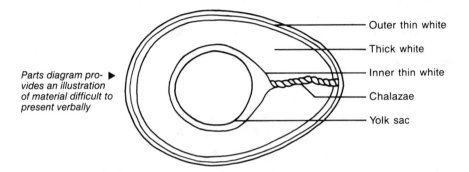

Figure 2: Parts of the white

The yolk sac and chalazae anchor the yolk and hold it in the center of the egg.

◀ *Detail on function also forms a transition to the next major part*

One-sentence definition contains nutritional information ▶ **The Yolk.** The egg yolk is a small, yellow sphere, composed roughly of half water and half solids, that provides most of the nutrients in the egg. Of the solids, about a third are proteins and two thirds are fats. The yolk also contains important minerals such as iron, thiamin, and riboflavin. The yellow color is a result of carotenoid pigments, good sources of vitamin A. However, the yolk also contains cholesterol, which many health professionals recommend reducing in the diet.

◀ *Details on composition give nutritional information, important to understanding the value of eggs in a diet*

Detail on composition gives a drawback of eggs in a diet

Emulsifiers are mentioned as theoretical background for understanding the egg's use in foods ▶ The fat, water, and protein are held together by emulsifiers, such as lecithin. An emulsifier is a substance that surrounds fat droplets and makes them soluble in water, thus making the yolk useful in stabilizing salad dressings such as mayonnaise.

The writer reviews his main points in his conclusion, to stress the importance of knowing about eggs

The parts of the egg, the white and the yolk, with their different nutrients and physical properties, have specific applications in preparing foods. Knowledge of these differences can help one make better and more nutritious food products.

Our second description is again of the hand auger. However, this description was written for a different audience from Mary's: the production section and management of Soiltest, Inc. The production section had had extensive experience with operating augers as well as with the materials involved and construction methods. Management was less familiar with operating augers but also knew materials and construction.

The hand auger described was a newly designed model being readied for production. The production section would use this description to construct a prototype, while management would use the description to understand the resulting product. Thus, the description is explanatory, not operational. (We have omitted the appendix in our example.)

DESCRIPTION 2
THE HAND AUGER

The hand auger is a soil-sampling tool used for taking disturbed samples to depths of 25 ft. In addition to sampling, the auger can also be used to advance a hole quickly to a desired depth, after which a split spoon or Shelby tube can be used to take an undisturbed sample. The auger is best operated by two technicians, but it can be used by one technician if the depth of the hole is less than 6 ft. The auger consists of a handle, one or several rods, and a cutting head.

The handle has a wooden cross-piece with a metal connector around the middle. These parts, and the remaining parts of the auger, can be seen in Figure 1, reproduced as an oversized diagram in the Appendix. The cross-piece (generally tubular in shape, contoured for hand comfort and grip) is 18 in. long, 1½ in. in diameter at its center, and 2 in. at each end. The cross-piece is made of hickory, shaped on a lathe, sanded smooth, and varnished.

The connector is a broad ring with a threaded nipple. The ring is made of ⅛-in. thick mild steel and fits tightly over the middle of the cross-piece. Externally, the ring is coated with anticorrosive paint. The nipple is 1 in. long, ¾ in. in outside diameter, and made of solid steel. This nipple, also coated with anticorrosive paint, is welded to the ring and threaded with standard pipe threads, to screw into the collar of the rod.

The rod consists of a metal tube with a collar at one end. The tube is 3 ft. long, with an outside diameter of ¾ in. and an inside diameter of ½ in., and is threaded at both ends. At one end, a standard ¾-in. plumbing collar of mild steel is fitted and tack-welded, leaving enough room on the collar to accommodate either the nipple on the handle or another rod. The tube is extruded mild steel with a smooth finish. The entire rod is coated with anticorrosive paint. The threaded end of the last rod screws into a collar on the cutting head.

The cutting head consists of a collar, a stirrup, and the blades. The collar is a standard ¾-in. plumbing collar of mild steel and is welded to the stirrup. The stirrup, a 1-ft. length of T-section mild steel, is bent into a horseshoe shape with the flange on the outside and the web on the inside. The flange is 1 in. wide and ¼ in. thick, while the web is ½ in. thick. The metal is rolled smooth, and the stirrup and collar are both painted. The prongs of the stirrup overlap the tops of the cutting blades by 1½ in. and are welded to the blades.

The blades can be visualized as a huge thimble from which alternate quarters are cut away, leaving two scoop-shaped blades opposite each other. Figure 2 shows a diagram of these blades.

Figure 2: Diagram of the auger blades
(The shaded area represents one of the quarters that is cut away.)

The material is ⅛-in. thick, tempered steel sheet. The diameter at the larger (top) end is 6 in., which tapers to a 4-in. diameter at the bottom, where the shape changes from a truncated cone to a hemisphere. (See Figure 3 for the shape of these blades.)

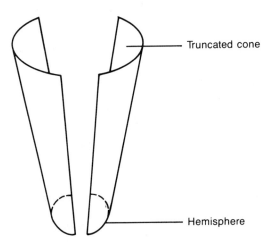

Figure 3: Diagram of the shape of the auger blades

The overall length of the blades is 1 ft. The lower end of each scoop is split from the tip upwards for a length of 6 in., and the leading half of each blade is bent upward until both cross. (See Figure 4 for a side view of the auger blade.)

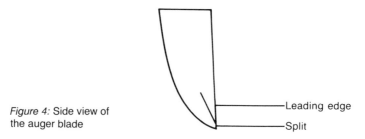

Figure 4: Side view of the auger blade

These leading edges will bite the soil first when the auger is rotated clockwise. They are beveled to cut the soil more easily but are not sharpened to a knife edge. The blades are not painted.

The hand auger can be disassembled into its major parts (handle, rod, cutting head), for ease of handling or storage, with the aid of two large plumber's wrenches. The length of the auger can also be adjusted during drilling by adding more rods.

APPENDIX

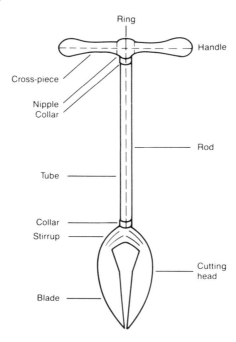

Figure 1: Parts of the hand auger

Exercises   **157**

EXERCISES

Topics for Discussion

1. Discuss the basis you would use to partition each of the following items. Are several bases possible? If so, do they result in the same list of major parts and the same order?

scissors	glove
ballpoint pen	door
watch	umbrella
pocket knife	book
shovel	paper clip
bucket	chair
tennis racket	key
newspaper	baseball diamond

2. List the parts for five of the items in Question 1. Then classify these parts into major parts and subparts. (Some items may have only major parts.)

Topics for Further Practice

1. Select an item from the list at the end of this question or an item from your field with which you are familiar. Then complete the following exercises:
   A. Use the reporter's formula to gather preliminary information about the item.
   B. Draw a diagram of the item and label its parts.
   C. Analyze a specific audience for a description of the item, your purposes in writing the description, and your audience's uses of it.
   D. Partition the item according to a basis and list the parts. Classify them into major parts and subparts.
   E. Plan your description.
      1) Using the list of possible details to include in the description of an item (page 143), select the details your audience requires. Arrange those details in a logical order.
      2) Insert the specific information about your item into your list to create an ordering device for your description.
   F. Write your description.

sieve set for soil gradation	Oswald viscometer
plumb bob	strain gauge
T square	balance sheet
lettering guide	air piston
C clamp	double-bitted axe
titration burette	insect's compound eye
pipette filler	fuel pin in a nuclear reactor
U-tube manometer	sanitary manhole
littan stethoscope	Sherman live-trap
blunt-nosed probe	integrated circuit package
nose twitch	single-effect evaporator

   (list continues on next page)

Pitot tube   picometer
Fortin barometer   spectrophotometer
Philadelphia rod   two-column paper
scalpel   slump cone
sling psychrometer   balloon catheter

2. Select a different audience from the one in Question 1 for a description of the same item. Give specific differences in content selected, arrangement, and style for a description you might write to this audience.

9

Descriptions
of Processes

THE NATURE OF PROCESS DESCRIPTION

Descriptions of processes are often found in technical documents. For example, a report recommending a plan for removing nitrogen from a plant's chemical wastes might include the procedure involved. A report on storage systems for by-products of nuclear reactors might include the process of radioactive decay as background for a discussion of various means of storage. Because process description is a common form of technical reporting, writing these descriptions is a skill you will often use.

If you compare these two examples, you will see that process descriptions differ, depending on reader use. Because readers of the first report would be instituting your plan, they would have to understand how to perform the process. Readers of the second report, however, would not be carrying out the process of radioactive decay, so they would merely need to understand how it occurs.

These different uses result in two types of process description. The first, for an audience that will perform the operation, is a set of instructions or an operating manual. The second, for readers who must understand but not duplicate the process, is a process narrative. These two kinds of process descriptions differ in content, arrangement (structure and format), and style.

However, for both descriptions, you should partition your process into steps, because you are detailing how a series of operations is done or occurs. In addition, the basis or rationale for division in both process descriptions is always chronology: The steps proceed in order of occurrence.

In the following sections we illustrate sets of instructions and process narratives. Our first illustration concerns Jim Gaunt, an analytical chemist and manager of the analytical services laboratory of the Engineering Research Institute at a major state

159

160   9 | Descriptions of Processes

university. Jim will be teaching a course next fall in methods of analyzing water and waste water.

The students enrolled in this course must learn to operate several complicated pieces of equipment in Jim's lab. Since these students will be performing laboratory tests on their own rather than under his direction, Jim has decided to provide sets of instructions for each piece of equipment they must use. The instructions will also be useful references for the five full-time employees at the laboratory and for advanced graduate students who may use the equipment in their research. In our example, we follow Jim through the writing of one set of instructions.

Our second example concerns Karen Christian, a senior in food technology at a major state university. During an advanced laboratory class, Karen had been making microscopic studies of bread structure using a microscopy technique described in her laboratory manual. As she became familiar with the process, she began to experiment and modify it to suit the needs of her work. When confronted with the task of writing an honors thesis for the University Honors Committee, she felt that her modified approach would be a worthy subject. In our example, we show Karen composing her thesis, a process narrative.

JIM'S INSTRUCTIONS FOR OPERATING AN ATOMIC ABSORPTION SPECTROPHOTOMETER

One of the pieces of equipment Jim's students must operate is an atomic absorption spectrophotometer. This instrument, which measures absorption by light of atoms in a flame, is used to detect metals in a solution.

Jim has titled his instructions ''Operating Procedure for the Perkin-Elmer Model 305B Atomic Absorption Spectrophotometer.'' The instructions are intended to describe the operation of a particular atomic absorption spectrophotometer — the Perkin-Elmer Model 305B — rather than atomic absorption spectrophotometers in general. Because these instruments are quite complex and differ from manufacturer to manufacturer and from model to model, operating procedures are written for each model.

Notice, therefore, that process descriptions may be of two types: a general description, where you consider a process as a whole (e.g., how to operate a spectrophotometer); or a particular description, where you consider a specific type of process within a larger group of processes' (e.g., how to operate the Perkin-Elmer Model 305B atomic absorption spectrophotometer). As you can see, Jim will be writing a particular process description.

Prewriting

Analyzing the Communication Context

Analyzing the Audience

Jim has one primary and three secondary audiences. His primary readers, the students in his class, are beginning graduate students with degrees in civil engineering, animal science, biological sciences, and chemistry. These students have a basic knowledge of laboratory techniques and are familiar with spectrophotometers in general, although they have not used the atomic absorption spectrophotometer.

His secondary audiences are the five employees in his lab, any new employees, and advanced graduate students who may need to operate the equipment in connection with their research. Although the current employees in the lab would have used the atomic absorption spectrophotometer, newly hired employees may or may not have had experience with it. Although they would have degrees in scientific disciplines and some chemistry training, their backgrounds might be in biology rather than in chemistry. Graduate students using the apparatus would be pursuing advanced degrees in sanitary engineering. They may have operated atomic absorption spectrophotometers before, but probably not this model.

Analyzing Purpose and Use

Your primary purpose in a set of instructions is always to instruct. Three of Jim's audiences must learn how to operate the atomic absorption spectrophotometer; the fourth might need a reference manual. This primary purpose depends on a secondary purpose: informing your readers about the steps in this procedure.

Although your readers will all employ sets of instructions to perform an action, their specific requirements and uses may differ. Jim's students and newly hired employees will use the operating procedure to gain the detailed instruction they require. His five employees, on the other hand, will use the procedure as a refresher, since they do not often operate the Perkin-Elmer, and as a reference for instructing newly hired employees. Advanced graduate students will use the procedure to learn specific details of the Perkin-Elmer Model 305B's operation.

Gathering Information

The reporter's formula can be used to gather information for a set of instructions:

1. *What* is the procedure?
   What are its steps?
2. *Who* performs it?
3. *Where* is it performed?
4. *When* is it performed?
5. *Why* is it performed?
6. *How* is it performed?

However, you will not necessarily apply every question to your procedure. For example, Jim finds that some questions do not generate useful information: who (students in my class); where (the analytical services laboratory); when (during class); why (to test water and waste-water samples). However, two other questions are relevant: *What* are the steps? *How* is the equipment operated? Answering these questions will provide him with much of the information he will need.

A second information-gathering technique for a set of instructions is to perform the procedure and list each step. This list of steps will answer the question What are the steps? in the reporter's formula and provide you with details for your step-by-step instructions. Be careful, however, to include *each* step in your list. Although *you* understand how to perform the procedure, your audience may not. Therefore, you must not omit details that nontechnical readers may require.

As an information-gathering technique, Jim does operate the spectrophotometer and list the steps he follows. This list will be further amplified and ordered when he composes his step-by-step description.

If your set of instructions concerns an apparatus, a diagram is a third information-gathering technique you may use. This diagram will help you remember the parts to mention in connection with specific steps.

If your process is complex, involving major steps with substeps, a flow chart of the process is a fourth information-gathering technique. Representing each step with

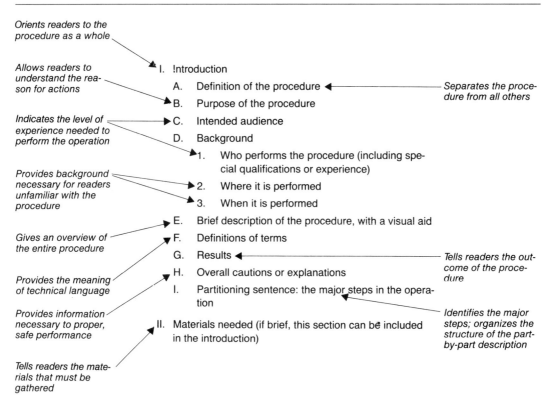

FIGURE 9.1
Checklist of possible details to include in a set of instructions

a symbol, often a block, and indicating the direction of the process with arrows drawn between each step will help you discover the relationship among steps.

A last information-gathering technique is a checklist of possible details to include in a set of instructions. We give this checklist in Figure 9.1, with annotations on the purpose of the details.

The introduction provides an overview of your procedure. A definition will identify the procedure for readers who are unfamiliar with it. The purpose of the procedure and background information, a brief description of the procedure, and a visual aid, if appropriate, will expand this definition. This visual aid is often a flow chart, e.g., a flow chart of the process for installing a washing machine, which gives the reader an immediate picture of the entire procedure. Results will then identify the outcome of the procedure.

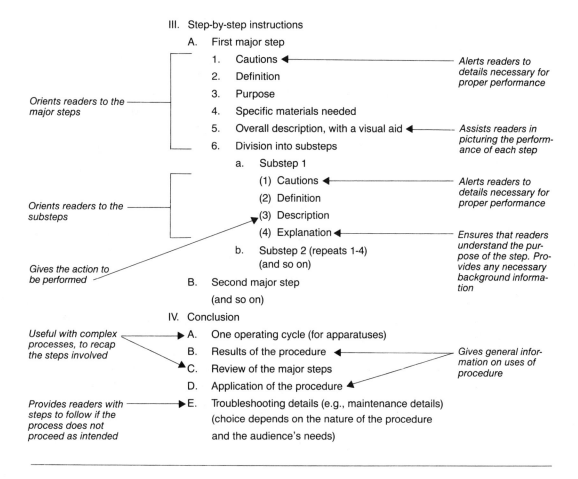

If you have words which readers must know before they can understand the instructions, you can place definitions for these terms in the introduction. Cautionary notes and explanations which pertain to the entire procedure can also be included in the introduction.

If your procedure requires few materials, you can include the materials needed in the introduction. (If your procedure requires many materials, these should be listed in a separate section after the introduction.) You then end the introduction with a partitioning sentence predicting the major steps in your procedure, as an organizing device for the reader and a transition to the step-by-step instructions.

The step-by-step instructions detail the actions the reader will perform. The conclusion then ends your set of instructions appropriately. If you are concluding with trouble-shooting details, you can present these details in a visual aid. For example, a chart of procedures to follow before calling a repair person for an automatic washing machine would organize this information for the reader.

Selecting and Arranging Content

Selection

The checklist of possible details to include in a set of instructions provides facts you might use: You then select the material your readers require. For example, Jim does not begin his procedure with a lengthy introduction because his readers are all familiar with experimental situations and spectrophotometers in general. Therefore, they do not require background. Moreover, they would find a general discussion of atomic absorption or of atomic absorption spectrophotometers interruptive rather than informative since their main concern is the Perkin-Elmer's operation. Jim also omits the ''materials-needed'' section, often in the form of a numbered list, because his readers will not be gathering any materials; all necessary equipment has been provided. (Our examples at the end of the chapter include a set of instructions with these details.)

To select your particular content, use the checklist of possible details to construct an outline of your information. This outline will assist you with arrangement.

Arrangement (Structure)

You may prefer to begin by arranging your step-by-step procedures, since they are the heart of a set of instructions. In order to arrange these steps effectively, you must first classify them into major steps and substeps. The major steps are the most comprehensive operations; the substeps are actions carrying out these operations. In addition, complex sets of instructions can have sub-substeps as well.

When deciding on the major steps in your procedure, you must be aware of two problems: providing too many major divisions or too few. Since the purpose of classifying steps is to organize the procedure for readers, too many or too few major steps will not effectively segment it. Therefore, your readers could become confused as they try to follow your instructions while performing the actions.

Jim's Instructions **165**

At first, Jim provided too many major steps. Figures 9.2a and 9.2b show Jim's information-gathering list of major steps and his restructured list after he had classified his major steps further.

FIGURE 9.2a
Jim's information-gathering list of major steps

Initial turn-on
Adjustments at turn-on
Lighting the flame
Adjusting the flame
Optimizing the signal output
Recording the signal output
Shutting down the spectrophotometer

FIGURE 9.2b
Jim's classified list of major steps

Turning on and adjusting the spectrophotometer
Lighting and adjusting the flame
Optimizing and recording the signal output
Shutting down the spectrophotometer

The clue to the need for further classification was the repeated words we have circled. These words indicated that some steps concerned similar topics and therefore should have been grouped.

You should also express the names of your major steps in parallel grammatical form. They will reappear as headings in your procedures and, since they are of equal level, must be identically worded. Notice that Jim has revised the names of his major steps to reflect this fact.

Once you have determined the major steps in your operation, list your substeps under these divisions. You should include only one or two actions in each substep, since lengthy substeps are difficult for readers to follow while performing the actions you describe. Then, if any of these substeps require division, list the sub-substeps as well. In Figure 9.3 on page 166, we present a portion of Jim's prewriting list for his first major step, illustrating these principles.

If the meaning and purpose of your major steps are not obvious to readers, you can begin them with an introduction. (See our checklist of possible details to include on page 163 for suggested facts.) However, Jim has not begun with introductory material because his readers do not require general orientation; the meaning and purpose of the steps are obvious.

FIGURE 9.3
Portion of Jim's prewriting list of major steps and substeps

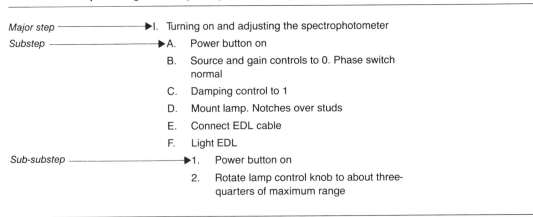

You should then plan your use of cautionary notes and explanations. Cautionary notes are very important because they alert readers to hazards or difficulties they may encounter. Therefore, these notes must be placed *before* the step or steps to which the notes refer. Otherwise readers may perform the actions incorrectly. Figures 9.4a and 9.4b, a portion of a set of instructions without and with a cautionary note, illustrate the importance of these notes.

Notice that the warning in Figure 9.4a is included in a step rather than being placed in a cautionary note before the step to which it refers. Therefore, readers are

FIGURE 9.4a
Portion of a set of instructions without a cautionary note

1. Set the operating dial to 4.
2. Turn on the power dial.
3. If the red light above the power dial flashes, turn the dial off.

FIGURE 9.4b
Portion of a set of instructions with a cautionary note

1. Set the operating dial to 4.
   CAUTION: IF THE RED LIGHT ABOVE THE POWER DIAL FLASHES, TURN THE DIAL TO OFF
2. Turn the power dial on.

not prepared to note the flashing red light and turn the dial off, as they would be with the warning in Figure 9.4b. Thus, well-planned cautionary notes may help avert errors and dangerous situations for readers performing an operation.

Explanations provide additional information readers need in order to understand a step, e.g., the purpose of a step or its result. Since readers are more likely to perform steps correctly if they understand the steps, explanations ensure proper performance of the procedure. In order to separate explanations from steps, explanations are placed *after* the steps to which they refer and are highlighted so reader attention is directed to them. (In Jim's draft, page 169, you will see illustrations of explanations.)

Lastly, you should plan your use of visual aids. Steps diagrams, which picture individual steps for readers, are frequently used in sets of instructions. These diagrams complement the text by conveying the same information visually and ensure that readers will understand actions hard to describe in words. (See our sample set of instructions, page 183, for an example of a steps diagram.)

In addition to arranging your step-by-step procedure, you must arrange the material for your introduction and conclusion. Your outline of the details your audience requires will assist you with this task. Jim has decided to include only a visual aid (a diagram of the Perkin-Elmer) and a partitioning sentence as introductory material, since his readers do not need further orientation. He uses the visual aid, because it will identify important parts of the Perkin-Elmer and serve to recall the apparatus to readers if they are reviewing the procedure away from the lab. The partitioning sentence will then predict the order of the instructions, thus assisting readers in following them. Jim has decided to conclude with troubleshooting details, because his readers must be aware of certain maintenance requirements if the Perkin-Elmer is to be operated correctly: keeping the air and gas tanks filled at all times to avoid a flashback explosion. Since these details are simple, he arranges them in a sentence list.

Arrangement (Format)

Format is important in a set of instructions to aid readers in following your directions while performing the actions.

Your major formatting decision involves the use of a list. Lists aid readers by segmenting each step, thus allowing quick retrieval of the information. Lists also help readers recall which step they are performing when they turn away to carry out the action. In order to segment your list, you should maintain adequate white space by single spacing steps and double spacing between steps.

A second formatting decision concerns highlighting your cautionary notes and explanations. Cautions should be highlighted by using capital letters, underlining, or marks of punctuation (dashes, asterisks), or by setting the notes off in boxes. Figure 9.5 on page 168 illustrates these possibilities.

Explanations should be highlighted by using punctuation marks, as shown below:

— This dial turns the power on and controls its flow.

*** This dial turns the power on and controls its flow.

168   9 | Descriptions of Processes

FIGURE 9.5
Highlighting for cautionary notes

CAUTION: TURN THE CONTROLS ALL THE WAY TO 4 BEFORE LIGHTING THE LAMP.

CAUTION: TURN THE CONTROLS ALL THE WAY TO 4 BEFORE LIGHTING THE LAMP.

CAUTION: TURN THE CONTROLS ALL THE WAY TO 4 BEFORE LIGHTING THE LAMP.

XXX CAUTION: TURN THE CONTROLS ALL THE WAY TO 4 BEFORE LIGHTING THE LAMP.

CAUTION: TURN THE CONTROLS ALL THE WAY TO 4 BEFORE LIGHTING THE LAMP.

In addition, both cautionary notes and explanations should be separated from your steps by adequate white space.

A third decision on format concerns placement of your visual aids. If you have decided to include a visual aid depicting the performance of each step, place the visual aid beside the step (if the visual aid is small) or immediately after the step (if the visual aid is large). This placement will assure that readers have the required information as the step is being performed. Surrounding these visual aids with adequate white space will also increase the readability of your set of instructions by reducing clutter and providing eye relief.

Jim does not include a visual aid for each step because none of the steps require complicated actions. He places his diagram of the Perkin-Elmer on page 2 of the instructions, because the visual aid is large.

Planning Style

Sets of instructions are always written in the imperative mood: commands, where the subject *you* is understood.

> *Mount* the lamp.
>
> *Light* the EDL.

This use of the imperative shortens each instruction, making the steps easier to follow. In addition, placing the verb first in each step assures that your readers know immediately the action they are to perform. You should also use short, simple sentences with only one or two ideas in each sentence regardless of your audience's level of education or technical experience. These short, simple sentences facilitate reading the steps while performing the action. However, you should avoid omitting the articles *a, an,* and *the* (e.g., ''Mount lamp''), because this omission creates an abrupt tone and an unnecessarily telegraphic style.

Jim's Instructions

169

Writing

We reproduce a portion of Jim's draft for major step 1 below, with his reasons for including and omitting certain details.

**OPERATING PROCEDURE
FOR THE
PERKIN-ELMER MODEL 305B
ATOMIC ABSORPTION SPECTROPHOTOMETER**

In this manual, you will learn the following techniques: turning on and adjusting the spectrophotometer, lighting and adjusting the flame, optimizing and recording the signal output, shutting down the spectrophotometer. (A diagram of the Perkin-Elmer is included on page 2 of this manual.)

TURNING ON AND ADJUSTING THE SPECTROPHOTOMETER

Capitalization emphasizes the name of a control, for reader ease

The audience is familiar with spectrophotometers, so definition is not required

1. Press the POWER BUTTON.
   - This button turns on the electronics.

Explanation necessary for Jim's students and newly hired employees

2. Turn the SOURCE AND GAIN CONTROL counterclockwise to 0.

Ensures proper operation

CAUTION: THE PHASE SWITCH MUST BE IN "NORMAL" POSITION.

Aids readers unfamiliar with the controls, thus saving reader time

Definition is not required

3. Set the DAMPING CONTROL.

4. Mount the electrodeless discharge lamp (EDL) in the lamp compartment.

Spelling out the lamp name ensures reader understanding of the abbreviation. Use of the abbreviation then shortens the instructions

Ensures proper alignment

   - All notches must be positioned over the alignment studs.

Ensures proper operation

CAUTION: TURN THE POWER SUPPLY OFF.

5. Connect the power-supply cable to the EDL.

Ensures proper operation

CAUTION: IF YOU CANNOT LIGHT THE EDL OR SET IT AT THE SPECIFIED CURRENT, TURN THE POWER SWITCH OFF AND SEEK ASSISTANCE.

6. Turn on and light the EDL.
   a. Turn the POWER SWITCH on.
   b. Rotate the LAMP CONTROL clockwise to approximately three-quarters of its maximum range.

Specifies the limit of operation, for proper performance

Rewriting

Your major criteria when revising a set of instructions are amount of detail, logical progression, and appropriate emphasis. You must include sufficient information to enable readers to perform the steps, arrange that information in effective steps and substeps, and emphasize important points. Otherwise, readers may have difficulty carrying out the procedure. Testing the instructions on an audience with the same level of expertise as your intended readers, to see if they can perform the operation, will aid you with this revision.

When Jim tests his procedure on a new employee, he finds that he has omitted a detail necessary for proper operation: the value at which to set the damping control. He rewrites that substep as follows:

3. Set the DAMPING CONTROL to 1.

He also discovers that his reader does not know where to find the alignment studs, so he adds that fact in the explanation:

□ All notches must be positioned over the alignment studs in the neck of the lamp socket.

When Jim rereads his steps, he feels that they are effectively classified. However, when he rereads for appropriate emphasis, he decides to revise one cautionary note. He feels the caution on inability to light the EDL, as phrased, does not stress the main point (turning off the power switch), because that point occurs at the end of the note. He revises the note as follows:

CAUTION: TURN THE POWER SWITCH OFF AND SEEK ASSISTANCE IF YOU CANNOT LIGHT OR SET THE EDL.

CHECKLIST FOR SETS OF INSTRUCTIONS

1. Have I used the following techniques to gather information?
   A. The reporter's formula
   B. A performance of the procedure and a list of steps
   C. A diagram of any apparatus involved
   D. A flow chart of the procedure
   E. A checklist of possible details to include in a set of instructions

2. Have I partitioned my procedure and classified the steps into major steps and substeps? Does the procedure include sub-substeps, if necessary? Have I included too many or too few major divisions?

3. Does my introduction orient readers effectively to the set of instructions?

4. Have I included a partitioning sentence at the end of the introduction, forecasting the major steps?

Karen's Process Narrative   **171**

5. In the step-by-step description, have I done the following?

   A. Introduced the steps, if necessary

   B. Used a numbered list to segment my steps

   C. Placed cautionary notes *before* the steps to which they refer and highlighted these notes

   D. Separated explanations from steps and highlighted the explanations

   E. Single spaced my steps and double spaced between steps

   F. Included visual aids, if appropriate

   G. Used short sentences and included no more than two operations in each substep

   H. Used imperative verbs, placed first in the sentence, and parallel grammatical form for each step

   I. Included the articles *a, an,* and *the*

6. Is my conclusion appropriate for my audience?

KAREN'S PROCESS NARRATIVE ON A TECHNIQUE FOR ANALYZING THE MICROSTRUCTURE OF BREAD

Unlike many process descriptions, Karen's process narrative does form a report in itself: *A Technique for Analyzing the Microstructure of Bread.* In this report, Karen discusses bread microscopy, describes the method she developed, and applies the technique to future research problems. In her process narrative, then, she talks about steps she has performed.

Process narratives can also be written about steps no one performs: for example, the stages in the growth of a plant from seed to maturity or the way a camera records an image. However, the structure of such process narratives is the same as the one Karen writes, even though no one performs these actions. (In our samples, we include a process narrative of this type.)

We now follow Karen through the writing of her report.

Prewriting

Analyzing the Communication Context

Analyzing the Audience

Karen's primary audience is the University Honors Committee (a group of six professors, one from each of the six colleges on campus: Sciences and Humanities, Design, Engineering, Industrial Education, Agriculture, and Veterinary Medicine) and her major professor. Since a copy of her report will be put on file in case future students in the Food Technology Department wish to review it, these students are her secondary audience.

The six members of the honors committee have had no experience with food technology in general or with microscopy techniques in particular. Future food technology students may have some general knowledge of microscopy. Karen's major professor is well acquainted with these techniques, although she has not worked specifically with bread.

Analyzing Purpose and Use

Karen's primary purpose with the honors committee and her major professor is to persuade them that she is an able researcher and knowledgeable in her field, so that they will decide to grant her a B.S. degree with honors. To accomplish this purpose, she must inform her readers adequately about the details of her technique, as a basis for decision making.

Her primary purpose with future food-technology students is to inform them about her technique, so they may consider it as an alternative to other techniques when deciding on research methods. Her secondary purpose is to instruct them in methods for carrying out her technique.

Gathering Information

Four of the information-gathering techniques you used for sets of instructions will also assist you with process narratives: the reporter's formula, a list of steps in the process, diagrams, and flow charts. Library and laboratory research are two other information-gathering techniques Karen employs. She must become familiar with standard documented microscopy procedures, to see if hers has merit. She must also develop, apply, and test her technique.

A checklist of possible details to include is a last information-gathering technique for a process narrative. We give this checklist in Figure 9.6.

Notice that this checklist is similar to the checklist for sets of instructions, with three omissions:

1. A ''materials-needed'' section, giving the materials that must be gathered

2. Cautionary notes

3. Troubleshooting details

Because readers will not perform the operation in a process narrative, these items of information are not necessary. The purposes of the remaining details are the same as the purposes for those in sets of instructions.

Selecting and Arranging Content

Selection

To select content for your process narrative, use the checklist of possible details to prepare an outline of material your readers require. This outline will assist you with arrangement.

Karen's Process Narrative

FIGURE 9.6
Checklist of possible details to include in a process narrative

I. Introduction
   A. Definition of the process
   B. Purpose of the process
   C. Intended audience
   D. Background
      1. Who performs the operation
      2. Where it is performed
      3. When it is performed
   E. Brief description with a visual aid
   F. Definitions of terms
   G. Materials and methods used
   H. Results
   I. Partitioning sentence: the major steps in the explanation

II. Step-by-step explanation
   A. First major step
      1. Definition
      2. Purpose
      3. Overall description with a visual aid
      4. Division into substeps
         a. Substep 1
            (1) Definition
            (2) Description
            (3) Explanation
         b. Substep 2 (Repeats 1–3)
            (and so on)
   B. Second major step (Repeats 1–4)
      (and so on)

III. Conclusion
   A. One operating cycle (for apparatuses)
   B. Results of the operation
   C. Review of the major steps
   D. Application of the process

Arrangement (Structure)

You may prefer to begin arranging a process narrative with the step-by-step explanation, because this portion is the most important part of the narrative.

To order content for this explanation, you must classify your information-gathering list of steps into major steps and substeps. In Figures 9.7a and 9.7b, we present Karen's information-gathering list and her classified list. (We have indicated the reasons for the classification.)

FIGURE 9.7
(a) Karen's information-gathering list of steps (b) Karen's classified list of major steps and substeps

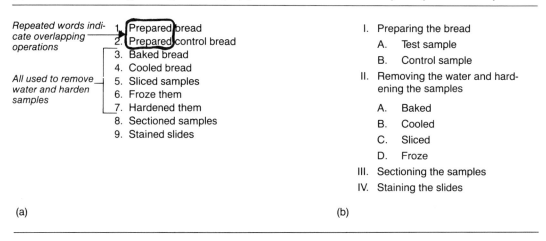

You then expand each section of your outline by including the details your readers require. In Figure 9.8, we present Karen's expanded outline.

After arranging your step-by-step explanation, you should arrange your introduction and conclusion. We list the details in Karen's introduction below, with her reasons for including some items of information and omitting others and for arranging her information in this order.

I. Definition of microscopy

   Six of Karen's readers may not know what microscopy is. They require a definition at the beginning of the introduction to understand the discussion.

II. Purpose of using microscopy in research on bread structure

   Karen's readers must understand the purpose of microscopy in order to appreciate the value of her research. This purpose would logically precede a description of the process, since readers must know the purpose to understand the description adequately.

III. Background: previous techniques used and their drawbacks

   Unless Karen's audience has some knowledge of previous techniques, they will not understand how her method differs.

FIGURE 9.8
Karen's expanded outline

    I. Preparing the bread
        A. Test sample
            1. Used straight-dough method
            2. Modified it
        B. Control sample
            1. Two variations
            2. Used as basis for comparison

    II. Removing the water (call it dehydrating?) and hardening the samples
        A. Baked
        B. Cooled   (Give purpose of each substep)
        C. Sliced
        D. Froze
            1. Cut small pieces from frozen slices
            2. Strengthened using paraffin embedding
              PERHAPS A VISUAL AID HERE: EMBEDDED SAMPLE

    III. Sectioning the samples
        A. Sliced samples thinly
        B. Affixed them to slides

    IV. Staining the slides
        A. Used iodine-based starch stain and a protein stain
        B. Developed stepwise staining procedure

    IV. Purpose of her technique
        The drawbacks of previous techniques logically lead to Karen's purpose for developing her technique.

    V. Partitioning sentence
        This sentence will predict the order of the step-by-step explanation for Karen's readers, inform them of the major steps, and form a transition to the step-by-step explanation.

Karen does not define the intended audience, since they are aware of this fact. She omits definitions of terms because she has decided she must write a nontechnical description of her process: Six of her seven primary readers are outside her field of study and have no knowledge of microscopy. She also omits materials and methods used. Since this information is highly technical, the same six readers would not understand it.

9 | Descriptions of Processes

You must also arrange an appropriate conclusion. Karen feels the results of her work are too technical for six of her primary readers. Her readers do not need a recapitulation of steps, since her process narrative will not be complex or involved. Therefore, she decides to discuss the application of her method to future research. These facts may convince the honors committee of the worth of the technique she has developed and suggest to future students further uses her method may have.

In Figure 9.9, we give Karen's outline for her conclusion.

FIGURE 9.9
Karen's outline for her conclusion

CONCLUSION

    I. Applied research

        A.   Additives

        B.   Ingredient variation

    II. Research on other baked goods

        A.   Doughs and batters

        B.   Before-and-after-baking comparisons

Arrangement (Format)

Your major formatting decision in a process narrative involves segmenting and highlighting your steps. Because readers will not be performing the action described, process narratives do not segment these steps as sharply as sets of instructions. The examples below illustrate this fact:

Set of Instructions	*Process Narrative*
A. Freeze bread slices overnight.	After baking and cooling, the bread was sliced and frozen overnight. Small pieces were cut from the frozen slices. . .
B. Cut small pieces. . .	

However, process narratives still segment and highlight major steps and (in complex processes) substeps.

You have several formatting techniques for highlighting these steps:

1. Headings
2. A numbering system
3. Paragraph indentation
4. White space

If your process narrative is short, white space and indentation may sufficiently segment your steps. However, if your narrative is long, headings — perhaps in conjunction with a numbering system — would be a more effective choice.

Because her process narrative is fairly long, Karen uses headings to segment and highlight her major steps, formatting the headings as follows:

Microscopy Technique Developed

Preparing the Bread

Dehydrating and Hardening the Samples

Sectioning the Samples

Staining the Slides

These headings will identify each step for her readers, thus guiding them through her discussion.

Planning Style

Unlike a set of instructions, a process narrative is written in paragraph form, as you will see in Karen's first draft.

Writing

We give Karen's first draft below, with annotations on the information she included.

One-sentence definition

Purpose of the process

▶ Microscopy is the analysis of the structural components of an object, using a microscope. Because the microscopic structure of a baked product greatly influences the macroscopic properties and quality of the product, microscopy has wide applicability to food-research problems and to the analysis of bread.

Previous techniques. These are not given in detail because the honors committee does not require a literature review

▶ Since the 1930s, bread microstructure has been studied and analyzed, and various microscopy techniques have been developed and discarded.

In spite of this ongoing research, a comprehensive, coherent body of sample-preparation procedures has yet to be compiled; researchers continue to follow a variety of procedures and hence achieve diverse results. More valid contributions to food-science research could be made if current techniques were evaluated and modifications devised.

Karen's purpose

▶ The objective of this research was to modify existing techniques and develop a coherent method. This method ◀— *Partitioning sentence: the major steps in the process* includes preparing the bread, dehydrating and hardening the samples, sectioning the samples, and staining the slides.

MICROSCOPY TECHNIQUE DEVELOPED

A brief introduction gives background for the steps → Features from past methods were combined with new procedures in preparing bread samples for microscope viewing.

Preparing the Bread

Bread was prepared following the straight-dough method outlined in the American Association of Cereal Chemists' Approved Methods (AACC, 1976): Specific modifications made on this method are described in Appendix A. A control bread (i.e., containing no substituted ingredients) was prepared; in addition, two variations were included to provide a basis for comparison during microscopic examination. The variations contained microcrystalline cellulose, an indigestible fiber product, substituted by weight for 5% and 10% of the bread flour.

Detail informs Karen's major professor and future students of the source of her preparatory method. If she had only nontechnical readers, she would omit the reference

Technical details are included in an appendix for her major professor and future students. Thus, these details will not interrupt the text for nontechnical readers

Two informal definitions define troublesome terms

Dehydrating and Hardening the Samples

This term is common enough that nontechnical readers will know it

Before the bread samples could be sliced to the thickness required by light microscopy (approximately 8 μm), the samples had to be dehydrated and hardened. After baking and cooling, the bread was sliced and frozen overnight. Small pieces were cut from the frozen slices; these pieces would eventually provide the cross sections for the microscope slides. A paraffin embedding was chosen to strengthen the bread samples. Thus, the dehydration steps served to remove water, which if present would have interfered with the paraffin as the wax infiltrated the sample. The product of the hardening procedure was a disk of paraffin, about 3.4 in. in diameter and 1/8 in. thick, in which the bread sample was centered. A representation of the disk is shown in Figure 1.

Exact dimensions are given for the technical audience but are subordinated, since these readers are not the primary audience

The purpose of cutting slices is explained for the nontechnical readers. Karen's major professor and students would know this fact

A visual aid, which her nonexpert readers require to picture the sample, is included. Expert readers would not require this visual aid. The dimensions of the disk, which expert readers would know, are also given

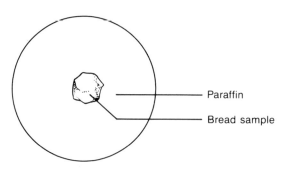

Figure 1: Paraffin disk with embedded bread sample (top view) ← *The perspective of the visual aid is given, so the reader knows how to view it*

Sectioning the Samples

The purpose of sectioning is given for nonexpert readers ▶

A requirement of light microscopy is that the sample be thin enough for light to pass through. Accordingly, the embedded bread samples were sliced into thin sections (8 μm thick). The sections, which appeared as a paper-thin version of the original disk with the bread at the center, were then affixed to microscope slides with an adhesive. (See Appendix D for sectioning details.)

The technical fact is subordinated but included for the benefit of expert readers ◀

◀ *An explanation her nontechnical readers may require, the appearance of sections, is included*

The method is mentioned but not described. The details are unnecessary for nontechnical readers

Staining the Slides

The purpose of staining, a fact her nontechnical readers will not know, is given

The final preparatory step involved applying stains to the slides; each stain would selectively color one fraction of the bread sample. Because starch and protein are the predominant structural components, an iodine-based starch stain and a protein stain were chosen. A staining procedure, ◀——— described in Appendix C, was developed to produce slides with clearly defined starch granules (blue) and protein matrix (red).

The purpose of using these stains, which her technical readers would know but her nontechnical ones ◀ *would not, is given*

—— *An explanation of the staining procedure is omitted. Nontechnical readers will not be reproducing it. Technical readers can follow it in the appendix*

APPLICATION OF TECHNIQUE TO FUTURE RESEARCH

While the technique described was used to produce a photographic study of basic bread microstructure, the procedure has value in applied research as well. That is, the effects of additives and ingredient variation on microstructure may be explored with the aid of the microscopy procedure detailed in the appendices. Disagreement still exists among researchers concerning certain structural interactions in bread; with thorough microstructure research and comparisons of results, an accurate, complete picture of bread microstructure can be assembled.

The usefulness of the microscopy technique is not limited to bread microstructure study; the technique can be adapted for the study of other baked goods. In particular, the procedure can be applied to the investigation of doughs and batters and to before-and-after-baking comparisons.

Rewriting

Your major criteria when revising a process narrative are amount and kind of detail and logical progression. You should include all of the information your readers require to understand your explanation. Moreover, your step-by-step explanation should proceed chronologically, and your narrative should be effectively segmented into steps. When Karen rereads for amount and kind of detail, she decides a visual aid would help her inexperienced readers picture the stained slides. Therefore, she includes the photograph you see below:

These starch granules and the protein matrix can clearly be seen in Figure 2.

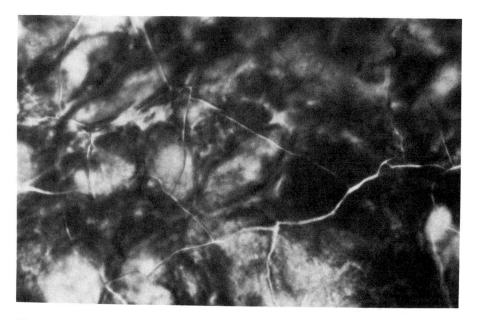

Figure 2: 5% MCC bread: stained protein (red) and starch (blue). 400x magnification.

Karen finds no changes she wishes to make in logical progression.

CHECKLIST FOR PROCESS NARRATIVES

1. Have I used the following techniques to gather information?
   A. The reporter's formula
   B. A list of steps
   C. A diagram of any apparatus involved
   D. A flow chart of the process
   E. A checklist of possible details to include in a process narrative
2. Have I partitioned my process and classified the steps into major steps and sub-steps?
3. Does my introduction orient my audience effectively to the process narrative? Have I included a partitioning sentence at the end of the introduction to forecast the major steps?

4. Have I described, in paragraph form, the substeps under each major step?

5. Have I highlighted these divisions by using headings, a numbering system, paragraph indentation, or white space?

6. Is my conclusion appropriate to my audience?

SAMPLE PROCESS DESCRIPTIONS

Our first example is a process narrative for a procedure no one performs: the coating of a telecommunications fiber. The narrative was included in an article on polymer protection for glass fibers, published in the Bell Laboratories *Record*. The primary purpose was to inform the audience, and the secondary purposes were to entertain and to persuade. As employees of Bell Laboratories, the readers would be interested in learning about Bell Labs' research projects. Readers would also be entertained by a scientific article and would be persuaded that Bell Laboratories was endeavoring to remain in the forefront of telecommunications research.

Notice that the narrative is brief, serving as an overview of the process rather than as a detailed description. Process narratives may be short, especially when they form parts of other documents. We annotate the narrative with the reasons for including certain information.

DESCRIPTION 1

Background on the apparatus used also identifies the start of the coating process → During the coating process, a cup-like reservoir fitted above a cone-shaped applicator or die holds the epoxy-acrylate prepolymer. The fiber is drawn from the molten preform—— *Steps are described in chronological order* through the prepolymer-filled reservoir and die. The movement of the fiber drags the liquid along with it and causes the prepolymer to circulate down along the fiber and up the walls of the reservoir. This motion, in turn, creates pressure in the die, which forces the epoxy-acrylate out of the applicator to coat the emerging fiber. Immediately after it leaves the coating applicator, the prepolymer-coated fiber passes through a hardening chamber. Coatings can be ◄—— *Conclusion gives the results of the process* applied to fibers in thicknesses of from about 10 to 100 micrometers or more using this technique.[1]

[1] © Bell Laboratories *Record*, December, 1979, pg. 317.

Process descriptions 2 and 3 are a set of instructions and a process narrative for the nitrate quick test on plant tissues. The set of instructions was written to be included in a sample kit, distributed to farmers by the county extension service. The

182 9 | Descriptions of Processes

process narrative was intended to be in a brochure, available at the extension office. The purpose of the brochure was to describe the test, so that farmers would decide to use it for checking the nitrate level of their plants.

DESCRIPTION 2
PROCEDURE FOR THE
NITRATE QUICK TEST ON TOMATO-PLANT TISSUE

The nitrate quick test on tomato-plant tissue is a speedy method of discovering the nitrate status of the plant. Farmers may perform this test to check the adequacy of the nitrate-fertilizer program they are currently using.

The best time to perform the test is during the rapid-growth stage, near maturity. During this time, nitrate intake is most likely to affect the plant's growth. Thus, the test results will be the most accurate.

Once these results are known, the fertilizer rate for the next planting season can be adjusted. However, fertilizer adjustment will be too late for this season's plants, as fertilizer should be applied long before maturity.

Although this process is described for tomato plants, it can also be applied to other crops.

MATERIALS NEEDED

> One tomato plant
>
> One razor blade
>
> One sheet of absorbent test paper
>
> One bottle of nitrate test powder
>
> One Nitrate Scale Reference Card

STEPS IN THE TEST PROCEDURE

The nitrate quick test consists of four major steps: selecting the plant tissue, squeezing the tissue, adding the test powder, and comparing the results with the card.

Selecting the Plant Tissue

Selecting the plant tissue involves choosing and removing a piece of the tomato plant for testing. The tomato plant consists of two major parts: the shoot (or stem) and the leaves, as shown in Figure 1 (page 2).

The plant tissue should be removed from the shoot.

1. Locate the shoot of the tomato plant.

    CAUTION: The section of shoot chosen should be free of foreign material (dirt, fertilizer, etc.) or the test results may be invalid.

2. Choose a succulent (not dry or brittle) area of the shoot halfway up the plant.

Sample Process Descriptions

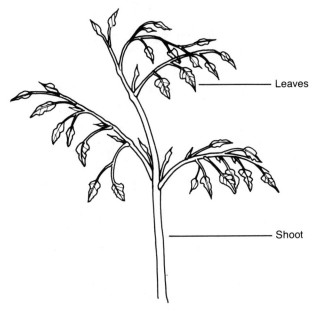

Figure 1: Parts of a tomato plant

3. Remove a ¼-in. section out of the shoot, using a razor blade.
4. Place the piece aside.

Squeezing the Tissue
Squeezing the plant tissue involves extracting the sap from the section of shoot obtained in the previous step. The squeezing step makes use of the absorbent test paper. This paper is composed of labeled parts: the nitrate, the potash, and the phosphate sections. For the purpose of this test, only the nitrate section is used.

1. Fold the nitrate corner of the absorbent test paper over, as shown in Figure 2.

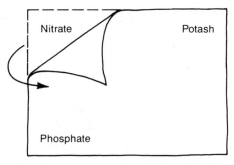

Figure 2: Folding the absorbent test paper

2. Unfold the paper and place the piece of shoot in the center of the crease.
3. Squeeze the piece of shoot firmly, using your thumb and index finger as shown in Figure 3.

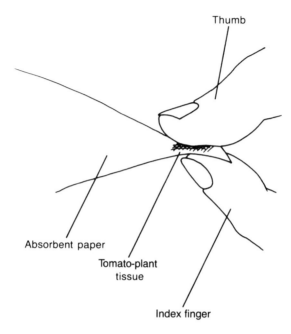

Figure 3: Squeezing the shoot piece

The paper should now have a moist circle around the plant tissue.

4. Leave the tissue in the crease.

Adding the Test Powder

This step involves adding the nitrate test powder, which indicates by color the level of nitrate present.

CAUTION: Avoid getting the nitrate test powder on your skin or clothes, as the powder is acidic and can cause allergic reactions. Wash your hands after you perform this step.

1. Open the absorbent paper and sprinkle a small amount of test powder on the tissue.
2. Fold the paper over again and gently blend the powder into the tissue, using your thumb and index finger.
3. Open the paper and let it stand for five minutes.
4. WASH HANDS!

Sample Process Descriptions

Comparing the Results

The results of the previous step are now used to determine the level of nitrate present in the plant. For this step, use the Nitrate Scale Reference Card, shown in Figure 4.

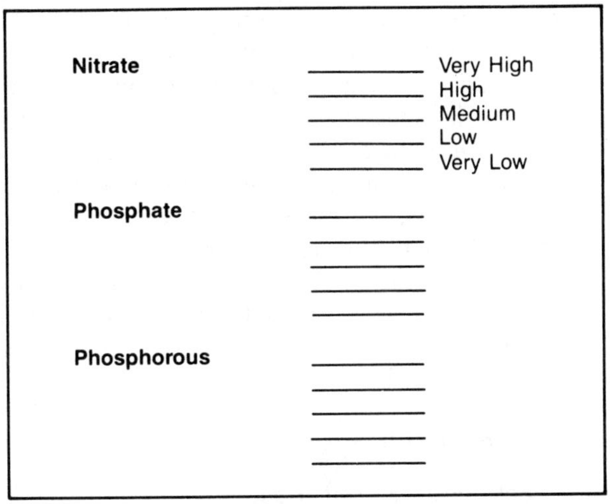

```
+-------------------------------------------------------+
|                                                       |
|   Nitrate               _____  Very High         |
|                         _____  High              |
|                         _____  Medium            |
|                         _____  Low               |
|                         _____  Very Low          |
|                                                       |
|   Phosphate             _____                    |
|                         _____                    |
|                         _____                    |
|                         _____                    |
|                                                       |
|   Phosphorous           _____                    |
|                         _____                    |
|                         _____                    |
|                         _____                    |
|                                                       |
+-------------------------------------------------------+
```

Figure 4: The Nitrate Scale Reference Card

The card is divided into three sections: Only the area marked "Nitrate" is applicable to this test. The colors in the nitrate section are various shades of red.

1. Lay the Nitrate Scale Reference Card next to the test area on the absorbent paper.
2. Find the shade of red that best matches the color obtained from the test.
3. Record your results (very low, low, moderate, high, or very high).

The nitrate quick test is a simple way of testing the level of nitrate present in a plant tissue. If the results of your test indicate a very low or low reading, you should increase the level of nitrate fertilizer you are applying to your tomato plants, in the next growing season. If your results indicate a high or very high reading, you should reduce this level. Finally, if the results indicate a medium reading, your present level of nitrate fertilization is sufficient.

DESCRIPTION 3
THE NITRATE QUICK TEST
ON PLANT TISSUES

The nitrate quick test on plant tissues is a speedy method of discovering the nitrate status of a plant. Farmers perform this test to check the adequacy of the nitrate-fertilizer program they are currently using.

The best time to perform the test is during the rapid-growth stage, near maturity. During this time, nitrate intake is most likely to affect the plant's growth. Thus, the test results are most accurate.

Once these results are known, the fertilizer rate for the next planting season can be adjusted. However, fertilizer adjustment will be too late for this season's plants, as fertilizer should be applied long before maturity.

The nitrate quick test is a simple process, involving only a plant, a razor blade, and the testing equipment: absorbent test paper, nitrate test powder, and a Nitrate Scale Reference Card. The steps involved are selecting the plant tissue, squeezing the tissue, adding the test powder, and comparing the results with the card.

STEPS IN THE PROCESS

Selecting the Plant Tissue

Selecting the plant tissue involves choosing and removing a piece of the plant for testing. A plant consists of two major parts: a shoot (or stem) and leaves, as seen in Figure 1.

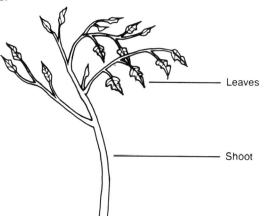

Figure 1: Parts of a plant

The plant tissue is removed from the shoot, approximately halfway up the plant on a succulent (not dry or brittle) area.

This tissue section varies in size. For small plants (e.g., tomatoes), the section may be approximately ¼ in. long, while for large plants (e.g., corn), the section may be approximately 1 in. long.

Squeezing the Tissue
Squeezing the plant tissue involves extracting the sap from the section of shoot. An absorbent test paper, seen in Figure 2, is used in this step.

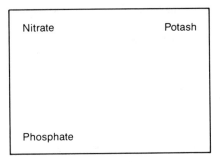

Figure 2: Absorbent test paper

Although the paper has several labeled sections, only the nitrate section is used in this test.

In the squeezing step, the nitrate corner is folded over the tissue and pressed firmly. The result is a moist circle around the tissue.

Adding the Test Powder
In this step, the test powder is sprinkled on the tissue. This powder indicates, by color, the level of nitrate present.

Comparing the Results
The results from adding the test powder are then compared with the Nitrate Scale Reference Card, seen in Figure 3. This card identifies levels of nitrate

Figure 3: Nitrate Scale Reference Card

(very low, low, moderate, high, very high) by shades of red. The color on the absorbent paper is simply matched to the appropriate shade on the card, and the results are noted.

The nitrate quick test is a simple way of testing the level of nitrate present in a plant tissue. If the results indicate a very low or low reading, the level of nitrate fertilizer should be increased. If the results indicate a high or very high reading, the level should be reduced. If the results indicate a moderate reading, the present level of nitrate fertilization is sufficient.

Kits for performing the nitrate quick test are available at the county extension office. Check the nearest office to obtain a kit.

EXERCISES

Topics for Discussion

1. In class, discuss the criteria for an effective set of instructions and for a process narrative. Consider content, arrangement (structure and format), and style.

2. In class, discuss the differences between a set of instructions and a process narrative. Consider content, arrangement, and style.

3. List the major steps you would include when partitioning the following common processes:

baking a cake	bathing a dog
tying a necktie	playing a musical instrument
playing solitaire	painting a picture
planting a tomato plant	mowing a lawn

Then list the substeps necessary for carrying out each major step.

4. Locate a set of instructions for performing a process.

   A. Describe the audiences for this set of instructions. Consider all the elements of an audience analysis: identification, classification, characterization.

   B. Label each section and detail in the instructions, using the checklist of possible details to include in a set of instructions (page 162). Give the author's rationale for including each item of information, based on the audience analysis you have conducted.

   C. Critique the set of instructions for content, arrangement, and style.

5. Locate a process narrative. Repeat all the activities in Question 4.

Topics for Further Practice

1. Rewrite the set of instructions in Question 4 as a process narrative, for the same audience. Describe the changes you have made in content, arrangement, and style.

Exercises   **189**

2. Rewrite the process narrative in Question 5 as a set of instructions, for the same audience. Describe the changes you have made in content, arrangement, and style.

3. Using the techniques in this chapter, write a set of instructions for a procedure your classmates could perform in the classroom. Test the set of instructions on your audience, then revise as necessary. Rewrite the instructions as a process narrative.

10

Technical Letters and Memorandums

THE NATURE OF TECHNICAL LETTERS AND MEMORANDUMS

Most jobs, whether in business, trade, or the professions, will require you to write letters and memorandums. These may be divided into two main groups: positive messages or negative messages, classified by likely reader reaction. If you feel a reader will react favorably to the letter or memorandum, it is a positive message. If you feel a reader will react unfavorably or be disappointed, it is a negative message.

Positive and negative messages may include several specific types: requests, which ask for information or action; responses, which answer readers' questions; announcements, which state information readers have not solicited. The principles we discuss in this chapter will assist you with writing these types.

In our illustrations, we first follow Jon Lang, a chemical engineer with Simms Corporation, through the process of composing a positive message: a request for information. Next we follow Terry Selton, a mechanical engineer in the Engineering Sales Division of A.L.T., Inc., through the process of composing a negative response to Jon's request. We then include a negative announcement in our samples.

JON'S POSITIVE MESSAGE: A REQUEST FOR INFORMATION

Jon Lang works in the process section of Simms Corporation, a producer of commercial bleaching agents. In one phase of producing the agents, sodium hypochlorite is recovered and cooled in a heat exchanger. Because sodium hypochlorite is highly corrosive, a problem has arisen with the cooling phase of the recovery process: Over the last four years, several exchangers made of corrosion-resistant materials have been destroyed by the sodium hypochlorite. Jon has been assigned the task of solving this problem.

No one at Simms has sufficient knowledge of material sciences to select a substance with adequate corrosion resistance and heat-transfer characteristics or to build an exchanger. However, in *Chemical Engineering* magazine, Jon sees an advertisement from A.L.T., Inc., specialists in producing heat-exchange devices for all applications. He decides to request information.

Prewriting

Analyzing the Communication Context

Analyzing the Audience

Unless your positive message is an in-house document or unless you have had previous contact with your reader, you may have very little information about the person you are addressing. In fact, you will often address your message to a person in a job role, e.g., Director of Sales or Personnel Manager. However, you can make some judgments about this audience.

First, the job role should indicate the probable technical knowledge of your audience. For example, Jon will write his letter of request to the head of engineering sales at A.L.T., Inc. Jon expects this reader to know heat-transfer terminology, although he or she may have to consult staff engineers for answers to specific questions.

Second, since readers of technical correspondence are busy professionals, your message should be written as effectively as possible, so that they can easily access the information. Jon will phrase his questions clearly and arrange them effectively to make his reader's task easy, thus increasing the chance that his reader will respond.

Analyzing Purpose and Use

The primary purpose of a positive message is to inform readers about your subject. For example, Jon must inform his reader of his questions about A.L.T.'s heat exchangers. Your secondary purpose is to persuade your reader to act as you desire: Jon must persuade the head of engineering sales to respond. If Jon does not adequately inform his reader about the information he requires, his reader will not be able to respond effectively and may decide not to reply.

Readers' primary use of a positive message is to gain knowledge about the subject of the letter or memorandum: The head of engineering sales will use Jon's letter as a guide to the material Jon requires. Therefore, a carefully written request will produce a more informed response. Second, your reader will decide whether or not to act: Jon's reader should decide to reply, since a potential sale might result.

Gathering Information

A list of questions will help you gather material for both positive and negative messages. We show this list in Figure 10.1, with Jon's answers to the questions.

FIGURE 10.1
List of questions for a positive or negative message, with Jon's answers

1. What is the situation giving rise to the message?
   a. We produce commercial bleaching agents, chlorine dioxide (ClO_2) and sodium bisulfite ($NaHSO_3$), and in the process recover sodium hypochlorite.
   b. Sodium hypochlorite is highly corrosive.
   c. Must reduce sodium hypochlorite temperature from 100°F to 75°F with a heat exchanger.

2. Are any special conditions or circumstances involved in this message?
   a. Has destroyed both glass and fluorinated-hydrocarbon exchangers.
   b. Value of destroyed exchangers: $37,000.
   c. Flow rate of sodium hypochlorite through exchanger = 100 gpm at 50 psi.
   d. Solution's viscosity = 1200 cp.
   e. Sodium hypochlorite is cooled by water.
   f. Water temperature = 62°F; flow rate = 165 gpm at 75 psi.

3. What is/are my specific request(s), response(s), announcement(s)?
   a. Does A.L.T. carry equipment to withstand sodium-hypochlorite corrosion?
   b. Can this equipment cool effectively?
   c. Predicted heat-transfer coefficient?
   d. Cost?

In addition, any previous correspondence you have had with your reader may provide information for your positive message. Since Jon has not previously been in contact with A.L.T., he cannot use this source of material.

Selecting and Arranging Content

Selection

Your reader may not require all the details in your prewriting list. Therefore, you should examine your list and omit any unnecessary details. For example, Jon crosses out the specific names of the bleaching agents, because sodium hypochlorite is the problem and not the bleaching agents. Thus, these chemical names and formulae could confuse the reader, who might waste time trying to figure out how these chemicals affect the equipment needed. Jon also crosses out the number of exchangers destroyed and their value, since this information will not help the reader, but he does not delete the materials from which they were made. The last detail will indicate that these materials are not adequate for this application, allowing the reader to eliminate exchangers made of these materials as possible solutions.

Arrangement (Structure)

When deciding on an arrangement for your message, you should first consider overall structure, then order the individual sections and the details within these sections.

Overall Structure. Your choice of an overall structure for your message depends on reader effect. Your reader will probably respond favorably to a positive message. Therefore, you should use a deductive structure, placing your main idea at the beginning of the message. Because readers will not resist your main point, you can immediately tell them the subject of your document: your specific request, response, or announcement.

Jon's reader should respond favorably to his request for information, since a potential sale may result. For this reason, Jon chooses a deductive structure. He will place his main idea (the fact that he is requesting information on heat exchangers) in the first paragraph of his letter.

Individual Sections. The standard reporting form for a positive message will assist you with arranging individual sections. Figure 10.2 gives this form, with annotations on the purpose of its content.

The purpose of the introduction in a positive message is to tell the reader your main idea as well as provide background on the reasons for the message and the choice of recipient. Readers who understand the causes for a message and the reason why they are receiving it are more likely to accept its contents favorably. Jon will

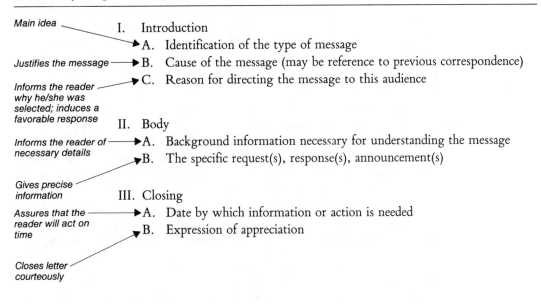

FIGURE 10.2
Standard reporting form for a positive message

identify his letter as a request for information, so his reader will immediately know how to use Jon's letter: as a guide to writing a letter of response. Jon will then inform his reader of Simms Corporation's problem with heat exchangers, making the cause of the message clear. He will also say that he selected A.L.T., Inc. on the basis of their advertisement. The reader will then know how much information Jon already has about A.L.T.'s heat exchangers and will realize that A.L.T. has been chosen as a potential seller.

The purpose of the body in a positive message is to provide background necessary for understanding the message and to give specific information on your request, response, or announcement. Here Jon will relate the history of Simms Corporation's problem with heat exchangers and its needs. He will then list his questions. The history and description of needs will give his reader general details necessary for selecting an exchanger to suit Simms' requirements. The questions will provide the reader with precise guidelines for answering the request.

The purpose of the closing is to persuade the reader to respond favorably to the message. Providing a date by which information is required indicates that you are convinced the reader will act and informs him or her of your deadline. An expression of appreciation then ends the message on a courteous note.

Details Within Sections. Your major decision on arranging details within sections concerns the order of your specific requests, responses, or announcements, if your message contains several. For example, Jon must decide on an order for his questions.

You can order your requests, responses, or announcements in several ways:

1. *Most to least important or least to most important* You could place your most important detail first, if the reader will not resist it or if you wish to emphasize it. You could place it last if the reader will resist it or if you wish to build to a point.

2. *Most to least complex or least to most complex* You could place your most complex detail first, so the reader will be fresh when reading it. You could place it last to ease the reader into your message.

3. *General to specific or specific to general* You could place your most general information first (e.g., Does A.L.T. have any heat exchangers meeting our specifications?). If the reader cannot respond positively to this information, he or she does not need to proceed to more specific points.

4. *Most important — least important — second in importance* You could place your least important point in the middle of your discussion. Other points would then receive emphasis while your least important point would be deemphasized.

Jon decides to order his questions from general to specific because his reader must respond positively to Jon's first question if his more specific questions are to be of use.

You should also classify your specific requests, responses, or announcements so similar information is grouped. For example, you might group all issues on cost (initial purchase price, delivery charges, maintenance fees) in one paragraph or as one major item in a list. However, since Jon's questions are simple, containing only one idea, he does not need to classify them.

Arrangement (Format)

Your major formatting decision in a positive message concerns emphasizing important information: The material in the body of your message should be highlighted so that the reader can easily access this information. Paragraphing or numbered lists can frequently be used, with indentation for additional white space.

Jon decides to number the questions in his request and to indent them five spaces from the left and right margins. In this way he will immediately call reader attention to the questions. His reader will also be able to use the questions as a checklist, ticking off answered ones as the information is gathered.

Planning Style

One major stylistic decision in a positive message concerns the material in the body: This material should be clear and precise to make your reader's task as easy as possible and persuade the reader to act as you desire. For example, Jon must avoid vague or all-encompassing questions, such as "Please send information on your heat exchangers." A request that does not contain specific questions does not adequately guide the reader. Therefore, because a reader would not know the type of information to gather, you might receive a great deal of useless data. Moreover, you would increase the possibility that you would have to write a second message because the first did not produce desired results.

A second stylistic decision concerning material in the body is the use of clear, direct sentences. If you are employing a list, you should include only one or two details for each item. Since you may not have a great deal of information about your reader, you cannot be sure of his or her technical expertise. Direct sentences will make accessing your information easier and will allow your reader to learn your most important information more quickly, thus increasing the changes of a favorable audience response.

A third stylistic decision involves tone, the approach you adopt toward your reader. Although your message conveys favorable news, you should still build goodwill with the reader. Therefore, you should avoid a demanding or condescending tone in order to accomplish your persuasive purpose.

Writing

Jon's first draft follows, with annotations on the content included.

SIMMS CORPORATION
3225 State Street
Salt Lake City, UT 84112

May 2, 1982

Head, Engineering Sales
A.L.T., Inc.
4343 Sand Road
Tonawanda, NY 14150

Dear Head of Sales:

Identifies the type of message — I am writing to inquire about the line of corrosion-resistant heat exchangers you advertised in *Chemical Engi-* ◄— *Reason for the choice of recipient*

Cause of message —► *neering,* because of Simms Corporation's corrosion problem with its current exchangers. Our firm produces bleaching agents and in the process recovers sodium hypochlorite. This solution, which is cooled in a heat exchanger, has ◄—— *Background* destroyed both glass and fluorinated-hydrocarbon exchangers. I am gathering information to recommend a change of equipment to my section supervisor.

The variables for this process are listed below.

Process fluid — sodium hypochlorite
—Inlet temperature = 100°F
—Outlet temperature = 75°F
—Pressure = 50 psi
—Flow rate = 100 gpm
—Viscosity = 1200 cp
Service fluid — water
—Inlet temperature = 62°F
—Pressure = 75 psi
—Flow rate = 165 gpm

Because of the corrosion problem, I would appreciate answers to the following questions, in light of the information given above.

Specific requests for ►
information

1. Does A.L.T. produce equipment that will withstand ◄ *List contains clear sentences, with one idea in each*
   sodium hypochlorite's corrosive quality?

2. If so, could the exchanger accomplish the cooling task?

3. What is the predicted heat-transfer coefficient of this process?

4. What is the cost of such an exchanger?

Checklist for Positive Messages   **197**

Time by which the ———▶ I hope to make my recommendation by June 1.
information is needed

       I greatly appreciate your time and look forward to ◀——— *Courteous close*
hearing from you soon.

Sincerely,

Jon Lang

Jon Lang

Process Engineer

Rewriting

Your major criteria for revising a positive message are amount and kind of detail, logical progression, and stylistic appropriateness. The description of the problem or situation should contain all the details your reader requires and the material in the body should be specific. This material should also develop logically, to make the reader's task of accessing information easier. Finally, your style should be appropriate, to increase the probability that the reader will respond as you desire.

As Jon reads his letter, he remembers one question he would like to add to make the letter complete. Since Simms Corporation has been expanding at a steady rate for the last several years, Jon would like information on A.L.T.'s exchangers with greater capacities. Therefore, he adds a fifth question "What exchangers with higher capacities does your company offer?" He places the question last because it does not concern his present problem.

Jon is satisfied with the remainder of his letter. He now types and edits it and sends it to A.L.T.

CHECKLIST FOR POSITIVE MESSAGES

1. Have I used a list of questions to gather information for my positive message? Have I used a previous communique, if available?

2. Have I omitted details from my prewriting list that my readers cannot use?

3. Have I arranged my positive message effectively?

    A. Have I used a deductive structure — main idea first?

    B. Have I used the standard reporting form for a positive message to sequence individual sections?

    C. Have I ordered my specific requests, responses, or announcements effectively? Have I classified and highlighted them?

198   10 | Technical Letters and Memorandums

4. Are my specific requests, responses, or announcements clear and precise? Are the sentences direct?

5. Is my tone appropriate to my message?

TERRY'S NEGATIVE MESSAGE: A RESPONSE TO A REQUEST

The head of the Engineering Sales Division at A.L.T., Inc., has given Terry Selton, a mechanical engineer in the division, the task of replying to Jon's request for information. Although A.L.T. does manufacture a heat exchanger that will fulfill Simms' needs, the exchanger is currently being tested to correct technical difficulties. Therefore, the model Simms requires is not immediately available. Since Terry must convey unfavorable news to Jon, Terry is writing a negative message.

Prewriting

Analyzing the Communication Context

Analyzing the Audience

As with a positive message, you may have very little specific information about the audience of your negative message, unless it is an in-house document or the result of previous contact. Again, you can use the reader's job role as a clue to his or her technical expertise. However, since Terry has received Jon's request for information, Terry uses this previous contact to analyze his audience.

From the letter, Terry learns that his reader is a process engineer. Therefore, Jon will understand all generic information about heat exchangers, although he may not be familiar with A.L.T.'s models. Terry also learns from Jon's questions that he is interested in performance of the equipment rather than in theory or design. These questions will guide Terry's response.

You should also note that your reader's reaction to your message might be unfavorable. Therefore, you will have to write your message effectively in order to retain the reader's good will.

Analyzing Purpose and Use

The primary purpose of a negative message is to inform your audience of disappointing information: your specific requests, responses, or announcements. However, in order to inform the reader effectively, you must persuade him or her to agree with or accept unfavorable material. For example, Terry must inform Jon that the heat exchanger Simms requires is not currently on the market because of technical problems. Terry must also persuade Jon that these difficulties are not major and that the heat exchanger will eventually meet Simms' standards and needs, so that Jon will continue to consider the exchanger as a solution to Simms' problem.

The reader of a negative message uses your communique to learn unfavorable information. If this information is presented effectively, he or she may be persuaded

Terry's Negative Message: A Response to a Request **199**

to act as you desire; despite the news Jon receives, he may still recommend purchasing A.L.T.'s heat exchanger when it becomes available.

Gathering Information

The list of questions we presented for the positive message will also assist with gathering information for a negative message. We give that list in Figure 10.3, with Terry's answers.

FIGURE 10.3
List of questions for a positive or negative message, with Terry's answers

---

1. What is the situation giving rise to the message?
   Simms Corporation needs information on corrosion-resistant heat exchangers. Sodium hypochlorite has destroyed current exchangers.

2. Are any special conditions or circumstances involved in this message?
   - ☐ Fluorinated-hydrocarbon and glass exchangers *not* possibilities.
   - ☐ Process fluid: in, 100°F; out, 75°F; 50 psi; 100 gpm; 1200 cp.
   - ☐ Service fluid, water: in, 62°F; 75 psi; 165 gpm.
   - ☐ Reader's questions.
   a. Do we carry exchangers resistant to sodium hypochlorite?
   b. Can our equipment accomplish the cooling?
   c. Heat-transfer coefficient?
   d. Cost?
   e. Range of capacities in our line?

3. What is/are my specific request(s), response(s), announcement(s)?
   a. We have an exchanger (the titanium-plated Maxflow) that would be suitable, but it is being tested for technical difficulties and is currently unavailable.
   b. It will accomplish the cooling.
   c. Heat-transfer coefficients have been as high as 1250 Btu/hr ft² °F.
   d. Require further data to calculate cost.
   e. Range includes exchangers to handle working temperatures to 450°F at 300 psi. Working fluid viscosities may be up to 30,000 cp. Heat-transfer area may be as large as 16,000 ft², and flow rates to 10,000 gpm.

If you have received a previous communique from your reader, you should use that document to help you gather material. As you see from Terry's answers, he has used Jon's questions as a guide to the information he will put in his response.

Selecting and Arranging Content

Selection

Examining your list of answers and omitting details your readers do not require will assist with content selection for a negative message. Because Terry has used Jon's request as a guide to gathering information, Terry does not find any details he wishes to omit.

Arrangement (Structure)

In a negative message, you must also consider overall structure, the order of individual sections, and the order of details within sections.

Overall Structure. Since your reader could respond unfavorably to a negative message, you should use an inductive structure, placing your main idea (the negative information) late in the message. This structure will allow you to prepare your reader for the negative information by presenting an explanation first. This structure will also allow you to place favorable information in the position of most emphasis, at the beginning of the message. You then deemphasize your negative information by placing it between more favorable details.

Since Terry's reader will be disappointed that the heat exchanger suiting Simms' needs is not available, Terry chooses an inductive overall structure for his letter. He will place his negative information late in the message, between more positive points.

Individual Sections. The standard reporting form for a negative message will assist you with arranging individual sections. We give this form in Figure 10.4, with annotations on the purpose of its content.

FIGURE 10.4
Standard reporting form for a negative message

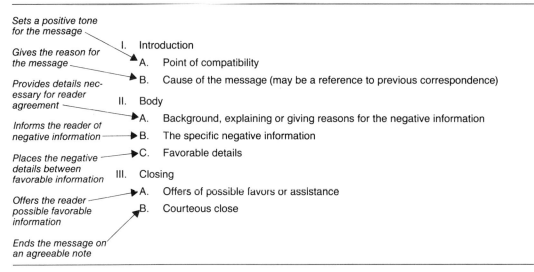

The purpose of the introduction in a negative message is to create a positive tone so that readers will be led into your message agreeably, as well as to inform them of the cause of the message. Thus, your introduction should establish a point of compatibility with the reader. The following is a list of possible points of compatibility, with examples of each:

1. Agreement
   I agree with you that Simms Corporation's problem is related to corrosion resistance.

2. Appreciation
   We appreciate your interest in A.L.T.'s heat exchangers.

3. Assurance
   We have carefully reviewed Simms' needs.

4. Compliment
   Simms Corporation has obviously studied the corrosion problem in depth.

5. Conviction
   A.L.T.'s heat exchangers are well suited to most industrial needs.

6. Cooperation
   We are happy to help solve Simms' corrosion problems.

7. Understanding
   We understand the costly nature of these problems.

Terry decides he will first thank Jon for inquiring about A.L.T.'s heat exchangers, thus establishing the cause of the message. Terry will then include a conviction: A.L.T.'s heat exchangers are well suited to most industrial needs. With this conviction, he hopes to convince Jon of the worth of A.L.T.'s product.

The purpose of the body in a negative message is to prepare the audience for the negative information, inform them of it, then deemphasize it by countering it with favorable information. Terry will first explain that A.L.T. has a heat exchanger meeting Simms' needs, but that this exchanger is currently undergoing tests. He will then state the negative information (the model is presently unavailable), after which he will provide more favorable information: the date by which tests will be completed, an offer to compute costs, and information on A.L.T.'s highest capacity heat exchanger, in case Simms should decide to order one of these instead of the Maxflow.

The purpose of the closing is to create goodwill. Offers of possible favors or assistance convince the reader that you have his or her interests at heart, even though you must convey negative information. These offers and your courteous close also end your message on a positive note, thus countering your negative information. Terry will include a brochure with his letter, so that Jon can see A.L.T.'s heat exchangers and have their specifications at his fingertips. Terry will also offer to install any exchanger free of charge as an incentive to Jon to recommend purchase.

Details Within Sections. Your major decision on arranging details is the order of your background and negative sections. Since the background section explains the reasons for the negative information, this section must develop logically. You should create a chain of reasoning leading the reader step-by-step to agree with the negative information you must present. Your negative information will then be the inevitable result of circumstances you detail. Your negative information should also be immediately countered with favorable details.

Terry will begin by informing Jon that the Maxflow, the exchanger meeting Simms' needs, is the newest model of heat exchanger on the market. He will then state that because of this fact minor design problems have arisen with the Maxflow's intake apparatus. He will not explain the problems in detail, because devoting additional space to them would emphasize them. Instead, he will announce the negative information: "Because further testing is necessary, the Maxflow is currently off the market." However, he will immediately counter this negative detail with positive information: "The Maxflow will be available when the tests are completed, at the end of the month."

In this way, Terry hopes to convince Jon that his negative information is necessary. He also hopes to convince Jon that the exchanger is a worthy product, so that Jon will still recommend purchase.

Planning Style

Your major stylistic decision in a negative message involves tone. Your tone should be agreeable. If you are abrupt in stating your negative information, you will not accomplish your persuasive purpose or create goodwill. Moreover, your tone should be confident even though you are conveying bad news. Choosing positive words — focusing on what you can or are doing rather than on what you cannot or are not doing — will assist in creating a confident tone. For example, Terry will state that the Maxflow *is* currently off the market rather than saying the exchanger *is not* on the market.

Writing

We give Terry's first draft below, with annotations on important details.

A.L.T., Inc.
4343 Sand Road
Tonawanda, NY 14150

May 10, 1982

Mr. Jon Lang, Process Engineer
Simms Corporation
3225 State Street
Salt Lake City, UT 84112

Courteous opening; refers to previous correspondence as the cause of the letter ──▶

Dear Mr. Lang:

Thank you for your letter of May 2d, inquiring about

Date allows the reader to locate previous communication in company files

Conviction as the point of compatibility — A.L.T.'s corrosion-resistant heat exchangers. As you know, these exchangers are well suited to most industrial needs.

Answers to the reader's first three questions — A.L.T. does produce a titanium-plated exchanger, the Maxflow, which has been used successfully to cool sodium hypochlorite and will meet the cooling requirements of your process. Heat-transfer coefficients have been as high as 1250 Btu/hr ft² °F.

Background explaining the negative information. — This heat exchanger is the newest model to be manufactured. Because of this fact, minor design problems have arisen with the exchanger's intake apparatus. Since solutions to these problems require further testing, the Maxflow is currently off the market. However, the exchanger will be available when tests are completed at the end of the month.

The negative information

Sufficient explanation is given to alert the reader to the difficulty, but details are omitted to deemphasize problems

Positive information countering the negative information

Answer to the reader's fourth question — We will be happy to quote a price for the exchanger, if you will send us the heat capacity of sodium hypochlorite. We require this information to calculate total heat-transfer area, the major determinant of cost.

Answer to the reader's fifth question — A.L.T. also produces higher capacity exchangers than the Maxflow. Our highest capacity exchanger has a total heat-transfer area of 16,000 ft². The following is a list of the *maximum* values for working fluid characteristics that our exchangers can handle.

The list aids the reader in accessing this information —

Temperature	450°F
Flow rate	10,000 gpm
Pressure	300 psi
Viscosity	30,000 cp

Offer of additional information and a favor — I am enclosing a brochure that will give you further information. Should Simms Corporation decide to purchase a Maxflow or another heat exchanger, A.L.T. will be pleased to install the equipment free of charge.

Courteous close — If I can be of further assistance, please feel free to contact me.

Sincerely,

Terry Selton

Terry Selton
Sales Engineer

TS/bn

204   10 | Technical Letters and Memorandums

Rewriting

Your major criteria for revising a negative message are amount and kind of detail, appropriate emphasis, logical progression, and stylistic appropriateness. You have to give your reader negative information, but you should not devote so much space to the explanation of this negative information that it receives additional stress. You should also lead the reader to agree with your negative information by a logically developed explanation and an agreeable, positive tone.

Terry finds no changes he wishes to make in amount and kind of detail or logical progression. However, he feels that his opening does not emphasize his interest in Jon's inquiry as strongly as he would like. He rewrites this opening to put greater emphasis on his concern for the reader:

> I was pleased to receive your letter of May 2d, inquiring about A.L.T.'s heat exchangers.

Terry now feels his negative message is ready to be typed, edited, and sent to Jon Lang.

CHECKLIST FOR NEGATIVE MESSAGES

1. Have I used a list of questions to gather information for my negative message? Have I used a previous communique, if available?

2. Have I omitted details from my prewriting list that my reader cannot use?

3. Have I arranged my negative message effectively?

   A. Have I used an inductive structure — main idea (the negative information) later in the message?

   B. Have I used the standard reporting form for a negative message to sequence individual sections?

   C. Have I ordered my background section so that the reader is led step-by-step to agree with the negative information?

   D. Have I immediately countered my negative information with positive details?

4. Have I devoted as little space as possible to the negative information?

5. Have I included favors or offers of assistance to create good will in my reader?

6. Have I projected a confident tone in my message? Have I chosen positive rather than negative words?

SAMPLE LETTERS AND MEMORANDUMS

Our first example is a negative message: a request for adjustment because of the delivery of defective merchandise. Bailey Hospital in Thomaston, Mass., ordered five microscopes, one of which proved to have technical defects. Although the firm from which Bailey purchased the microscopes, Dartmouth, Inc., serviced it once, the

Sample Letters and Memorandums **205**

microscope has continued to perform unsatisfactorily. Thus, the supervisor of laboratory facilities at Bailey must write a request for adjustment. Because the writer is requesting a replacement free of charge, the message contains negative information.

5003 Arnold Drive
Thomaston, MA 10023
November 16, 1983

Ms. Elizabeth Conklyn
Director, Product Sales
Dartmouth, Inc.
16 Nash Street
Franklyn, CT 20361

Dear Ms. Conklyn:

On October 1st, Bailey Hospital ordered five F647 microscopes from Dartmouth, Inc. Although the model has always proven satisfactory in the past, one of the new microscopes failed to produce the necessary resolution.

Your service department subsequently checked the microscope, but its performance is still unsatisfactory. Because our laboratory work requires instruments in the best condition, we feel we must request a replacement for the F647.

Dartmouth, Inc., has always served us well, and we have been pleased with your laboratory equipment. Therefore, we feel you will want to send a new F647. Of course, we will package and return our defective microscope. We would like to have the new microscope within a week, as our laboratory work requires a full line of equipment.

Thank you for your consideration.

Sincerely,
BAILEY HOSPITAL

Jan Whiting

Ms. Jan Whiting
Supervisor, Laboratory Facilities

JW/ml

Our second example is a positive message: a response to a request. The director of student affairs at Clark University in Centerville, Ala., had sent a memorandum to selected departments requesting their participation in a student tutorial program. The

206   10 | Technical Letters and Memorandums

following memorandum was the Physics Department's reply. Because the Physics Department is happy to assist, the memorandum contains positive information.

MEMORANDUM
Clark University
Centerville, AL 70512

TO: Robert Grant
         Director, Student Affairs

FROM: Barbara Winter *BW*
         Chair, Physics Department

DATE: October 1, 1983

SUBJECT: Student Tutorial Program

The Physics Department is happy to participate in a student tutorial program, as outlined in your memorandum of September 15, 1983.

We will send two faculty persons to the three tutorials scheduled for our introductory physics course (104). These faculty persons will have prepared study questions on the material covered in lecture and will be available for student questions and discussion.

We hope these tutorials will assist beginning students with the task of mastering physics as well as help with their adjustment to college life.

Our third example is a negative message: an announcement of a change in procedure. Management of Taylor, Inc., a precision tool-making firm, recently instituted a computer billing system that will necessitate a new method of customer payment. Because the method could involve some alterations in customers' accounting systems, the request for the change contains negative information.

---

150 Brownlee Road
Clareville, MO 50103
January 5, 1983

Manager, Accounting Department
D & O Tool Company
1605 Ring Road
Saylorsville, MO 50631

Dear Manager:

As you know, computer billing systems are being instituted industrywide to save both customer and company funds. Because Taylor, Inc., is interested in reducing costs, we have replaced our manual billing system with computers.

Under the manual system, bills were compiled at the close of each month and sent on the tenth day of the next month. Because of the computer's speed, this delay in billing is no longer necessary. We will now compile our bills and send them to customers on the *last day* of each month. The bills will then be *due* on the 10th.

This operation will allow us to realize considerable savings, which we intend to pass on to our customers. As an example, large-scale orders (15 items or more) will now be discounted 10 percent.

We are sure you will find our computer billing system accurate as well as cost effective. If you have any questions about the new procedure, we will be happy to answer them.

Sincerely yours,
TAYLOR, INC.

John Singer
Manager, Accounting Department

JS/tw

EXERCISES

Topics for Further Practice

1. Choose a familiar product used in your field. (You might want to select a piece of equipment used in one of your laboratory courses.) Write a positive request for information about the equipment to an audience you devise.

2. Write a negative response to the request you wrote for Question 1. Assume that the equipment is no longer available.

3. You work for a firm that has had contract work increase 19 percent per year for the last three years. Your company now has an outdated copier that only runs one page every eight seconds. The copier is also inefficient and costly in terms of labor, because pages must be placed, one at a time, on the copying surface and duplicates must be sorted by hand. Since much of the copying done by your firm is multiple copies of documents that range in length from 50 to 150 pages, this system is no longer workable. Also, oversized graphics that must be reduced are currently sent out for duplication. You have been assigned the task of gathering information about copiers and making a recommendation on purchasing a new one. Write a positive request to the Zan Duplicator Company. Assume you saw an advertisement from this company in the November 1981 edition of *Office Equipment* magazine. The ad described Zan Duplicator Company's "Multi-Function Series" photocopy line, which is of particular interest to you.

4. Write a positive response to the request in Question 3. You have copiers available that will fulfill the reader's needs.

5. As a graduating senior in _____, you feel a change in your department's requirements for a major would be beneficial. However, this change would be somewhat controversial. Write a negative request to the head of your department asking that the department institute the change. (You may stipulate the change.)

6. Write a negative response to the personnel department of a company where you have previously worked. Assume they have offered you a position for the summer, but you have chosen to decline. You have made this decision because you have been offered a more challenging job with another firm.

7. Write a positive announcement to the undergraduates in your department, stating that you are founding a majors' club and asking for their participation.

11

Resumes and
Letters of Application

THE PURPOSES OF RESUMES
AND LETTERS OF APPLICATION

Compiling and writing an effective resume and letters of application are essential parts of a job search, whether you are looking for a first job or for a better position later in your career. In either case, the primary purpose of these documents is to persuade a potential employer that you are the best candidate for the position. This primary purpose, however, depends on a secondary aim: informing your reader of your skills. Your success in effecting your primary purpose will depend on how well you fulfill your secondary aim.

Even though resumes and letters of application have the same general purpose, these documents differ in reader use. Your reader will use your resume to decide whether or not you suit the position, based on facts drawn from your total background. Therefore, the resume presents a broader range of detail than the letter of application, which the reader uses to decide if you suit the particular position, based on specific details you emphasize. The letter of application, then, never simply repeats your resume. Instead, the letter highlights and expands on facts you choose.

Even though the information given in resumes is more nearly complete than that in the letters they accompany, your resume does not necessarily list every detail of your background. You must still select appropriate details for your resume from the data you know about yourself. For example, as a biomedical engineering major, you may have taken courses in chemistry and biology as well as those in your major discipline. However, you would not list all these courses on your resume, perhaps because this list would be too long or because some of the courses are not important require-

209

ments of the job for which you are applying. Instead, you would select from all the courses you have taken only those indicating your particular strengths for the job in question. In fact, selection in resumes is so important that at times you might compose several resumes, each with a different emphasis and each containing a different selection of details.

In this chapter, we illustrate resumes and letters of application, using the example of Dave Wells, a senior who will graduate in May with a double major in sociology and psychology.

DAVE'S RESUMES

Prewriting

Analyzing the Communication Context

Knowing the field you would like to enter — the people in it are your audience — will help you select and arrange details for your resume. Dave would prefer to work with the developmentally disabled (those with emotional, behavioral, or physical difficulties). However, since he is unsure he will be able to find a position of this type, he must consider other areas, particularly recreational programs for cities and day-care programs. Because his audiences differ, his best strategy is to construct two resumes, tailoring each to suit the group addressed.

The primary audience for one resume will be the directors of institutions for the developmentally disabled. These directors must be informed of his training and work experience in developmental education. The primary audience for the second resume will be the directors of programs in recreation and heads of day-care centers. Directors of programs in recreation need to know his background and experience in recreational leadership and leisure studies. Heads of day-care centers will be particularly interested in his work with children. These readers will all use Dave's resumes to decide whether or not to interview him.

Possible secondary audiences include members of the boards of directors for the institutions, city officials, persons holding the position for which Dave is applying, and other employees. These audiences may read the resumes in order to approve the directors' decisions or to advise them on Dave's capabilities.

Gathering Information

You must know a great deal about yourself and your interests to construct an effective resume. Completing a personal inventory will help you gather this information. Examining sample resumes will then indicate possibilities for your resume.

Completing a Personal Inventory

A personal inventory asks you to assess your characteristics in terms of the employment you would like to obtain. As such, the inventory will be useful at various points in the job-hunting process. Perhaps you are certain of both the field you wish to enter and the availability of jobs. In this case, the personal inventory will help you gather the information you need to construct your resume. On the other hand,

Dave's Resumes   **211**

you may have a firm idea of the field you would prefer to enter but must consider
other areas because positions you would like might not be available, or you may have
very little idea of the type of position you would find interesting. The inventory will
also assist you in these cases, by asking you to assess details you know about yourself.

We present a personal inventory in Figure 11.1, then discuss the purpose of each
section.

FIGURE 11.1
Personal inventory

Personal Inventory

I. Personal data
    A. Name
    B. Address(es)
       1. Permanent
       2. Temporary
    C. Telephone number(s)
       1. Permanent
       2. Temporary
    D. Personal description
       1. Height
       2. Weight
       3. Health
       4. Birth date/Age
       5. Marital status
       6. Service record
       7. Hobbies and interests

II. Education
    A. College
       1. Institution
       2. Dates attended
       3. Degree obtained
       4. G.P.A. (grade point average)
          a. Major
          b. Overall
       5. Relevant course work
          a. Major
          b. Minor
          c. Other
       6. Awards/Honors
       7. Activities
    B. Graduate school
       1. Institution
       2. Dates attended
       3. Degree obtained
       4. G.P.A.
       5. Relevant course work
       6. Awards/Honors
       7. Activities

FIGURE 11.1 continued

    C. Training programs
        1. Title of program
        2. Dates attended
        3. Certificate obtained
        4. Description of program/Skills attained

III. Work experience
    A. Positions held
        1. Position
        2. Date held
        3. Employer
        4. Supervisor
        5. Description of duties
    B. Additional relevant information

IV. References
    A. Name
    B. Title/Job role
    C. Address
    D. Telephone number

V. Skills and personal qualities/Interests
    A. Assess yourself according to the following hierarchy of skills[1] by circling those that apply to you.

Data-Related	People-Related	Thing-Related
Synthesizing	Advising	Precision working
Innovating	Negotiating	Operating/Controlling
Analyzing	Supervising	Manipulating/Handling
Coordinating	Instructing	Tending
Computing/Compiling	Entertaining	
Copying	Helping/Serving	

Decreasingly prescribed (left axis)   *Increasingly prescribed* (right axis)

        1. Are your skills largely data-, people-, or thing-related?
        2. List your strongest skills.

    B. Assess yourself according to the following grid of personal qualities.

Attribute	Well-Developed	Under-Developed
Work-Related Characteristics		
1. Adaptability		
2. Communication skills		
3. Cooperation		
4. Decision-making ability		
5. Dependability		
6. Determination		
7. Flexibility		
8. Independence		
9. Initiative		
10. Judgment		

Attribute	Well-Developed	Under-Developed
11. Leadership 12. Logical thinking 13. Organization		
Personal Characteristics 1. Confidence 2. Emotional stability 3. Energy 4. Health 5. Maturity 6. Neatness 7. Poise 8. Tact		

    1. Assess your strongest qualities.
    2. Assess your weakest qualities.

C. List your interests in the following areas, then assess these interests.

The most interesting job/activity you have performed

The most interesting job/activity you can imagine

The least interesting job/activity you have performed

The least interesting job/activity you can imagine

    1. Assess the kinds of employment you would find interesting.
    2. Assess the kinds of employment you would not find interesting.
    3. Assess the hobbies/activities you find most interesting.
    4. Assess the hobbies/activities you find least interesting.

D. List, in order from most to least appealing, five specific jobs for which you might apply.

E. State your goals.
    1. Short-range
    2. Long-range

[1]We are indebted to Richard Bolles, *What Color is Your Parachute?* (Berkeley: Calif.: Ten Speed Press, 1984), p. 78, for this adaptation of the hierarchy-of-skills chart.

In the first three sections of this personal inventory, you gather the facts that form the basis of your resume: personal data, education, and work experience. An effective resume contains information in each of these areas.

In section IV, you list three or four references, taking care to choose people who know your skills well enough to write favorable letters and whose backgrounds best support your case. If you are a student, you will usually list professors, but you should also include a variety of references, perhaps a part-time or full-time employer. Moreover, be certain to contact these people before including their names, to obtain their permission.

In section V, you consider your skills, personal qualities, and interests in order to discover kinds of positions you might like to hold and your suitability for those you find interesting.

Hierarchy of Skills. This hierarchy asks you to consider your skills according to the following categories:

Skill	*Duty*	*Example*
Data-related	Working with information	Computer analyst Accountant
People-related	Working with others	Clinical psychologist Teacher
Thing-related	Working with objects	Mechanical engineer Civil engineer

Of course, these three categories may overlap. For example, an accountant must also work with people, and a mechanical or civil engineer must work with facts and figures. However, separating these skills into three categories allows you to ascertain the pattern of your talents.

In addition, these skills are arranged hierarchically, from most prescribed at the base of each list to least prescribed at the top. Prescription refers to the amount of freedom in a job using that particular skill. For example, consider a data-related job involving copying: feeding documents into a copier. Persons holding such a position would have very little freedom in defining and carrying out their duties. These duties would be very heavily prescribed. On the other hand, a data-related job involving synthesis — computer programming, for example — would be largely nonprescribed. Computer programmers would decide precisely how they would construct each program and exactly what it would contain.

If you prefer a job with more freedom of decision, you should concentrate your search on positions involving your nonprescribed skills. If you prefer a more prescribed position, you should look for a job requiring those skills. For example, Dave prefers jobs with more freedom of decision, so he will consider positions using his people-related, less prescribed skills: advising, negotiating, and supervising.

Grid of Personal Qualities. The grid of personal qualities asks you to rate yourself according to characteristics employers find important. This grid allows you to see patterns, which you then assess in relation to employment you would prefer. You may choose jobs using your strongest qualities, or you may develop the qualities your profession requires. For example, Dave will select jobs using his strongest characteristics: communication, cooperation, and leadership.

Assessment of Interests. The purpose of the assessment of interests is to help you rate previous activities as areas you might or might not like to enter and to suggest new directions. For example, Dave realizes he does not enjoy competitive activities, so he will avoid jobs involving competition.

List of Possible Jobs. This list of possible jobs is designed to stimulate thoughts about employment you would enjoy, if you are unsure of the field you would like to

Dave's Resumes

enter. Listing these jobs may also reveal that you know little about your job possibilities and thus need to research further.

Dave lists the jobs he would prefer: working with the developmentally disabled, working in recreational programs and in day-care centers. He also lists positions as a teacher's aide and a hospital orderly as further possibilities, in case he cannot obtain the jobs he would prefer.

Short-Range and Long-Range Goals. Your short-range goals are your immediate career objectives. Your long-range goals are your future objectives: the directions you would like your career to take. For example, your short-range goal may be to obtain an entry-level accounting position. Your long-range goals may be to obtain your C.P.A. rating, further your schooling, and move into a managerial position. Dave's short-range goal is to supervise, teach, or counsel the developmentally disabled. His long-range goal is to direct social programs.

You should specify your goals as precisely as possible, in order to be certain you are sure of them and can express them effectively on your resume.

Examining Resumes

If you are unfamiliar with your options in terms of the content, arrangement, and style of resumes, examining as many sample resumes as you can will help you discover possibilities. You should obtain these resumes from people in your field, because resume styles vary among disciplines. Your college placement office may be able to provide sample resumes in addition to other job-hunting information.

Selecting and Arranging Content

Selection

A resume contains personal data, and information on education and work experience. Some of this information is obligatory; other details are optional.

Personal Data. The obligatory personal data include your name, address(es) (permanent and temporary), and telephone number(s). Other personal details are optional: They should be included only if they support your candidacy. For example, a single marital status may indicate a willingness to relocate, while hobbies may reveal your range of interests.

Although a career objective is also optional, including this objective is often useful. The objective indicates your interests, so potential employers can determine if you fit their needs. The objective also allows you to stress your personal goals and illustrates your goal-directedness. You may include only short-range goals if you are unsure of your future direction or if the job for which you are applying does not fulfill your long-range goals. However, including both short-range and long-range goals in your career objective better informs your audience of your desires.

Education. The name of your college or university and the degree or expected degree with its date are obligatory educational details. (Data on high school are usually omitted when you are a college graduate.) You may include your G.P.A. (grade point average) if this is strong (3.0 or above). If your overall G.P.A. is lower than that in your major, you might list only the latter, or include them both, clearly identifying each. If you include course work, you should select only the most relevant courses for the positions you want; omit general courses an employer would assume you had taken. Your honors should be listed, and activities may also be included if they indicate your particular interests and strengths. However, avoid a lengthy list of activities; select only the relevant ones and the ones where you held an office. You should also explain, in parentheses, any honors or activities with which your audience would be unfamiliar, e.g., chairperson, VEISHEA (Iowa State University spring festival).

Work Experience. You should include all important jobs you have held, whether or not they were in your area of expertise. Jobs in your area indicate your experience in this field. Other jobs indicate your general capabilities. However, you should omit unimportant positions (e.g., the paper route you held at age 12), unless you have had no other work experience. In this case, include all jobs to favorably impress employers with the fact that you have worked.

List the position, the employer, the dates, and a description of duties for each job. This description of duties will allow you to indicate your responsibilities in the position, stress your suitability for the positions for which you are applying, and give a sense of your capability.

References. Here you list three or four references or indicate that references are available on request. The latter method eases the burden of letter-writing for your references and allows you to select those you would prefer for a particular position.

Dave decides to include his hobbies for optional personal data, because they indicate his active nature — e.g., jogging, backpacking, slow-pitch softball. Thus, his hobbies will support his candidacy for physically demanding jobs and for those in recreational leadership. He will also include career objectives on each resume because phrasing these differently will help him suit each resume to its specific readership.

Under education, Dave decides to include a list of course work, because he can then select different courses for each resume and emphasize different aspects of his background. For activities, he chooses only his participation in cross-country track, again supporting his active nature.

Under work experience, Dave will include his last three jobs (resident assistant in a dormitory, assistant house supervisor at Storr County Home for the Developmentally Disabled, camp counselor at Rockville YMCA Day Camp). One specifically relates to his experience in working with the developmentally disabled; the other two indicate his interpersonal skills, important for all the jobs he would like. He will omit one job he has held (liaison between newspaper carriers and the management of a newspaper), because this position did not involve skills he wishes to stress and because he has had other, more significant, work experience.

Arrangement (Structure)

The overall sequence of information in a resume is very important, because position determines the amount of emphasis details will receive. The pattern of most to least important is the most effective order.

Obligatory personal data (name, address, telephone number) begin a resume since audiences first need to know who you are and where to contact you. The remaining personal data, if included, are usually subordinated by being placed at the end of the resume because other information is more important in presenting your capabilities.

Your career objective usually follows your obligatory personal data, so that potential employers can immediately determine if you suit their needs. You then have a choice of information to put after the objective: education or work experience. (Some resumes, called skills resumes, are organized around the applicant's skills rather than education and work experience. In our samples, we include a resume of this type.) If you are a student, your strongest qualification may be education. However, if you have had significant work experience in the field for which you are applying, work experience can be placed before education, so that your work experience receives appropriate emphasis.

You should arrange both your education and your work experience from most to least recent because prospective employers want to see the most recent information first. However, you have a choice of organizing principles for the order of information *within* entries. You might construct a chronological resume, putting the dates of your education or work experience first, in which case the date you attended the school or held the job is emphasized. You might also construct a functional resume, beginning each entry under education with the school attended or the degree attained and each entry under work experience with the employer or the position held. In this case, these data would be stressed rather than the dates, which you subordinate by placing them later in the series. Of these two resumes, the functional resume is generally more useful than the chronological because other items of information are more impressive than the dates you attended school or held a job.

If you include a selection of course work under education, you should classify these courses and arrange them from most to least important: An undifferentiated, random list does not allow you to stress specific items. For example, you might group your courses as major and minor or classify them into areas of emphasis, e.g., sociology, psychology. You would then place the most important group first.

You should also classify your honors and activities, placing the most persuasive first. Honors are usually more impressive than activities, unless your activities involve skills important to your field.

Optional personal data, if included, then follow the sections on education and work experience. Since personal data are less important than education and work experience, you subordinate them by this placement. The last section of the resume concerns references. This information *is* important, since readers will use it to obtain others' opinions of you, but it alone will not persuade prospective employers of your capabilities, so it should be placed at the end of the resume.

Dave decides to place his work experience before education in resume 1 because of his position as assistant house supervisor at Storr County Home, a job in the field of developmental education. Placing work experience first will allow him to emphasize these data. In resume 2, he will place education before work experience, because his jobs have not been directly related to recreational leadership or day care. However, he has taken coursework in the former, which this placement will allow him to emphasize. He will group these courses according to the areas of his double major, placing psychology first on resume 1 and sociology first on resume 2, to suit his readers' differing interests.

Dave also decides to construct a functional resume, beginning his entries with the name of his university and with the position he has held. He places these facts first in the series because he wants to stress them. His university has the reputation for an excellent intern program in developmental education. Moreover, he held a fairly responsible position at Storr County Home.

Arrangement (Format)

An effective format will allow readers to access the information on your resume easily. Moreover, format may also affect their opinion of you. A poorly formatted resume may suggest a disorganized or careless worker.

Position on the page, capitalization, and underlining may be combined to emphasize or subordinate information appropriately. In addition, headings will signal the classification of data in your resume (e.g., education, work experiences, personal details) and call reader attention to these groups. However, if you have several levels of headings, be sure to format those on the same level consistently. Lists may also be used under education and work experience to segment information for your reader.

Effective use of white space is especially important in a resume. Incorporate vertical white space by single-spacing entries and double-spacing between entries. Incorporate horizontal white space by grouping or blocking material and indenting material you wish to subordinate — e.g., coursework you list under education. This indentation will differentiate the levels of your information, providing organization and contributing to ease of reading. In addition, maintain adequate margins all around your resume.

Dave decides to center his obligatory personal data and to capitalize his name so it will stand out. He will use headings — left margined, capitalized, and underlined — for his career objective and the remaining sections of his resume. He chooses to left margin these headings because centering them would increase the length of his resume; they would have to be raised above the information to which they refer. He will list the information under these heads, being careful to incorporate white space by indenting and blocking.

Planning Style

Phrases instead of sentences are usual in resumes because sentences make accessing information difficult and increase the resume's length. For parallelism, however, you should use phrases throughout the resume.

You should also carefully consider your choice of words, particularly in your career objective and description of duties under work experience. The words in your career objective shoud be precise, since a very general career objective may not provide your reader with the needed information. Compare Dave's first statement and the revision below:

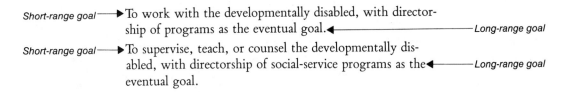

Statement 1 does not suggest the kind of short-range position Dave would prefer or the type of programs he would like to direct. Therefore, his readers cannot use the statement to decide if he fits their needs. The word choice in statement 2, on the other hand, specifies these facts. Moreover, statement 2 has a more definite tone, thus stressing Dave's goal-directedness more effectively.

Your description of duties under work experience should be phrased to include precise action verbs. These action words stress your energy and capability and pinpoint for your reader the exact kind of experience you have had. Consider the following examples:

1. *Had experience with* campers with behavioral and emotional difficulties.

    Instructed and supervised campers with behavioral and emotional difficulties.

2. *Worked with* educational and recreational programs.

    Planned and directed educational and recreational programs.

In each case, the second sentence is more forceful and more specifically informs the reader about the nature of Dave's work experience.

Writing

We give Dave's first drafts of his resumes on the next two pages, with annotations pointing out the differences between them. The first resume was prepared for directors of institutions for the developmentally disabled. The second was to be sent to directors of programs in recreation and to heads of day-care centers.

Rewriting

Revision of your resume is very important, to ensure an effective document. Finding other readers to critique the resume will aid you in the revision task.

You should select competent readers who are acquainted with the principles of constructing resumes and with the styles of resumes common in your field. The

Resume 1

DAVID J. WELLS
1133 Beedle Street #9
Westville, WI 40632
(515) 294-8591

CAREER
OBJECTIVE

To supervise, teach, or counsel the developmentally disabled, with directorship of social-service programs as the eventual goal.

WORK
EXPERIENCE

Moore House, University of West Wisconsin, Westville, WI 40632. 1980–82.

Work experience is placed first for emphasis.

The date is subordinated in a functional resume.

Position: Resident Assistant
Duties: Supervised 20 men on a dormitory floor.

Precise action verbs stress Dave's capabilities.

Storr County Home, Saylorville, IL 45910. May–Aug., 1981.

Indentation indicates levels of information.

Position: Assistant House Supervisor
Duties: Lived in as counselor on a dormitory floor. Took responsibility for 10 boys, ages 4–10. Directed leisure activities for 30 clients of all ages.

Rockville YMCA Day Camp, RR#1, Rockville, IL 41781. May–Aug., 1980, 1979.

Positions: Assistant Director
Counselor
Duties: Instructed campers with behavioral and emotional difficulties. Planned and directed educational and recreational programs for all campers. Supervised campers of mixed ages and abilities.

EDUCATION

University of West Wisconsin, Westville, WI 40632.
B.A., Psychology/Sociology, expected May, 1985.
G.P.A.: 3.7 overall
3.8 major

Course work is arranged to stress Dave's background in psychology.

Course work: Developmental Psychology
Counseling Theories and Techniques

Small Group Dynamics
Leisure Studies
Recreational Leadership

Blocking information contributes to reading ease.

Honors/Activities: Dean's List (3 years)
President's Scholar
Cross-Country Team (2 years)

HOBBIES

Jogging, backpacking, slow-pitch softball, gourmet cooking, ceramics.

REFERENCES

Available on request.

Resume 2

DAVID J. WELLS
1133 Beedle Street #9
Westville, WI 40632
(515) 294-8591

CAREER
OBJECTIVE

To oversee educational and recreational activities for citizens of all ages and abilities, with the eventual goal of directing social programs.

The career objective is more general about the kind of position desired, the people with whom Dave will work, and his long-term goal, to interest readers with varying needs.

EDUCATION

University of West Wisconsin, Westville, WI 40632.
B.A., Sociology/Psychology, expected May, 1985.
G.P.A.: 3.7 overall
       3.8 major

Education has been placed first, so it receives more stress.

The majors have been reversed to stress Dave's background in sociology.

Courses have been rearranged to emphasize Dave's preparation for recreational leadership.

Course work: Recreational
           Leadership
           Leisure Studies
           Small Group
           Dynamics

           Counseling
           Theories and
           Techniques
           Developmental
           Psychology

Honors/Activities: Dean's List (3 years)
                 President's Scholar
                 Cross-Country Team (2 years)

The sequence of courses has been rearranged to emphasize Dave's preparation for work with people of all abilities.

WORK
EXPERIENCE

Moore House, University of West Wisconsin, Westville, WI 40632. 1980–82.
    Position: Resident Assistant
    Duties: Supervised 20 men on a dormitory floor.

Storr County Home, Saylorville, IL 45910. May–Aug., 1981.
    Position: Assistant House Supervisor
    Duties: Lived in as counselor on a dormitory floor. Took responsibility for 10 boys, ages 4–10. Directed leisure activities for 30 clients of all ages.

Rockville YMCA Day Camp, RR#1, Rockville, IL 41781. May–Aug., 1980, 1979.
    Positions: Assistant Director
               Counselor
    Duties: Planned and directed educational and recreational programs for all campers. Supervised campers of mixed ages and abilities. Instructed campers with behavioral and emotional difficulties.

These duties have been rearranged to emphasize Dave's work with children of all ages and abilities.

HOBBIES

Jogging, backpacking, slow-pitch softball, gourmet cooking, ceramics.

REFERENCES

Available on request.

director of placement at your college or professionals who have conducted interviews for their companies are excellent choices as mock audiences. You should then review your resume, using as criteria amount and kind of detail, appropriate emphasis, logical progression (including format), stylistic appropriateness, and mechanical accuracy.

When Dave reviews his resume, he decides to add a detail: an explanation of President's Scholar (senior award), because his audience will not know the significance of this honor. To achieve appropriate emphasis, he rearranges his G.P.A.'s so that his major G.P.A. appears first. Since it is stronger than his overall G.P.A., he wishes to stress the major G.P.A. as the more persuasive detail.

After carefully reviewing your resume, you should edit it for mechanical accuracy. Then have it copied on standard 8½" by 11" paper. Plain white is usually preferred, and a high-quality photo offset of your typed copy is better, in most cases, than a photocopy. You can also have your resume professionally printed. However, professional printing tends to standardize your resume and may suggest to an employer that it has also been professionally prepared.

Dave has photo offsets made. His resumes are now ready to be included in letters of application.

CHECKLIST FOR RESUMES

1. Have I used the following techniques to gather information?

   A. Conducting a personal inventory

   B. Examining sample resumes

2. Have I selected appropriate details from my personal inventory on education, work experience, and personal data? Have I included too many or too few details?

3. Have I considered the arrangement of details for appropriate emphasis?

   A. Have I constructed the most appropriate form of resume: chronological or functional?

   B. Have I followed the overall sequence of most to least important?

   C. Have I arranged details within sections so that the most important details precede those of lesser importance?

4. Have I used an effective format?

   A. Are my headings highlighted? If I have used several levels, are they formatted in parallel ways?

   B. Have I used white space effectively?

      1. Have I maintained adequate margins?

      2. Have I indented and blocked appropriate material?

      3. Have I single-spaced entries and double-spaced between entries?

   C. Have I highlighted appropriate material by position, capitalization, underlining, and white space?

Dave's Letter of Application   **223**

5. Have I written an effective career objective? Do I want to include both short-range and long-range, or only short-range, goals?

6. Have I used phrases throughout the resume? Have I used precise language in my career objective and action verbs in my job descriptions?

7. Have I carefully edited my resume?

DAVE'S LETTER OF APPLICATION

Prewriting

Analyzing the Communication Context

You must gather all the information you can about the audience of your letter of application. If you are answering an advertisement, it will often state an audience. For example, Dave is writing his letter of application in answer to the following advertisement:

> *House Supervisor:* Black County Home for the Developmentally Disabled. Position requires B.A. degree in one of the following areas: psychology, sociology, education. Black County Home is a live-in facility. Candidate will be responsible for overseeing live-in clients and directing programs in education and recreation for additional clients. He or she will also supervise three assistants. Excellent communication skills required. Send resume and letter of application to
>
> Ms. Camille Monard, Director
> Black County Home for the
> Developmentally Disabled
> 386 Goucher Road
> Stanton, MN 47981

The director of Black County Home, Ms. Camille Monard, is Dave's primary audience. She will use the letter, in conjunction with his resume, to decide whether or not to interview Dave. He must inform her effectively of his strengths for the position so she can make this decision.

Secondary audiences may include the board of directors, who would ratify Ms. Monard's decisions, and other employees at the home: the current supervisor and assistant house supervisors. These readers would use the letter to advise Ms. Monard for or against granting an interview.

If you cannot discover the name of the person to whom you are addressing your letter, you can address it to a job role (e.g., Director of Personnel). You should then gather all the information you can about the company, in preparation for writing your letter. You should also research the job for which you are applying, so that you know the issues readers will expect you to address. This research may also help you discover alternative jobs, in case you cannot obtain the ones you would prefer.

224   11 | Resumes and Letters of Application

Gathering Information

Consulting Information Sources

Your college placement office and library contain many sources for information on jobs. The following are some of these sources. The starred entries will give information on specific companies.

1. Books

   Richard Bolles. *What Color Is Your Parachute?* Berkeley: Ten-Speed Press, 1984.

   Career Planning Handbook: A Guide to Career Fields and Opportunities. U.S. Civil Service Commission. Washington, D.C.: Government Printing Office.

   Muriel Lederer. *Guide to Career Education.* New York: Quadrangle/New York Times Book Co., 1974.

   *Occupational Outlook Handbook. U.S. Bureau of Labor Statistics. Washington, D.C.: Government Printing Office.

   SRA Occupational Briefs. Chicago: Science Research Associates.

2. Magazines

   Business Periodicals Index contains a listing of magazines you might consult.

*3. Newspapers

   Classified advertisements will suggest possible positions and expected qualifications as well as actual openings. The *Wall Street Journal* and the *New York Times* are particularly useful sources.

4. Publications

   *College Placement Annual. Bethlehem, PA: College Placement Council, annually.

   Fortune Magazine's Directory of the 500 Largest Industrial Corporations. New York: Time, Inc., annually.

   *Middle Market Directory. New York: Dun & Bradstreet, annually.

   *Million Dollar Directory. New York: Dun & Bradstreet, annually.

   *Reference Book of Manufacturers. New York: Dun & Bradstreet, annually.

   Standard & Poor's Registry of Corporations, Directors, and Executives. New York: Standard & Poor's Corp., annually.

   *Standard Corporation Records. New York: Standard & Poor's Corp., monthly.

   Thomas Register of American Manufacturers. New York: Thomas Pub. Co., annually.

5. Professional journals

   These journals frequently contain advertisements of positions available. You should consult those journals in your field.

6. Company annual reports, brochures, magazines, and newsletters

   Many companies will send these upon request.

7. Lists of positions

   Most placement offices maintain lists of available positions.

Conducting a Job-Analysis Inventory

A job-analysis inventory asks you to evaluate your skills and abilities in relation to an actual position. By using this inventory, you systematically examine your strengths and weaknesses in terms of the characteristics your audience finds most desirable. You then use this information in your letters of application.

We give this job-analysis inventory in Figure 11.2 on page 226.

Notice that this job-analysis inventory contains the same categories of information as your personal inventory: personal data, education, and work experience. However, the job-analysis inventory differs from the personal inventory in that you are determining the requirements of a specific position, then selecting information from your personal inventory that indicates your suitability or unsuitability for that position. Thus, the job-analysis inventory assists you in selecting content for your letter of application by directing you to points you will want to stress or deemphasize.

Selecting and Arranging Content

Selection

If you are answering a specific advertisement, you should use it as an aid in selecting content. For example, Dave's advertisement lists several responsibilities of the job as house supervisor: overseeing clients, directing programs in education and recreation, supervising assistants. The advertisement also stresses the necessity for excellent communication skills. Such information is valuable because it gives the topics you must address. You must support the fact that you can perform the duties required or have the skills the job demands. In addition, such an advertisement frequently follows a hierarchical order, with the most important duty or skill placed first. Therefore, you may want to arrange your letter according to the order the advertisement suggests, if that order is effective.

If you are not answering a specific ad, you should determine from the facts you have learned about the company or institution and from section II of your job-analysis inventory (qualifications in relation to position) the topics your letter must address. In either case, your task in selecting content is to tailor your letter as effectively as possible to the needs of your audience, then to choose details from your background to support your points.

A checklist of possible details to include in the first and last paragraphs of a letter of application is another aid in selecting content. We give this list in Figure 11.3 on page 237, then discuss the details further.

The first paragraph of your letter should contain two items of factual information: (1) the specific position for which you are applying, since a company may be

FIGURE 11.2
Job-Analysis Inventory

Job-Analysis Inventory

I. Position
   A. Title/Job role
   B. Company/Institution
      1. Name
      2. Description
   C. Job description

II. Qualifications in relation to position
   A. Personal data
      1. Assess yourself in relation to skills required
         a. Data-related
            (1) Skills required
            (2) Assessment
         b. People-related
            (1) Skills required
            (2) Assessment
         c. Thing-related
            (1) Skills required
            (2) Assessment
      2. Assess yourself in relation to personal qualities required
         a. Major characteristics required (from most to least important)
         b. Assessment
      3. Assess your interests and hobbies in terms of this job
         a. The most interesting aspect of this job
         b. The least interesting aspect of this job
         c. Use of your hobbies or activities
   B. Education
      1. Describe the educational background this job requires
         a. Degree/Discipline
         b. Additional training
         c. Course work
      2. Assess your educational background in terms of this job
         a. Degree/Discipline
         b. Additional training
         c. Course work
   C. Work experience
      1. Describe the previous work experience this job would demand
      2. Assess your work experience in terms of this job

III. Preferences
   A. Salary
   B. Geography
   C. Size of organization
   D. Fellow workers
   E. Work environment

FIGURE 11.3

Checklist of possible details to include in the first and last paragraphs of a letter of application

First Paragraph
1. Position applied for
2. Source of information
3. Attention-getting opening

Last Paragraph
1. Interview request
2. Dates of availability
3. Inclusion of telephone number and times to be reached

advertising for several openings, and (2) the source of your information about the opening, which tells your reader how much you know about the position. However, this beginning paragraph should also capture your reader's attention. Several openings will help you accomplish this task[1]:

1. *A name opening* Used if you know the person you are addressing or someone with whom he or she would be familiar

2. *A news-item opening* Used if the company or institution, or someone connected with it, has recently appeared in the news

3. *A personal opening* Used if you have a particular interest or hold a belief shared by your audience: for example, Dave's desire to work with the developmentally disabled or his philosophy concerning this group of people

4. *A summary opening* Used as a summary of your background in relation to the job, in a forecasting or predicting sentence. This sentence appears at the end of the first paragraph or, if lengthy, in the second paragraph, and sets up the order of information in the letter. The summary opening is effective because it points out your strengths, which you reiterate in the topic sentence of each supporting paragraph, and relates your background to the position: the purpose of any letter of application

These openings may be used singly, or you may combine them for a more effective paragraph.

Dave investigates back issues of the *Tribune*, where he discovers a feature story on Black County Home's Special Olympics Day. In the story, Ms. Monard articulates her philosophy on the developmentally disabled, which Dave shares. He decides he will use a combination of openings: the news-item and the personal opening, then

---

[1]We are indebted to Herta Murphy and Herbert Hildebrandt for this information. See *Effective Business Communications* (New York: McGraw-Hill Book Company, 1984), pp. 364–65.

the summary as a forecasting sentence to identify his qualifications, link them to the position, predict the topics of his letter, and form a transition to the remainder of the letter.

In the last paragraph of a letter of application, you should conclude on a courteous but positive note by requesting an interview, the appropriate "action closing" to a letter of application. You should also make the reader's response easy by stating the dates you will be available and giving your telephone number and the times when you are likely to be home.

Arrangement (Structure)

Your major structural decision in a letter of application concerns the middle paragraphs of the letter, where you cover your qualifications for the position in detail. Here you may include information on the three areas of any job-related document: personal data, education, work experience. However, you have a choice of overall pattern for these paragraphs:

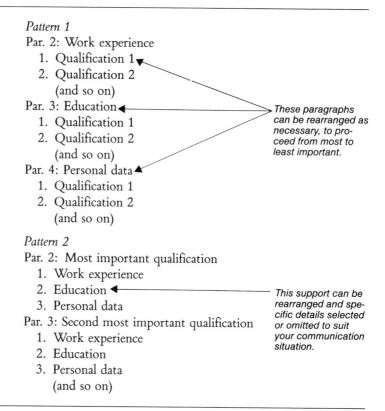

Dave's Letter of Application   **229**

Although either pattern may be used, pattern 2 results in a more effective letter, because it allows you to stress selling points: your qualifications for the job, arranged from most to least important. Thus, you direct your reader's attention to the specific points you wish to emphasize, as given in your summary sentence, by stating each point in a topic sentence and supporting that sentence with concrete examples and details. In this way, you tailor your letter to audience needs. Dave decides to use pattern 2 in his letter and structure it around the duties and requirements of the job, as given in the advertisement he is answering.

Once you have determined the topics you will address and the overall pattern you will use, you should outline your content for the body of the letter of application. We give Dave's outline in Figure 11.4, with annotations on important points.

FIGURE 11.4
Dave's outline of content for the body of his letter

I. Overseeing live-in clients ◄——————————— *This information is placed first because the advertisement listed this duty first. Dave continues to follow the order given in the ad.*

This support is arranged from most to least important, in terms of Dave's audience. They will be most interested in his work with the developmentally disabled and then in his live-in experience as resident assistant. ——► A. Work experience
   1. Directed living experiences of clients at Storr County Home
   2. Supervised dorm floor members as resident assistant

II. Directing programs in education and recreation
   A. Work experience ◄——————————————— *Dave omits education and personal data as support, because work experience in the field best indicates his actual skills. Education and personal data only indicate his interest.*
      1. Directed leisure activities for clients and taught living skills at Storr County Home
      2. Directed educational and recreational programs at Rockville YMCA Day Camp
      3. Taught campers recreational skills

Only courses that pertain specifically to this requirement of the job are given. ——► B. Education
      1. Leisure Studies
      2. Recreational Leadership

This paragraph is placed third because Dave has had the least experience in supervision. Thus, he places the paragraph between two stronger qualifications, to de-emphasize it. ——► III. Supervising assistants
   A. Work experience
      1. Supervised clients and campers
      2. Led peers as resident assistant in dorm

IV. Communication skills
   A. Work experience
      1. Wrote individual program plans and assessments of individual clients at Storr County Home
      2. Wrote semiannual reports as resident assistant
   B. Education
      1. Wrote a report as part of my *Counseling Theories and Techniques* practicum: a case study of a client, with suggested therapy
      2. Course: *Writing Professional Papers and Reports*

---

230   11 | Resumes and Letters of Application

Writing

We give the first draft of Dave's letter of application, with annotations.

133 Beedle Street #9
Westville, WI 40632
October 24, 1984

Ms. Camille Monard, Director
Black County Home for the
Developmentally Disabled
386 Goucher Road
Stanton, MN 47981

Dear Ms. Monard:

News-item opening ▶   I read the *Minneapolis Tribune's* recent article on Black County Home's Special Olympics Day with considerable interest. The philosophy you indicated
Personal opening ——— concerning the developmentally disabled — that they are merely people with
special needs — is mine as well. Therefore, I would like to apply for the open-
Position applied for ——— ing as house supervisor that you advertised in the *Tribune.*
Source of information ———
            You indicated that overseeing live-in clients, directing educational and rec-
reational programs, supervising assistants, and communicating well would be
important considerations. I feel well-prepared in these areas.

Topic sentence: first ▶   I have had considerable experience overseeing live-in clients, both at Storr
point in summary County Home and in the university residence halls. My duties at the home
opening involved responsibility for the living experiences of my group of developmen-
Specific examples ◁ tally disabled boys, from 6:00 p.m. to 8:00 a.m. Because I was considered par-
and concrete details ticularly able, I was given the rank of weekend supervisor. In addition, as
given as support resident assistant, I encouraged an agreeable living atmosphere by instituting a
new procedure for handling grievances.

Topic sentence: sec- ▶   I directed educational and recreational programs in many of my jobs as well.
ond point in summary At Storr County Home, I set up new leisure activities in fitness and outdoor
opening education, and taught living skills. At Rockville YMCA Day Camp, I directed
programs for all campers and taught developmentally disabled campers volley-
ball, swimming, and crafts. My course work in leisure studies and recreational
leadership prepared me for these positions.

Topic sentence: third ▶   My supervisory experience is best shown through my work with clients,
point in summary campers, and students. For example, I effectively supervised dormitory parties
opening by turning away uninvited guests in ways that avoided disruptive scenes.

Topic sentence: ——▶   My jobs have all involved communication skills. I have written individual
fourth point in sum- program plans, assessments of clients, and semiannual reports. As a student, I
mary opening

have also written an extensive case study for *Counseling Theories and Techniques* and have taken a course in writing professional papers and reports.

Mention of resume alerts the reader to it ▶ When you have reviewed my credentials and the enclosed resume, I would *Request for interview* ▶ be interested in further discussing my qualifications for the position of house supervisor. I will be in the Minneapolis area at Christmas and could visit Black *Time of availability* ——— County Home at that time. You may reach me at (515) 294-8591 any evening *Telephone number and time to be reached* ——— after 6:00 p.m.

Sincerely,

David J. Wells

David J. Wells

Encl: resume

Notice that the letter of application does not simply reproduce information given on the resume. Instead, it expands on selected details from the resume. For example, Dave adds the facts that he was responsible for his group of boys at Storr County Home from 6:00 p.m. to 8:00 a.m. and the rank he was given. He omits the boys' ages, as that information is on his resume.

Rewriting

Your major criteria for revising a letter of application are amount and kind of detail, appropriate emphasis, and logical progression. You should include enough concrete examples to support your candidacy and arrange your points so they are effectively stressed.

When Dave rereads for amount of detail, he feels that the simple mention of his course work in paragraph 4 does not provide adequate support, since readers could obtain that information from his resume. He adds the following details to the sentence:

My course work . . . prepared me for these positions by showing me the skills necessary for effective direction and giving me practice in planning and scheduling.

Since Dave has followed the order of the requirements given in his advertisement, he finds no changes he wishes to make in logical progression. He also feels he has given details appropriate emphasis by placing his weakest area (supervising) between two stronger ones.

He now types his letter and edits it carefully, knowing that any mechanical errors will seriously interfere with his persuasive purpose.

CHECKLIST FOR LETTERS OF APPLICATION

1. Have I used the following techniques to analyze my audience and gather information for my letter?

   A. Examining an advertisement

   B. Consulting information sources

   C. Conducting a job-analysis inventory

2. Have I given the position I am applying for and the source of my information in the first paragraph?

3. Have I chosen an appropriate opening? Should I use several?

4. Does my summary sentence include the important requirements of the job, as determined from an advertisement or from my job-analysis inventory?

5. Is the order of the requirements in the summary sentence effective? In my paragraphs, do I take up each requirement in the order given and announce the requirement in my topic sentence?

6 Have I included sufficient concrete examples as support? Do the examples extend the information given in my resume rather than simply repeat data?

7. Have I requested an interview, given dates of availability, and provided a telephone number and times to be reached in my closing?

8. Have I carefully edited my letter?

SAMPLE RESUMES AND LETTERS OF APPLICATION

Resumes

Our first sample was written to accompany a letter of application to the Farmers' Home Administration for a permanent job. The writer has placed education first, despite her summer work experience with FHA, because a B.S. or B.A. degree is a requirement of the position she would like.

Resume 1

ANN MURPHY
207 Fifth Street # 5
Lehigh, PA 60154
(614) 281-7634

CAREER OBJECTIVE	To assist farmer clientele in the area of agricultural financing, with the eventual goal of specializing in loans.
EDUCATION	Iowa State University, Ames, IA. B.S. degree in Agricultural Business, May, 1983.

Sample Resumes and Letters of Application   **233**

Specialization: Public policy, with a concentration in agricultural operations, production, finance, and law.
G.P.A.: 3.85/4.0.
All college expenses were financed by work, scholarships, grants, and loans.

WORK
EXPERIENCE

Farmers' Home Administration, Le Mars, IA. Student trainee, summer, 1980–81.
Responsibilities: writing and analyzing loans, analyzing cash flow, appraising real estate, inspecting developments and securities.
Le Mars Daily Sentinel, Le Mars, IA. Staff member, summer, 1978–79.
Responsibilities: laying out and designing ads, proofreading, writing sports and feature articles.
Part-time while in school: Iowa State University, cleaner and mail clerk.

HONORS AND
ACTIVITIES

Phi Kappa Phi honorary
Gamma Sigma honorary (agriculture)
Phi Eta Sigma (freshman honorary)
National Merit scholarship

National Agri Marketing Association (member)
Ag Business Club (vice president)
Residence Hall House Cabinet

REFERENCES

Available on request.

Our second sample is a skills resume, where the categories of education and work experience are not used to present the material. Instead, the resume is organized around skills or areas of expertise. Skills resumes are constructed by persons who have been out of school a number of years and have accumulated significant on-the-job experience. Notice that the skills are arranged from most to least important. Also notice that obligatory personal data still begin the resume.

Resume 2

THOMAS ALDEN
16 Lincoln Drive
Roland, MI 60125
(618) 429–7601

OBJECTIVE

A joint position in administration and teaching with responsibilities in media production and/or graphic design.

EMPLOYMENT

Media Resource Center   State University   1980–present
                          Riverside, MI

234   11 | Resumes and Letters of Application

Resume 2 (continued)

Station KXI-TV	Brook Haven, MI	1979–80
Graphic Arts Department	Northridge University Binghamton, NY	1974–79

REVIEW OF WORK EXPERIENCE

Media Production

☐ Held position of writer/director, in charge of shooting, processing, and editing materials for university and private clients.

☐ Experienced in all phases of producing filmstrips, films, and videotapes with accompanying dialogue.

☐ Regularly responsible for videotaping university sports events.

☐ Produced an award-winning film for a campus department. (Midwest Media Production competition)

☐ Have produced approximately 50 filmstrips, films, and video tapes in a four-year period.

Graphic Design

☐ Designed brochures, pamphlets, and posters for state and campus organizations.

☐ Regularly responsible for all advertisements for Northridge Center, a fine-arts, entertainment, and conference complex.

☐ Designed By-Ride signs for city transportation system.

☐ Illustrated numerous manuals for clients.

☐ Produced a cartoon map of the city for a promotional campaign.

Animation

☐ Drew cells for animated children's films shown on public television.

☐ Responsible for animation of the Lollypup series, shown on a five-state network.

Photography

☐ As consultant, shot, developed, and enlarged photographs for private clients.

☐ Experienced in use of all darkroom equipment.

EXHIBITIONS AND FILMS

Still-Life Photography (One-man show)	First National Bank Riverside, MI
Wildlife of Minnesota (Graphic designs and photographs)	Center for the Arts Riverside, MI
October Moon (Blue ribbon winner, Photography)	State Fair Minneapolis, MI
Penny Carr: Running for Life (Award-winning film)	Food and Nutrition Department State University Riverside, MI
Excellence in the '80s (Fund-raising film)	State University Riverside, MI

| EDUCATION | M.A. degree in Fine Arts | State University Riverside, MI | 1974 |
| PORTFOLIO | Includes photographs and graphic-design work. Available on request. | | |

Letters of Application

Our sample letter of application was written by a senior English major who was applying for a position as a publisher's book representative. The person holding such a job would visit various colleges and universities to market texts. Therefore, excellent communication skills would be a major requirement of the position.

148 Moore Street
Sewanee, IA 50321
November 5, 1984

Ms. Lydia Blake
Personnel Manager
Terence Publishing Co.
1671 East Street
Chicago, IL 43271

Dear Ms. Blake:

A circular posted in the Sewanee State Placement Office advertised your opening for a book representative. I would like to be considered for the position, because of my communication skills and sales experience.

College has helped me develop communication skills, but my earlier years contributed as well. I was a reader before I entered grade school. In high school, I debated in country and state competitions and entered speech tournaments at the local and state levels. These experiences gave me self-confidence and poise in communication situations, important qualities in a person who must meet and talk with professors on a daily basis.

I continued to develop these qualities in my college years. I debated for two of those years. I also joined the creative-writing club and, by my senior year, had been appointed editor of *Sketch,* our club magazine. Through this experience, I learned how to communicate better through writing. But I also learned how to direct others and how to cooperate with them, because producing a magazine involves both managerial skills and collaborative efforts.

My sales experiences involved similar skills. As assistant sporting-goods manager in a local chain store, I had to translate my superior's directives to my fellow workers, after which we would cooperatively carry these orders out. I received this position in my third month of employment because of my leadership abilities and skill at communicating. I then received additional raises every six months, based on amount of goods sold. When I left the store at

236　11 | Resumes and Letters of Application

Letter of Application (continued)

graduation, our department's sales record was the best in the state, in part because of my colleagues and in part because of my sales abilities.

I will be in the Chicago area next month and would like to make an appointment with you to discuss my credentials. After reviewing the enclosed resume, you can reach me at (504) 231-8605 any evening. Thank you for your consideration.

Sincerely,

Michael Forbes

Michael Forbes

Encl: resume

EXERCISES

Topics for Further Practice

1. Complete a personal inventory in preparation for constructing a resume.

2. On the basis of your personal inventory, select one career area you feel you might like to enter.
   A. Analyze your audiences for a resume. Include both primary and secondary readers.
   B. Analyze the purposes and uses of your resume.
   C. Plan your resume. Include content, arrangement (structure and format), and style.
   D. Compose your resume.

3. Choose a second area from your list of five possibilities as the basis for a resume.
   A. Discuss specific differences between the audiences for this resume and the audiences in Question 2.
   B. Discuss specific differences between your purposes and the uses of your two resumes.
   C. Discuss specific differences in content, arrangement, and style, based on your audiences, purposes, and the uses of your resumes.

4. Locate a specific job for which you might apply. (If possible, provide an advertisement.)
   A. Complete a job-analysis inventory in preparation for writing a letter of application.
   B. Research the job and the affiliation in the sources we have listed.
   C. Analyze the audiences for your letter of application. Include both primary and secondary audiences.
   D. Analyze the purposes and uses of your letter.
   E. Outline your letter.
   F. Write your letter.

5. Locate a second job. Discuss specific differences between the letter of application in Question 4 and one you might write for this position.

12

Proposals

THE NATURE OF PROPOSALS

A proposal is a written offer to solve a problem. At times, a company may find that a problem cannot be solved from within, causing the company to seek outside help. At other times, an employee of a company may discern that a change in procedure would be advisable, or an agency such as the Department of Energy, the National Institute of Health, or the National Science Foundation may fund research projects to discover possible solutions to large-scale problems. Each of these situations might initiate a proposal.

Proposals are extremely important. They allow you to offer your services to a client and compete in the marketplace, or present ideas to improve your firm. Proposals also allow you to seek research funds that might otherwise be unavailable. Therefore, writing proposals is an important skill.

Proposals vary greatly in length and format. A short in-house proposal might be written as a memorandum, while a brief out-of-house proposal might be submitted as a letter. A proposal to a funding agency or a long out-of-house proposal might be written in full report form.

However, all proposals can be classified as either solicited or unsolicited, depending on whether or not the reader has requested the proposal. The solicited proposal is written in response to a company's Request for Proposals (RFP) or a funding agency's announced intention to review research projects. In each case, the RFP or the funding agency's guidelines provide a statement of the problem to be solved or the area of research interest. For example, the state of Utah might publish an RFP for designing and building a bridge to provide access to a developing town. The National Science Foundation might announce its intent to review proposals on the subject of

237

corrosion control. (Some funding agencies, such as the National Endowment for the Humanities, accept proposals on many topics, but only at stated times.)

The unsolicited proposal is initiated by the writer rather than by the reader. For example, you may have discerned that using rental carriers would be more cost effective for your company than maintaining a fleet of trucks. In this case, you might write an unsolicited proposal to your superior, requesting that you be allowed to study the problem.

In our illustration, we follow Arlen Hon, a mechanical engineer, through the process of composing a solicited proposal to solve a company's problem. Our samples at the end of the chapter then include a solicited proposal for a program of research and an unsolicited in-house proposal.

Arlen works for UOL Empiricists, Inc., an independent testing laboratory that competes for consulting contracts concerning fluid flow. Arlen has been assigned the task of writing a proposal in response to an RFP from Amnex Plumbing Manufacturers. Amnex has just begun to produce several smaller-sized pipes and fittings and would like to be a supplier for the Chemical Process Industry. However, before Amnex can enter this market, the company must become a certified equipment supplier. To do this, the Chemical Process Industry's review board requires that an independent laboratory test the company's pipes for pressure drop per unit length and friction factor, and the fittings for equivalent length. Arlen's supervisor feels the problem is one that UOL can solve. Therefore, he asks Arlen to write a proposal.

Before you begin a proposal, you must prepare a schedule for your task. This schedule is necessary because most proposals have a firm deadline. That is, they *must* be received at or before a stipulated time if they are to be considered at all. Your schedule will help ensure that you submit on time.

While making out the schedule, you should consider the time necessary for your technical inquiry and for writing the proposal, and also time for revising, constructing visual aids, typing, proofreading, and mailing. In addition, you should leave a generous amount of time between the projected delivery date and the deadline. Since your task may not always proceed as planned, this cushion will still allow you to meet your deadline.

FIGURE 12.1
Arlen's schedule for proposal preparation

January 4:   Analyzing the RFP and beginning the technical inquiry

January 5–18:   Continuing the technical inquiry

January 11–12: Consulting with management about selecting personnel for the project and estimating cost

January 18–19: Writing the draft and delivering sketches to the Graphics Department

January 20–21: Revising

January 22:   Typing, proofing, and duplicating

January 25:   Mailing by noon

                Deadline: February 1

To construct this schedule, begin with the submission date and proceed backward to the beginning of your work, to ensure that you do not exceed your time limits. In Figure 12.1, we show Arlen's schedule for preparing the proposal.

ARLEN'S PROPOSAL

Prewriting

Analyzing the Communication Context

Analyzing the Audience

The audience of an in-house unsolicited proposal may be clear (e.g., your supervisor or upper management of your company). In this case, you should gather all the information you can about your readers and their reactions to your project. In particular, you should note whether readers are likely to perceive the need for your project and whether they will react to it positively or negatively. For example, if your project involves changes in procedures that have existed for some time or if it eliminates the need for certain positions in your company, your readers may react negatively to the proposal. Knowing this fact will help you answer readers' questions and counter readers' objections.

The audience of an out-of-house solicited proposal, unlike the audience of an in-house proposal, may be difficult to identify and characterize.

Checking your company's files for records of previous work with this client, contacting members of your staff who might have dealt with the client, or telephoning appropriate officials at the funding agency may provide useful information. The RFP or the funding agency's guidelines may also assist you. For example, the RFP may state the names and job roles of the persons you should contact, giving you some information on their interests and levels of technical expertise. The funding agency's guidelines may give the members of the review board who will read your proposal and the level of expertise at which you should write.

Because UOL has not had any association with Amnex, Arlen will have to rely on the descriptions of the readers given in the RFP. Two audiences are listed: upper-division management, including Amnex's vice president, James Lolly, and engineers in the Design Department.

Since upper-division management and Mr. Lolly will make the final decision on awarding the contract, they are one primary audience. However, the RFP also states that the technical proposal, which will primarily determine the capabilities of a firm to perform the project, will be evaluated by design engineers. Therefore, this group is also a primary audience.

Arlen assumes James Lolly is college educated in the area of business. He may or may not have an engineering background. (In general, assuming the reader of a proposal knows less is preferable to assuming he or she knows more.) However, as vice president of a company that designs and produces plumbing equipment, he will have read reports and will have had at least some experience with projects measuring fluid-flow characteristics. He may even have some knowledge of the theory behind this type of testing. Certainly he will be well able to evaluate all cost details, which will interest him greatly. These details will also interest the other members of upper-division management.

The design engineers will be knowledgeable about fluid-flow characteristics and how they affect the design of plumbing fixtures. These readers will also have extensive knowledge of the theory, design, and operation of testing programs in this field. They will be most interested in the technical details of the proposed tests.

Analyzing Purpose and Use

The primary purpose of a proposal is to persuade readers to accept your solution to the problem. For example, Arlen's primary purpose is to convince Amnex that UOL's capabilities outweigh those of competitors, so that Amnex will award UOL the contract. In addition, you may need to persuade readers that a need exists for the project you propose and that the long-term benefits will outweigh any short-term, seemingly negative effects. This primary purpose of persuasion depends on a secondary aim: informing your readers effectively of any pertinent technical, managerial, and financial details. If readers do not understand these details, they will not grant

FIGURE 12.2
The Amnex RFP with Arlen's annotations

Request for Proposals

The bidder's proposal should be submitted in one volume, directed to Mr. James Lolly, vice president, and should include the following sections:

- ☐ Introductory Summary
- ☐ Technical Proposal
- ☐ Management Proposal
- ☐ Cost Proposal
- ☐ Bidder's Capabilities

Definition of the problem

Objective

Summary of solution

Product of project

Scope of project

Introductory Summary

This opening section of the bidder's proposal should define and/or analyze the problem he/she is proposing to solve and state the objective of the project. A summary of the proposed method of solution and possible problems should be provided, along with a statement of the product of the project. The scope of the project is extremely important. This brief summary will provide an overview of the proposal for upper-division management.

Technical Proposal

Since the bidder's technical proposal will primarily determine the capabilities of his/her organization to participate in this procurement, the technical proposal should be complete in every detail. The proposal should be practical and straightforward, providing a concise delineation of capabilities to satisfactorily perform the contract being sought.

Description of methods

Project's phases

Other methods considered but not chosen, and why

Methods rejected, and why

The proposal should contain an outline of the investigative methods, approaches, and plans to solve the stated problem; the phases or steps into which this project might logically be divided; and other information considered pertinent to the problem. *The bidder should outline the actual work proposed as specifically as possible.* Engineers in our Design Department will evaluate this section of the proposal.

In addition, methods or approaches considered but not chosen should be described. If they are feasible, explain why they have been relegated to a secondary position. Describe briefly any methods that have been disqualified and explain why.

Management Proposal

This section is divided into two subsections: an identification of personnel responsibilities and a program plan.

The identification of personnel responsibilities should include the position of each person working on the proposed project and how he/she fits into the hierarchy of responsibilities and into the corporate structure of the bidder's company. The latter should be presented in the form of an organizational chart. In addition, care should be taken so that each task is identified for personnel accountability.

The program plan should divide the entire project into tasks. Each task should have projected starting and finishing times. Include the time required for report preparation, and represent the relation of all these elements by a milestone chart. Also in this section, include the proposed duration of the entire project; a contract will be awarded on February 8, 1982.

Hierarchy of responsibilities of each person working on the project

Visual aid: organizational chart

Personnel accountability for tasks

Tasks, including report preparation

Visual aid: milestone chart showing relationship of tasks

Length of time for entire project

Cost Proposal

This contract will be awarded on a cost-plus-fixed-fee basis. The cost proposal should include labor costs broken down by task and total labor costs broken down by workers' levels of expertise. Charges for equipment should be included, if applicable. Finally, a fixed fee and a total for the complete project with a statement of method of payment should be included in this section.

Labor costs by task

Labor costs by levels of expertise

Equipment charge

Fixed fee and total cost

Method of payment

Bidder's Capabilities

This section gives the bidder an opportunity to outline the resources his/her company possesses that are applicable to the project. Qualifications of all key personnel (up to and including the project engineer) should be included in the form of resumes.

In addition to these human resources, the bidder should describe the facilities or equipment his/her company possesses that are, or may be, useful in the proposed project. The bidder may also note special capabilities that his/her company has available.

Resumes of all personnel directly involved in the project

Facilities we have that will be used on the project

Special capabilities

---

your proposed request. Therefore, Arlen must inform Amnex about the testing program in order to accomplish his persuasive purpose.

Readers primarily use the proposal to decide whether or not to accept the solution. This decision is based on the information they gather from the proposal. Thus, James Lolly and the design engineers will decide whether or not to retain UOL on the basis of the technical, managerial, and financial data presented.

Gathering Information

When gathering information for a solicited proposal, you should study the RFP or funding agency's guidelines closely, since these documents will often indicate desired information. You should then annotate the RFP or the guidelines, noting details you will include in your proposal. In Figure 12.2, we show the Amnex RFP, with Arlen's annotations.[1]

---

[1] This RFP is based in part on the Department of Energy's Solicitation DE-RP19-81BC10435, issuing date June 9, 1981.

242   12 | Proposals

After completing your annotations on the information requested, you should make a list of details to include in the proposal and begin your technical investigation.

Arlen's RFP is very detailed and clear. However, you may often have to gather information for a solicited proposal with a vague RFP, or for an unsolicited proposal. In Figure 12.3, we give a checklist of possible details to include in any proposal, with annotations on the purpose of the details. We then discuss the details further.

The purpose of the technical section is to identify the problem and the proposed solution, and persuade the reader that you can carry out the solution. A strong technical section is essential to any proposal: If you do not convince a reader of the need for your project, and of your technical knowledge and abilities, your proposal will

---

FIGURE 12.3
Checklist of possible details to include in a proposal

1. Technical section

Identifies the problem → a. Statement of the problem — *Helps the reader understand the problem*

b. Background clarifying the problem ←

Creates the need for the solution → c. Need for research

d. Objective of the project or the specific solution proposed ← *Identifies the proposed solution*

Gives reasons for undertaking the project → e. Justification for undertaking the project

(1) Relation of proposed research to other research in the area

(2) Earlier work in the area by the proposer

(3) Significance of the project

(4) Benefits from the project

Alerts the reader to the use and necessity of funding →

(5) Future applications of results from the project

f. General statement of why funds are necessary and how they will be spent

Indicates the probability that the project will be successful → g. Feasibility of the project or the solution — *Delineates the project's boundaries*

h. Scope of or limitations on the project ←

Identifies the items to be delivered at the project's close → i. Product of the project (may be a written report) — *Describes methodology*

j. Methods to be used (may be step-by-step procedures) ←

k. Summary of methods not chosen, and why ← *Indicates a thorough consideration of the approach to the problem*

2. Managerial section

a. Personnel

Gives personnel accountability → (1) Assignment of responsibilities for tasks

(2) Hierarchy of responsibilities ← *Gives levels of responsibility for managing tasks*

Summarizes personnel qualifications → (3) Qualifications of personnel (may be presented as resumes)

Gives an overview of the project's schedule → b. Tasks/time schedule (may be a visual aid) — *Alerts the reader to the availability of facilities and equipment necessary for the project*

c. Facilities and equipment available ←

Indicates previous experience in the area → d. Description of work previously completed on similar projects

e. Descriptive literature ← *Provides information on your organization*

3. Financial section

Suggests when and how payment will be made → a. Budget for the project ← *Describes all expenditures*

b. Method of payment

Arlen's Proposal

not succeed. Therefore, clear statements of the problem with any necessary background, the need for research, and the objective of the research or the specific solution proposed are extremely important. Creating the need for research is especially important if the proposal is unsolicited, since the reader may not have realized a need exists.

When considering the technical section of your proposal, you should express the problem you are investigating in a sentence or two. Avoid generalizations, such as the following:

Farm-implement dealers are experiencing financial difficulties.

Factory processes are not keeping pace with money-saving technological advances.

General statements do not provide sufficient direction for your project. Moreover, if you include an overly general problem statement in your proposal, your reader may feel you have not carefully thought out your project and analyzed the need it is to answer.

Instead, you should state your problem as precisely and accurately as possible:

Subsidization programs may be related to farm-implement dealers' financial difficulties.

The nonautomated factory lines in our company's Manufacturing Division are outmoded and expensive.

You should also classify related problems. For example, congestion, noise, and ineffective routing of traffic in your office may all be due to inefficient layout of office space. An effective problem statement for this situation would classify these difficulties rather than listing each separately and would name the root cause:

Root Cause ——————— Inefficient allocation of space in our office has resulted in congestion, noise, and poorly routed traffic.

Once you have defined the problem, you should state the precise objective of your research or the specific solution proposed:

I will investigate the income of farm-implement dealers before and after subsidization, to identify trends and possible causal relationships.

I will study robotics (using robots) as a solution to the problem of nonautomation.

Redesigning the existing space will alleviate these difficulties.

A precise statement of the objective or solution will also convince the reader you have carefully considered your project.

The remaining details in the technical section justify your project and its cost, and persuade readers you can carry out the project and deliver the stated product. Here you should answer any questions readers may have and counter possible negative reactions. For example, relating your research to that of others in the field or to your earlier work indicates you are thoroughly familiar with the subject. Pointing out the significance of the project or its benefits may help overcome reader objections to any short-term effects. Stating the scope shows that you have considered the limitations of your project, such as using one particular method or confining your study to certain topics. Providing methodology convinces readers you know how you will proceed, and discussing alternatives suggests you have deliberated about options for conducting your project, from which you have chosen the best one. Stating the product to be delivered assures readers you have considered the project to its end and helps persuade them that you can complete the study.

The purpose of the managerial section is to identify the resources that will be devoted to the project — personnel, time, facilities, and equipment — and persuade readers that these resources are adequate for the project and will be efficiently managed. For example, the clear designation of tasks in relation to personnel and time, and the hierarchy of responsibilities convince readers that your research procedure is well planned. If readers feel their problem will not receive adequate attention, or if they are unsure of precisely how the project will proceed, they will not grant your request. The description of special facilities or equipment available may point out your superiority over competitors. The qualifications of personnel, description of work completed on similar projects, and descriptive literature provide information on your abilities, indicate the project is within your range of capability, and illustrate your advantages over other proposers.

The purpose of the financial section is to state clearly all costs involved in the project as well as the method of payment. Even though a contract usually follows a proposal's acceptance, the proposal is also a legal document. Therefore, the proposed budget must be carefully prepared. In fact, the budget is so important that financial experts may be asked to compile it from cost data you have gathered.

Selecting and Arranging Content

Selection and Arrangement (Structure)

If you are writing a solicited proposal, the RFP or the funding agency's guidelines will often dictate the content you should include and a structure for that content. You should follow these dictates as closely as possible, varying them only for specific purposes, since your readers will expect the information they have requested in the order they have given. Listing your annotations from the RFP or funding agency's guidelines will provide you with a skeletal outline of your proposal. In addition, you may note the audiences who will be most interested in particular sections of the proposal, if this information is available. You can then alter the writing in these sections to accommodate audience needs. In Figure 12.4, we show Arlen's skeletal outline and audience notes.

If you are writing an unsolicited proposal, you should select the details your audiences require from the checklist of possible details to include, in Figure 12.3, and annotate your outline with audience notes. That list of possible details also suggests a general order. For example, your readers will first require knowledge of the situation

FIGURE 12.4
Arlen's skeletal outline and audience notes

LOLLY, UPPER
MANAGEMENT

  I. Introductory summary
     A. Definition of the problem
     B. Objective of the project
     C. Summary of the solution
     D. Product of the project
     E. Scope of the project

ENGINEERS IN
DESIGN

  II. Technical Proposal
     A. Description of the method for solving the problem, with the project's phases
     B. Description of other methods considered but not chosen, and why
     C. Description of methods rejected outright, and why

LOLLY

  III. Management proposal
     A. Hierarchy of responsibilities of each person working on the project
     B. Organizational chart
     C. Personnel accountability for tasks
     D. Tasks involved in the project
     E. Milestone chart
     F. Length of time required for the entire project

LOLLY

  IV. Cost proposal
     A. Labor costs by task and by workers' levels of expertise
     B. Equipment costs
     C. Fixed fee and total cost
     D. Method of payment

LOLLY, ENGINEERS
IN DESIGN

  V. Bidder's capabilities
     A. Resumes of all personnel directly involved in the project
     B. Special facilities or equipment that may be used on the project
     C. Special capabilities

as it currently is (a statement of the problem and background on it), since they may be unaware that a problem exists, and a statement of need, pointing out the consequences of the problem. Once you have created the need, your readers should be prepared for your ideas of what ought to be (your solution to the problem as an answer to the need) along with the methods you will use for carrying out the project. You may then provide motivation for accepting your proposal by justifying your project and discussing its feasibility. After motivating readers, you might discuss how you would manage the project in terms of personnel, schedule, and equipment, and give the project's cost.

After you have created a skeletal outline of your proposal and added audience notes, you should expand your outline by placing specific details under each entry. We show a portion of Arlen's expanded outline in Figure 12.5.

You should also plan your tasks/time schedule for inclusion in your proposal. To construct this schedule, you should first classify the separate activities involved in

FIGURE 12.5
A portion of Arlen's expanded outline

I. Introductory summary
   A. Definition of the problem
      Amnex needs an independent laboratory to determine the flow characteristics of new plumbing fixtures, in order to be a certified supplier for the Chemical Process Industry.
   B. Objective of the project
      The project will produce the pressure drops and friction factors for 1-in., 3/4-in., 1/2-in., and 3/8-in. pipe. Also equivalent lengths for 1-in. and 3/4-in. gate valves, 1-in. tee fittings, and 1-in. 90° elbow fittings will be determined.
   C. Summary of the solution
      To obtain these characteristics, a flow-measurement device will be used to determine the mass flow rate (mass/unit time) of liquid through the system. At the same time, the pressure drops through the pipes and fittings will be measured. Several flow rates will be used to ensure accuracy. From these figures, the friction factors will be calculated.

your study into major tasks. For example, Arlen must set up and calibrate several apparatuses. However, he classifies these tasks as one major task: setting up and calibrating equipment. His other major tasks will be collecting data on pipes, analyzing data, determining equivalent lengths of fittings, and preparing the report. You should then schedule each of these tasks as to time, according to their expected duration. You should also schedule your reporting activities: the dates when progress reports and the final product or report are due. In this way, your reader receives an overview of your project and can immediately discern the relationship of activities.

If your project involves few tasks, you may state your tasks/time schedule in words. (See sample proposal 2 at the end of this chapter for an example of a tasks/

FIGURE 12.6
Time line picturing a tasks/time schedule

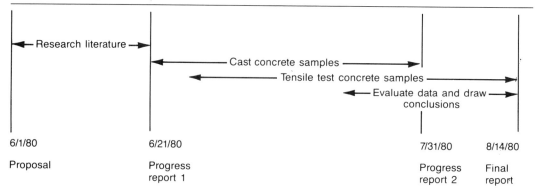

time schedule stated in words.) However, if your project involves several tasks, a visual aid will picture your tasks/time schedule more usefully for the reader.

Several types of visual aids can be used to picture your schedule. If your tasks are sequential or slightly overlapping, a time line is a useful visual aid. Your research tasks can be plotted on a horizontal line, and your reporting dates can be scheduled by vertical lines, as you see in Figure 12.6, a time line for the tasks involved in testing concrete.

If your tasks flow into or follow from each other, a flow chart is an appropriate visual aid. In Figure 12.7, we show a flow chart for the tasks involved in designing a highway bypass.

FIGURE 12.7
Flow chart picturing a tasks/time schedule

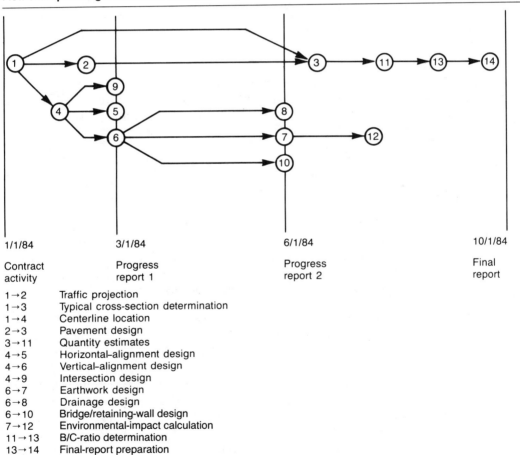

FIGURE 12.8
Milestone chart picturing a tasks/time schedule

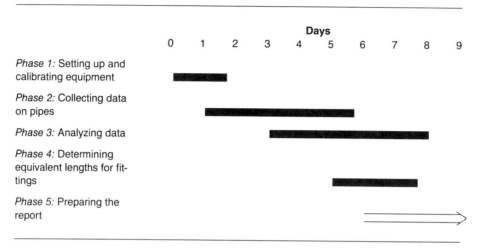

If your tasks begin and end at different times or if you are involved in many tasks, a milestone chart is a useful visual aid. Dates for your project are placed along the top of the chart, and bars or arrows are drawn on the chart to indicate the duration of each task. In Figure 12.8, we show Arlen's milestone chart, requested in his RFP.

Notice that the reporting phase is signaled by a double arrow (⇒) to highlight this task and indicate that it is not a research activity. Progress reports might be highlighted in the same way. Since Arlen's project is brief, he will not be submitting progress reports.

When constructing your tasks/time schedule, be sure to phrase your tasks in parallel grammatical form (e.g., sett*ing* up and calibrat*ing* equipment, collect*ing* data on pipes). Also be sure your wording is precise, so your readers will have a firm idea of your project. Such tasks as "conduct research" do not adequately inform the reader because they are too general and might suit anyone's project.

Arrangement (Format)

Your major formatting decision for a proposal concerns the use of headings to guide readers: The divisions given in your RFP or funding agency guidelines, the sections in your list of possible details (technical, managerial, and financial), or other sections you wish to highlight should be signaled by first-level headings. If your RFP or agency guidelines provide titles, you should use them as headings, because of reader expectations. Therefore, Arlen will head his sections "Introductory Summary," "Technical Proposal," "Management Proposal," "Cost Proposal," and "Bidder's Capabilities," as given in his RFP. If you do not have suggested titles, the three we have provided ("Technical Section," "Managerial Section," "Financial Section") may be used, or you may use headings you devise.

In addition to first-level headings, you should use second-level headings if your sections are long. These second-level headings may be suggested by the RFP or by the content of the section. For instance, in his management proposal, Arlen decides to use the second-level headings ("Personnel Responsibilities" and "Project Plan") suggested by the RFP. He also decides to use second-level headings in the technical proposal, though none are given in the RFP, because this section is long. The names of his tasks are appropriate headings, as the section concerns the phases or steps in the project.

Planning Style

When planning sentence construction and word choice for your proposal, you should consider the audience notes on your outline: The style of each section of the proposal should suit the readers who will be most interested in that section. For example, Arlen will use simpler sentence constructions and nontechnical language in the introductory summary, because James Lolly and upper-division management may not be familiar with the material presented or the engineering terms relating to the proposed testing program. However, Arlen will write the technical proposal for the expert because the readers judging this section are experienced engineers.

In addition, you should consider voice and tone in your proposal. Industrial and research proposals are frequently written in the passive voice because the actor may not be as important as the action. For example, Arlen will use the passive voice in his project description because he wishes to emphasize what will be done rather than who will do the work. However, brief unsolicited proposals may be written largely in the active voice, as in sample proposal 2 at the end of this chapter.

You will also want to project a tone of confidence in your proposal because you wish to impress readers with your certainty about your capabilities. Therefore, you should avoid the use of such terms as *seems, probably,* or *might,* which communicate uncertainty, in favor of more positive, definite words. You will also want to establish a formal tone, since this tone reflects professionalism and helps gain your reader's respect.

Writing

The first draft of Arlen's proposal follows, with annotations on important points.

INTRODUCTORY SUMMARY

Definition of the prob- ▶Amnex needs an independent laboratory to determine the
lem   flow characteristics of new plumbing fixtures, in order to
             be a certified supplier for the Chemical Process Industry.   *Objective of the pro-*
             The project will produce the pressure drops and friction   ◀ *ject*

factors for 1-in., ¾-in., ½-in., and ⅜-in. pipe. Also, equivalent lengths for 1-in. and ¾-in. gate valves; 1-in. tee fittings; and 1-in., 90° elbow fittings will be determined.

Summary of the solution. This summary is brief, to serve Lolly's and the upper management's needs. The technical proposal then expands this summary.

A flow-measurement device will be used to determine mass flow rate (mass/unit time) of liquid through the system. At the same time, the pressure drops through the pipes and fittings will be measured. Several flow rates will be used to ensure accuracy. From these figures, the friction factors and equivalent lengths will be calculated. This information will be presented to Amnex in a report.

◄ *Product of the project*

Scope of the project

Water at 70°F will be the only fluid tested in the pipes and fittings. However, even with this limitation, the testing will produce sufficient information for the certification review board. In addition, the limitation will significantly reduce the cost of the project.

◄ *Favorable cost detail serves to motivate the reader.*

TECHNICAL PROPOSAL

Description of the procedure. The forecasting sentence predicts the phases discussed and highlights them as tasks in the project.

The testing program will be carried out in four phases: setting up and calibrating equipment, collecting data on pipes, analyzing data, and determining equivalent lengths of fittings.

Phase 1: Setting Up and Calibrating Equipment

The project will begin with setting up and calibrating the fluid-flow apparatus. Mass flow rate and pressure drop will

A technical process description supplies information the design engineers require.

be measured using a manometer positioned across a venturi meter and a device that will collect water supplied at constant pressure for 20 seconds. The water will be collected and weighed to determine mass flow rate. Collecting and weighing will be done at 15 different supply pressures, and a log-log, linear-regression chart will be drawn plotting the flow rate versus pressure drop.

— *Positive, definite language inspires confidence.*

Phase 2: Collecting Data on Pipes

In this phase of the project, the different pipes will be installed in the flow loop of the apparatus. The pressure drop across each pipe will be measured with a manometer. Mass flow rates indicated by the venturi meter reading will be determined, using the linear-regression chart constructed in phase 1. The pressure drop and the flow rate for each run will be recorded.

Arlen's Proposal

Phase 3: Analyzing Data

The data collected in phase 2 will be analyzed to determine the friction factors for the pipes. From the mass flow rates, the velocity of the water will be determined, using the following equation:

$$v = \dot{m} / A \, (\rho) \tag{1}$$

where

v = velocity,

$\dot{m}$ = mass flow rate,

A = cross-sectional area of the pipe,

(ρ) = density of the water.

Using the velocity of the water, the Reynolds number for the flow will be calculated with the equation

$$Re = \frac{vD\rho}{\mu} \tag{2}$$

where

Re = Reynolds number,

D = diameter of the pipe,

μ = dynamic viscosity of water.

With these calculations complete, the friction factor will be determined from the equation

$$\frac{1}{\sqrt{f}} = 4.0 \log (RE\sqrt{f}) - 0.40 \tag{3}$$

where

f = friction factor.

Other equations are also available, depending on the value of the Reynolds number. Tables and graphs may be consulted if the equations do not give a representative friction factor.

Phase 4: Determining Equivalent Lengths of Fittings

This phase will end the testing program of the project. Each fitting will be installed in the flow loop of the flow-measurement device. The manometer will be connected to measure the pressure drop across the fitting being tested. The mass flow rate through the fitting will be used to determine the water's velocity by the same procedure described in phase 3. (See equation 1.) After recording the pressure drop at a specific velocity, the friction factor for each fitting will be calculated, as described in phase 3. (See equations 2 and 3.) The equivalent length will then be determined using the equation

$$L_{eq} = \Delta P(D)(g)/2f\mu(v^2) \tag{4}$$

where

g = gravity,

f = friction factor,

μ = dynamic viscosity.

Mention of other testing procedures considered but not chosen, and why. Since none are feasible, they are not described.

At this point, all information needed by the certification board for Amnex's new fittings will be complete. No other substantially different testing program appears to be feasible. No other testing program was considered and rejected.

MANAGEMENT PROPOSAL

The proposed plan for managing the project is described below in two parts: (1) personnel responsibilities and (2) program plan. In the first, all tasks in the project and those responsible for the tasks are identified, and an organizational chart is provided. In the second, the tasks involved in the project are given in a milestone chart, with the time required for the entire project.

◀ Forecasting sentence predicts the sections of the management proposal.

Personnel Responsibilities

Hierarchy of responsibilities and personnel accountability

The work for this project will be done by a UOL group led by project engineer Arlen Hon, who will supervise all phases of the project and write the final report. Two group engineers under him, James Conty and Fred Lam, will set up and calibrate the equipment, and collect and analyze the data. Mr. Hon will also have the services of contract supervisor John Rado if any problems arise. An organizational

Organizational chart ▶ chart of UOL's Liquid-Flow Division is shown in Figure 1.

Project Plan

The project will comprise four phases of testing, plus report preparation. The length of time required for each phase is shown in the milestone chart in Figure 2.

Arlen's Proposal

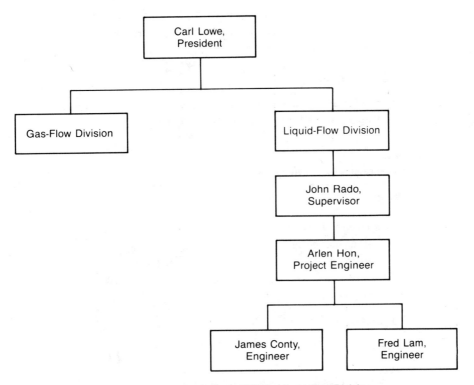

Figure 1: Organizational chart of UOL's Liquid-Flow Division

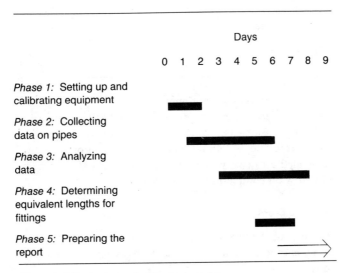

Figure 2: Milestone chart of the project plan

If a contract is awarded on February 8, 1982, as scheduled, work will begin on February 10. The final report will be delivered to Amnex on February 23, 1982.

COST PROPOSAL

Labor costs by task ▶ Labor costs broken down by task and levels of workers'
and workers' levels expertise are shown below in Tables 1 and 2, respectively.
of expertise

TABLE 1:
Labor Cost Breakdown by Task

Phase	Cost ($)
1: Setting up and calibrating equipment	180.00
2: Collecting data on pipes	1,540.00
3: Analyzing data	1,320.00
4: Determing equivalent lengths for fittings	780.00
5: Preparing the report	480.00
TOTAL LABOR COST	4,300.00

TABLE 2
Labor Cost Breakdown by Levels of Expertise

Level	Cost ($)
Project engineer	2,440.00
Group engineers	2,860.00
TOTAL LABOR COST	4,300.00

Equipment charges ▶ Equipment charges are not applicable.

Fixed fee and total ——▶ The fixed fee for this project will be UOL's standard
cost 15% of project costs, bringing the total cost of the proposed project to $4,945.00. Payment for the project would
Method of payment ▶ be divided into two parts: 50% on the award of a contract and 50% on completion of the project.

BIDDER'S CAPABILITIES

Resumes included ▶ Resumes of the project engineer, Arlen Hon, and the two group engineers, James Conty and Fred Lam, are attached to this proposal.

Arlen's Proposal   **255**

Special equipment ———▶ UOL will use their Liquid-Flow System, Model 1000,
available, and capa- as the flow-measurement device in this project. This device
bilities is specifically designed for testing programs such as the one
proposed. Because UOL owns the device, no equipment ◀—— *Cost detail serves to*
charges will be necessary.   *motivate acceptance*
                                                            of the proposal.

Rewriting

Your major criteria for revising a proposal are amount and kind of detail, appropriate
emphasis, logical development, and stylistic appropriateness. You must include all the
information requested in the RFP or funding agency's guidelines. If the proposal is
unsolicited, you must include sufficient detail to convince your readers that the project should be undertaken and that your proposed work will be effective. You must
also arrange these details so they receive appropriate emphasis and proceed in logical
order, and you must express them appropriately.

To revise for amount and kind of detail, first check your list of details from the
RFP or funding agency's guidelines, or your list of details to include: Each listed
detail should appear in your proposal. Then reread all sections of your proposal, comparing what your readers might ask with the information you have supplied.

To revise for appropriate emphasis and logical progression, consider the order in
which the reader has requested information: You should have followed this order,
unless you varied it to stress an important point. If you are writing an unsolicited
proposal, reconsider the general order we discuss on pages 244–245 and any variations you have made for appropriate emphasis.

Arlen checks his list of details from the RFP. He feels his proposal is complete
and progresses logically. However, he finds a problem of emphasis: The advantage
UOL has in owning the specialized fluid-flow device is not emphasized strongly
enough because the device is described in the concluding section of the proposal,
along with the mention of resumes for personnel. Moving the description to a section
of its own, earlier in the proposal, would emphasize this advantage more strongly
and fulfill Arlen's persuasive purpose better than the order of information requested
in the RFP. Arlen decides the new section would most logically precede the technical
proposal, since this placement will allow readers to understand the mechanism so central to the testing program described in the next section.

Because he has added a section to the proposal, Arlen must also select a first-level
heading. He decides *SPECIAL EQUIPMENT* is most precise and best stresses
UOL's advantage over other firms. In Figure 12.9, we show the section Arlen has
added, with annotations on the purpose of the details.

To revise for stylistic appropriateness, you should check to ensure that your language is precise, particularly in your statements of the problem and the solution or
objective, and that you have suited the style of each section to its reader. You should

FIGURE 12.9
Arlen's section on special equipment

Added details on the capabilities of the equipment more clearly describe why it is important ——▶

Added detail emphasizes the time savings involved in owning the device ——▶

SPECIAL EQUIPMENT

The testing program outlined in the next section of this proposal requires determining mass flow rates through the new pipes and fittings. UOL will use their Liquid-Flow System, Model 1000, as the flow-measurement device. This device is specifically designed for testing programs such as the one proposed. The Model 1000 allows for pipes and fittings to be installed in its flow loop and is equipped to measure mass flow rates quickly and accurately. Pressure taps built into the loop facilitate measuring pressure drops within pipes and fittings. Because UOL owns the Model 1000, no equipment charges will be necessary. Also, because this special equipment is currently available, testing can begin immediately.

also reread for positive, definite language and a formal and confident, but not boastful, tone.

Arlen does find a problem with stylistic appropriateness. Under "Personnel Responsibilities," he has said the contract supervisor would be available "if any problems arise." Arlen now feels that this sentence does not project a tone of confidence: Since the sentence suggests the possibility of difficulties with the project, the statement is not positive. He changes it to read as follows:

> Mr. Hon will also have the services of contract supervisor, John Rado, for consultation purposes.

Arlen now feels his proposal is ready to be typed, edited, and sent to Amnex.

CHECKLIST FOR PROPOSALS

1. Have I made out a schedule for preparing my proposal?

2. Have I discovered all the information I can about the audience of my proposal? Have I considered the audiences listed in an RFP or funding agency's guidelines?

3. Have I used the RFP or funding agency's guidelines to gather information, by noting the details these documents request? If the proposal is unsolicited, have I selected appropriate information from the checklist of possible details to include in a proposal (Figure 12.3)?

4. Have I created a skeletal outline of my proposal, with audience notations? Have I expanded that outline with specific details?

5. Have I written a precise statement of the problem and objective of the project?

6. Have I classified my activities into tasks, and scheduled these tasks and reporting dates as to time? Would an appropriate visual aid help me picture this tasks/time schedule? Are my tasks expressed in parallel grammatical form and are they worded precisely?

Sample Proposals   257

7. Have I used first-level headings to identify the major divisions given in the RFP or funding agency's guidelines, or to signal the technical, managerial, and financial sections of my proposal? Have I used second-level headings as necessary, to guide the reader?

8. Have I suited the style of each section of the proposal to the audience most interested in it?

9. Is my language positive and definite? Is my tone formal and confident?

SAMPLE PROPOSALS

Our first sample, which we have annotated, is a solicited research proposal submitted to the National Institute of Health. The sections of the proposal reflect those given in the funding agency's guidelines. Though this proposal will be reviewed by peers of the writer (others in the field), the writer is careful not to assume they will be aware of the significance of the proposed research. Thus, the section dealing with background and significance is long, although the experimental plan is simple.

Numbers enclosed in parentheses in the proposal are references to published papers on research in the area of the proposal. We have not included the budget or the cited literature for this proposal.

PROPOSAL 1

PROPOSAL TO THE NATIONAL INSTITUTE OF HEALTH

Aim of the Research

Statement of the problem, with documentation as background → Incidence of cryptococcosis has been increasing in immunologically compromised or defective individuals (1–4). The disease presents a problem in these patients, because they are not responsive to drug therapy (5,6). Gentry and Remington (7) suggested this lack of efficiency of chemotherapy may be due to a failure of the host defense mechanisms. If more were known concerning

Aim of the research and future applications of results ▶ the mechanisms of resistance to *Cryptococcis neoformans* infections, a means of treating the immunologically deficient host through reconstituting the proper immunological components might be devised.

Significance of the Research

Related research in the area, as background ▶ The humoral response has been studied by many workers (for example, 8–14), and it appears to play only a minor role in the control of cryptococcal infection (13,14). Some indirect evidence suggests that cellular mechanisms provide the major defense against *C. neoformans* (13, 14, 7). Recently, we have studied the

Earlier research in the area by the writer development of delayed-type hypersensitivity in parallel with the numbers of viable *C. neoformans* cells in the tissue of mice previously injected with *C. neoformans* (15). We found that the total numbers of *C. neoformans* cells per mouse began to increase immediately after inoculation. However, as the delayed-type hypersensitivity response developed, the total numbers of *C.*

neoformans cells per animal declined. This result indicates that cell-mediated immunity could be responsible for the reduction of cryptococcal cells.

Relationship of proposed research to writer's previous work

The primary goal of this proposed study is to determine how mice that lack thymus-dependent lymphocytes and thus lack the ability to develop delayed-type hypersensitivity manage a *C. neoformans* infection as compared to mice that are immunologically normal.

Mice homozygous for the mutation "nude" (nu/nu) are athymic and lack thymus-dependent lymphocytes (16); therefore, they make an excellent model for studying the growth of *C. neoformans* in an immunodeficient host. Heterozygous (nu/+) litter-mates of nude mice are phenotypically and immunologically normal, making them useful as normal control animals. If nude mice are

Significance of the study

unable to control the *C. neoformans* infection, the result would indicate that host defense against *C. neoformans* is dependent on T-cell function. On the other hand, if nude mice respond to *C. neoformans* infections in the same manner as heterozygous litter-mates, defense mechanisms other than thymic-dependent ones are functioning.

Relation of specific aims to longer-term objectives

This information would identify the major defense mechanism against a cryptococcal infection and would add to our knowledge and understanding of the disease process. The information would also allow for planning experiments on immunological reconstitution to enhance defense against cryptococcal infections. Conceivably, this research could provide the background for developing protective therapy for patients with cryptococcosis.

Experimental Design and Methods

Specific plan of research

Forty-five nude mice and 45 heterozygous litter-mates will be given 10^4 *C. neoformans* cells intraperitoneally. At weekly intervals, 5 mice from each group will be footpad tested for delayed hypersensitivity, then sacrificed for determinations of numbers of viable *C. neoformans* cells in the liver, lungs, spleen, and brain. Sera will be collected for antigen- and antibody-level determinations. Data will be collected over a 63-day period.

Sample Proposals 259

Our second sample is an unsolicited in-house proposal within a small agribusiness firm. The writer feels she can improve the efficiency of a computer-based system that records data on clients' flocks of sheep and individual sheep of the flock. The project to improve efficiency is not part of the writer's daily work. Therefore, she must persuade her supervisor to approve undertaking a project that will require two weeks of her time.

PROPOSAL 2

MEMORANDUM

To: David Simmons, Supervisor
From: Vera Langley, Flock Analyst
Date: March 16, 1983
Subject: Proposal to improve efficiency of the flock-records program

As you know, the increase in types of information and calls for specific information from our flock records have caused us to split the flock-records program into two parts: individual and group programs. Updates of flock summaries require transferring data from the individual to the group program in three steps:

1. All desired individual summaries are saved on disk.

2. The group-records program is called up and put on the screen.

3. The individual summaries are retrieved from disk and integrated into the group record.

Step 3 is time-consuming, because summaries are retrieved singly and may need to be retrieved often in a single update. The time required to perform a typical update has now grown from three hours to nine.

To save time, a third program is needed, consolidating several summaries into one. Thus, only the consolidated summary would be retrieved, thereby cutting the time required for the typical update in half.

Proposal

I propose writing a "utility" program that will perform the consolidation task quickly and easily. I will also write an instruction manual to accompany the program.

Design

When designed, the program will have two parts:

1. The selection routine (selects summaries from a disk)

2. The consolidation routine (performs the consolidation task)

260   12 | Proposals

Schedule

Work on the program could start immediately and should take about two weeks. Each routine would require four days, and the instruction manual would require two.

The end product of three complementary programs will significantly reduce the time necessary to update flock records, thus increasing efficiency and saving employee wages.

If you have any questions regarding this proposal, I will be happy to answer them.

EXERCISES

Topics for Class Discussion

1. As a basic for class discussion, locate an RFP or a funding agency's guidelines. Annotate the RFP or guidelines by noting details to be included in proposals. Also note the audiences to which proposals will be addressed.

2. In class, create a skeletal outline of a proposal in response to the RFP or guidelines in Exercise 1. Include audience notations of readers most interested in particular sections.

3. Classify the following activities into major tasks. Then discuss appropriate visual aides for a tasks/time schedule, to be included in a proposal. If time permits, create the visual aid.
   A. Instituting a new interview sign-up procedure
   B. Setting up a majors' club in your department
   C. Altering your school's grading system
   D. Improving residence-hall living
   E. Changing the available recreational opportunities at your school
   F. Adding a fall break similar to spring vacation
   G. Instituting a professional field trip for graduating seniors in your department

Topics for Further Practice

1. Formulate a problem from material you have learned in one of your courses. Write a brief problem statement.
   A. Select two members of your department. Consider them and your department chair as readers of a proposal for a research project to solve the problem you have described. Perform audience analyses for the readers.
   B. Using the checklist of possible details to include in a proposal (Figure 12.3), write a brief outline of a proposal to solve the hypothetical problem you have described. (All entries in the list may not apply to the project you are proposing.) Note the audiences most interested in each section of the proposal.
   C. Devise a hypothetical tasks/time schedule for the project and construct an appropriate visual aid for your schedule.

Exercises   **261**

      D. Select a member of a department other than your own. Consider this person an additional reader of the proposal to solve the problem. After performing an audience analysis, list the changes in your brief outline that might be necessary to accommodate the new reader.

      E. Write a draft of a proposal to solve your problem. Assume you have four readers: the three from your department and the one from outside your department.

2. Devise audiences for one of the proposed subjects in Topics for Class Discussion, Question 3. Create a skeletal outline of a proposal on the subject, noting audience interests. Expand that outline, then write the proposal.

3. Select a research problem in your field of study. After reading about projects completed by others, select a limited topic of research. Using the checklist of possible details to include in a proposal, write a skeletal and a detailed outline of a proposal for the project. Consider the director of undergraduate studies in your department as your audience. Also assume that the proposal will not require funds but fulfills a requirement for a senior research project in your major. Write the proposal.

4. After surveying the equipment available in your courses, write a proposal to the chairperson of your department, proposing the acquisition of a new piece of equipment that will enhance the learning opportunities for students in the department.

13

Progress Reports

THE NATURE OF PROGRESS REPORTS

Progress reports are written to inform supervisors or clients about work that has been done and work that remains on an unfinished project. This information is useful to supervisors who must coordinate many projects. Moreover, clients are assured that work is being accomplished and are given the opportunity to evaluate the direction of the project. Because these activities are essential, you will often be asked to write progress reports.

Progress reports may be submitted in different forms: as a memorandum, if the report is a brief in-house document; as a letter, if the report is a short out-of-house document; as a full report, if the document is lengthy. The progress report, however, always concerns a project that has been started but is not complete. This report may be one of several on the status of a project or the only report on status before completion.

In our illustration, we follow Don Murphy, a field geologist with Simplex Geochemical, Inc., through the process of composing the first progress report on a project his company has been contracted to perform. We then include his second progress report in our samples at the end of this chapter.

Don has been assigned to a field crew working in the Galena Hill area of the Yukon Territory in Canada. The crew is performing a geochemical stream survey for Merax Mining Company, to locate lead, zinc, or silver mineralization. Simplex was awarded a contract on June 1, 1982, to perform the testing. They must submit this first progress report on June 12th and a second on June 24th. Don has been assigned to write these reports.

DON'S PROGRESS REPORT

Prewriting

Analyzing the Communication Context

Analyzing the Audience

The primary audience for your progress report will frequently be those who hired you to perform the project or those who recommended that you be retained. Therefore, this primary audience may be the same as the primary audience of your proposal, if one was submitted.

The progress report may also have secondary readers: people who are not directly involved in the project but who wish to check on its status, or future readers who might locate the report in company files to review what was done. You must carefully consider all these audiences when writing your progress report.

Don's primary audience is José Varga, a mining engineer who coordinates exploration and new mining operations for Merax and who contracted to have Simplex perform this work. Mr. Varga is familiar with geological exploration as it relates to mining heavy metals and with interpreting results from geochemical surveys. He has little interest in the theory underlying the survey project but is extremely interested in the findings of the team.

Don's secondary audience includes upper-management people of Merax, who routinely receive copies of all progress reports. These readers have a general knowledge of geological exploration but would not be able to interpret the raw data Simplex has collected. Therefore, Don will interpret these in his report. His interpretations will also be useful to future readers who may review the report.

Analyzing Purpose and Use

Your primary purpose in a progress report is to inform your audience of work completed and work to be completed. Your secondary purpose is to persuade your audience that the job is being effectively accomplished. Fulfilling this purpose may be easy if your project is on schedule but becomes more difficult if it is not.

In addition, you may need to recommend a change in the project plan. Perhaps you require support services or more time to perform a certain task. In these cases, you must inform your audience effectively about the need for these changes, so they will be persuaded to agree with your request.

Don's primary purpose is to inform José Varga of Simplex's activities, and his secondary purpose is to persuade Mr. Varga of Simplex's abilities. Don will not recommend any changes in the project plan because none are necessary.

Readers use progress reports primarily to gain information on the status of the project. However, they may also use them to make decisions. Perhaps the project has produced results that may dictate changes in future work. Readers would use the information provided to decide on these changes. For example, Mr. Varga may use results Don will give to decide whether or not to schedule additional exploration of the Galena Hill area.

Gathering Information

You can rely on your memory when gathering information for a progress report, recalling work you have completed and work left to be done on a project. In addition, written sources can be of help. If a proposal has preceded a project, that document includes a description of the project, the project's tasks, and a time schedule, all of which can provide information for your progress report. However, you must consider changes from what was proposed to what has actually been done.

Other types of routine writings can also supply information. Daily laboratory or field reports provide a record of work completed, as do journals containing lab or field notes. In fact, when you know you will eventually be asked to write a progress report, keeping a detailed record of activities will make your future writing task easier.

An earlier progress report (or reports) on the same project is another source of information. Since this report describes work completed and work planned, it suggests topics you must address in later progress reports.

Don uses the proposal Simplex submitted and his field notes as sources of information for his progress report. The proposal described the project and listed three tasks: preliminary surveys of the water and sediment (1) in the main creeks and (2) in the tributaries, and (3) a detailed survey of water and sediment in tributaries that preliminary work showed might lead to ore bodies. These tasks provide Don with the major topics of his progress report: He must report on what has been accomplished and what remains to be done in these areas. His field notes then give specific details.

Selecting and Arranging Content

Selection

Your first decision when selecting content for your progress report involves the details you provide for work accomplished and projected: your specific activities. If you have submitted a proposal for your project, you should use the tasks listed in the tasks/time schedule to classify your daily activities. If you have not submitted a proposal, you must classify daily work on your project into major tasks (e.g., data analysis; computer simulation), then list activities you have accomplished and your projected activities under each task.

Don classifies his activities using the tasks delineated in Simplex's proposal. Figure 13.1 shows his list of activities.

A second decision on content selection involves the results you will include. Although results of a project are also given in the final report, including some findings in progress reports may allow the reader to redirect a project or change its focus. These findings may also convince the reader that work is being effectively accomplished. However, including too many results will alter the focus of the progress report, which stresses completed and projected *work*. Therefore, you must include only a careful selection of results.

For example, Don and his crew have completed the water-and-sediment survey of Christal Creek. In his field notes, Don has 24 readings of heavy-metal concentra-

FIGURE 13.1
Don's classified list of activities

I. Preliminary survey of water and sediment in main creeks
  A. Survey of Christal Creek has been completed.
  B. Twenty-four readings of heavy-metal concentration were taken.
    1. Sample spacing was five per mile.
    2. Results of water survey showed an anomaly downstream from Silver Gulch.
    3. Result may be due to an active silver mine:
       Tributaries will be surveyed to clarify the data.
  C. Two creeks remain to be surveyed: Flat Creek and Duncan Creek.

II. Preliminary survey of water and sediment in tributaries
  A. Six tributaries are scheduled to be surveyed: Silver Gulch, Betty Creek, North Star Creek, Little Ripple Creek, Parent Creek, and Alma Creek.
  B. Some tributaries may be dropped, if preliminary surveys of the main creeks indicate ore is not present.
  C. Sample spacing will be five per mile.
  D. Results will be mapped and interpreted.

III. Detailed survey of water and sediment in tributaries that preliminary surveys showed might lead to ore bodies
  A. The promising tributaries in (II) will be surveyed in detail.
  B. Sample spacing will be determined on-site.
  C. Results will be mapped and interpreted.

tion for each survey. At the moment, however, José Varga's interest is in trends, not raw data, because knowing these trends will allow him to redirect the project, if necessary. Therefore, Don decides that including all the readings in the progress report would not be useful, although they might be included in his final report. Instead, Don will interpret the results in a summary statement. He will also map the results, so that the location of each reading will be clear, and he will append the maps for Mr. Varga's reference. He can then use this interpretation and the maps to make preliminary decisions on further exploration.

A third decision on content selection involves details in the introduction, preliminary assessment of findings, and conclusion of your progress report. We give a checklist of possible details to include in Figure 13.2, with annotations on the purpose of these details.

The purpose of the introduction to a progress report is to recall the circumstances of a project for the reader. Therefore, the introduction is usually brief, recapping information already given in a proposal or contract. Figure 13.3 shows Don's list of details for his introduction.

The purpose of the preliminary assessment of findings is to provide the reader with tentative conclusions drawn from results to date, in order to give useful information or convince the reader work is proceeding satisfactorily. This assessment may

FIGURE 13.2
Checklist of possible details to include in the introduction, preliminary assessment of findings, and conclusion of a progress report

Provides context for the report → 1. Introduction
    A. Reference to previous communication (e.g., a proposal, contract, or authorization)
Recalls circumstances of the study → B. Statement of the problem on which the project focuses
→ C. Brief summary of the proposed solution
Locates this report in the overall schedule → D. Tasks/time schedule for the project
→ E. Statement of the purpose of this report
Tells the reader how to use this report
      (1) To describe work accomplished and projected
      (2) To recommend a change in the project plan

Interprets results to date → 2. Preliminary assessment of findings (optional)

. . .

3. BODY

. . .

4. Conclusion
Tells the reader if the project is on schedule → A. Statement of the status of project
    B. Preliminary assessment of findings (if necessary and not included after the introduction)
Closes the report courteously → C. Offer to answer questions

FIGURE 13.3
Don's list of details for his introduction

I. Introduction
    A. Refer to June 1, 1982, contract date
    B. Preliminary survey of the Galena Hill area needed so Merax can determine potential for mining lead, zinc, or silver
    C. Simplex to perform a geochemical survey
    D. Include time line as visual aid picturing tasks/time schedule
    E. Purpose of report: to describe status

be placed after the introduction, if you wish to emphasize the results. The assessment may also be placed in the conclusion, where the assessment will receive less emphasis, or may be omitted if you cannot draw valid conclusions at this time.

Don has obtained favorable results; therefore, he decides to include an assessment after his introduction. He will state that preliminary findings show anomalies indicating the presence of ore, but the ore may be due to an already existing mine. Later surveys should clarify this situation.

The purpose of the conclusion is to inform the reader about the project's status and convince him or her that work is being effectively accomplished. If the project is on schedule, this fact may simply be stated. However, if the project is behind schedule, you should provide reasons justifying the delay. You may also briefly assess your preliminary findings here.

Since Don's project is on schedule, he will state that fact and offer to answer any questions José Varga may have.

Arrangement (Structure)

Your first decision when arranging a progress report involves the overall structure of the body. You can organize the body in two ways: by time or by tasks.

Time. If your major tasks are sequential or slightly overlapping, you should organize the body of your progress report by time. We show this organization in Figure 13.4.

FIGURE 13.4
Body of a progress report organized by time

1. Summary of work in past periods (included in second and subsequent progress reports)
2. Detailed description of work in the period just past
3. Detailed description of work in the next period
4. Summary of work in subsequent periods

Notice that the amount of detail included in sections of the body differs, depending on the reporting period being described. Work completed in the period just past and work projected for the next period are always described in detail. Work prior to the period just past and work after the next period are summarized, because this work has either already been reported or will be scheduled in detail in later progress reports.

If you are involved in several tasks for each time period, these tasks can be used as second-level headings under your time designations, as you see in Figure 13.5.

FIGURE 13.5
Body of a progress report involving several tasks in each time period

1. Summary of work in past periods
   A. Task 1 (optional, depending on length)
   B. Task 2
2. Detailed description of work in the period just past
   A. Task 1
   B. Task 2
3. Detailed description of work in the next period
   A. Task 2
   B. Task 3
4. Summary of work in subsequent periods
   A. Task 3 (optional, depending on length)
   (and so on)

Notice that these tasks overlap. However, because the overlap is slight, organization by time is still most useful.

Tasks. If your tasks overlap greatly or are performed concurrently, you should organize the body of your progress report by tasks. We show this organization in Figure 13.6.

FIGURE 13.6
Body of a progress report organized by task

Task 1
1. Summary of work in past periods
2. Detailed description of work in the period just past
3. Detailed description of work in the next period
4. Summary of work in subsequent periods

Task 2
1. Summary of work in past periods
2. Detailed description of work in the period just past
3. Detailed description of work in the next period
4. Summary of work in subsequent periods

Task 3
1. Detailed description of work for the next period
2. Summary of work in subsequent periods

Don's Progress Report   **269**

Notice that these tasks overlap greatly: Only task 3 has not yet begun, while tasks 1 and 2 have been concurrent. Therefore, organization by task is most useful because this organization gives a reader the whole picture on each task at once.

Because Don's tasks overlap only slightly, he will organize the body of his report by time. Because he is involved in several tasks for period 2, he will also use tasks as second-level headings.

Your second decision when arranging a progress report involves sequencing the activities included under each task. Entries are always given chronologically, from the first activity to the last. However, specific day-to-day activities are grouped such that each activity is represented by only one entry in the report. For instance, the following is a strictly chronological list of activities:

December 1–3	Literature search
December 4	Equipment procurement
December 5	Literature search
December 8	Equipment setup
December 9	Literature search
December 10–11	Equipment calibration
December 12	Additional equipment procurement

This list must be reorganized before being incorporated into a progress report: You would group these activities, then present them in general chronological order. The following list shows the reorganized material as it might appear in a progress report:

1. Literature search
2. Equipment procurement
3. Equipment setup
4. Equipment calibration

Don will group his activities as follows: conducting a water survey, conducting a sediment survey, mapping and interpreting results.

Arrangement (Format)

Your major formatting decisions in a progress report involve the use of headings and lists. Since readers often scan progress reports for material of interest, headings make retrieving this information simpler. Lists then highlight and segment the individual activities you have accomplished or plan to accomplish.

First-level headings for a progress report organized by time are chronological designations: e.g., ''Work in Past Periods,'' ''Work in the Period Just Past.'' Second-level headings may then be the names of tasks. First-level headings for a progress report organized by task are the names of your tasks. Second-level headings may then be chronological designations. In addition, if your introduction is long, you may use second-level heads such as "Background," "Description of the Project," and "Purpose of the Report" to guide the reader. You may also want to use a first-level head such as "Preliminary Assessment" for the section assessing results. However, if the introduction and assessment are single paragraphs, headings are unnecessary.

Since Don's introduction is short, he omits second-level headings. He will, however, use a first-level heading to highlight his assessment because he has obtained favorable results. We give his selected headings in Figure 13.7.

FIGURE 13.7
The headings for Don's progress report

Preliminary Assessment of Results
  Work Completed
    Preliminary Survey of Main Creeks
  Work to be Completed by June 24, 1982
    Preliminary Survey of Main Creeks
    Preliminary Survey of Tributaries
  Work Projected
    Detailed Survey of Tributaries

Don will then use a list to highlight and segment the activities under his second-level headings.

Planning Style

One stylistic decision in a progress report involves word choice: phrasing your work statements precisely so that they emphasize actions rather than facts or results. For example, the following sentence informs the reader of a fact:

> If the project is to be undertaken, the cost of mining the ore must be offset by the price obtained for the metals.

If included in a progress report, this sentence should focus on an action rather than on a fact:

> We *will determine* the cost of mining the ore and *compare* this cost to the price obtained for the metals.

Notice that action sentences contain verbs identifying the work you will do or have done, thus retaining the appropriate focus for a progress report.

A second stylistic decision in a progress report involves tone. A businesslike, direct tone will project confidence in your work, convincing the reader that the project is being successfully carried out. Using positive, definite language (e.g., "We surveyed . . .") will help create this tone.

Writing

Following is the first draft of Don's progress report, with annotations on important details. (The progress report is in letter form because the report is a brief out-of-house document.)

P.O. Box 2544
Denver, CO 80222

June 12, 1982

José Varga
Coordinator, Mining Operations
Merax Mining
343 Elm Street
Bozeman, MT 59715

Subject: Progress Report 1 — Preliminary Geological Exploration of the Galena Hill Area

Dear Mr. Varga:

On June 1, 1982, Simplex was contracted to perform a preliminary geological exploration of the Galena Hill area, Yukon Territory, Canada, for your company. Merax needed a preliminary survey of this area in order to determine its potential as a region for mining lead, zinc, or silver. Simplex is now conducting surveys of water and sediment in the main creeks and tributaries, according to the following schedule:

◀ *Reference to previous correspondence*

◀ *Statement of the problem*

◀ *Summary of the solution*

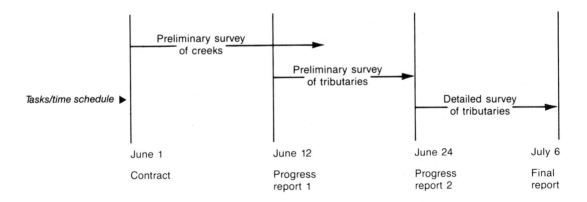

Statement of the purpose of the report ▶ This report describes preliminary findings and progress on the survey, as of June 12th.

PRELIMINARY ASSESSMENT OF RESULTS

The water survey indicated an anomaly just downstream from Silver Gulch. However, the sediment survey did not indicate an anomaly. These findings may result from an active silver mine located near Silver Gulch, but the tributaries in the area will be surveyed to clarify the finding.

WORK COMPLETED ◄─────────────────────── *Work in the period just past: detailed*

Preliminary Survey of Creeks

Sample spacing is included to verify Simplex's methodology.

1. We have completed the water survey of Christal Creek at a sample spacing of five per mile and have discovered only one anomaly, located at Silver Gulch. This anomaly did not persist downstream. (See appended Map A for the location of these findings.) ◄─┐ *The raw data are interpreted to show trends. Only the interpretations are included.*

2. We have also completed the sediment survey of Christal Creek and have found no significant anomalies. (See appended Map B for the location of these findings.)

WORK TO BE COMPLETED BY JUNE 24, 1982 ◄─────── *Work in the next period: detailed*

Preliminary Survey of Creeks

1. We will conduct preliminary stream-water and sediment surveys of Flat and Duncan Creeks to discover if anomalies exist. Sample spacing will be five per mile.

2. We will map and interpret the results of these surveys.

Preliminary Survey of Tributaries

1. We will survey Silver Gulch and other tributaries we identify as possibly leading to ore bodies. Tributaries that may be scheduled are Betty, North Star, Little Ripple, Parent, and Alma. Sample spacing will be five per mile.

2. We will map and interpret the results.

3. We will select promising tributaries for a detailed survey.

WORK PROJECTED ◄─────────────────────── *Work in the subsequent period: summarized*

Detailed Survey of Tributaries

1. We will conduct detailed water and sediment surveys of tributaries that show anomalies in the preliminary survey.

Don's Progress Report 273

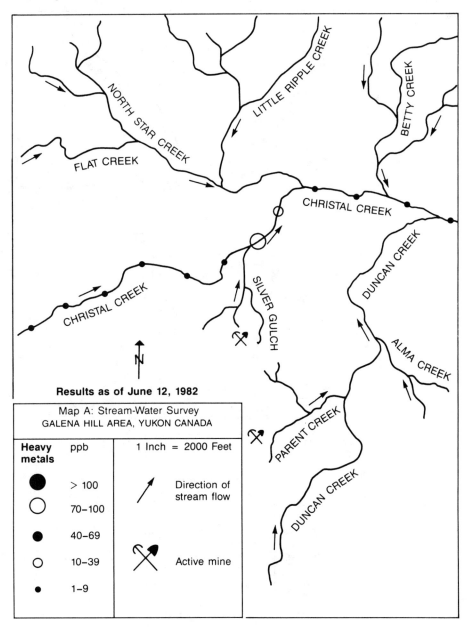

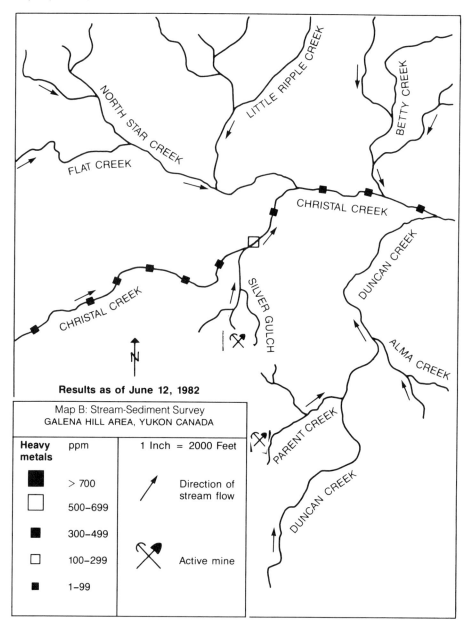

Checklist for Progress Reports   **275**

2. We will map and interpret the results.

3. We will submit the findings in a formal report.

If no serious problems arise, the project will be completed ◄ *Statement of the status of the project*
on schedule and the final report delivered by July 6, 1982.
If you have any questions about this report, please feel free ◄ *Offer to answer questions*
to contact me through our main office.

Sincerely yours,
SIMPLEX GEOCHEMICAL

Don Murphy

Don Murphy
Geologist

DM:sc

Enclosures (2): Maps
cc: Mr. Tom Kinsey

Rewriting

Your major criteria for revising a progress report are amount and kind of detail, and logical progression. When reporting results, you must include only the details your reader can use, saving the remainder for your final report. You must also be certain you have included all your activities, classified them, and given them in chronological order.

When Don rereads for amount and kind of detail, he notices he has omitted a very important activity in summarizing projected work: He will draw conclusions from all the surveys as part of the final report. This activity will certainly be of interest to Mr. Varga, so Don adds another entry. "We will draw conclusions from the results of all surveys." He places this entry as 3 under "Work Projected," renumbering the entry on submitting the final report to 4, in order to retain chronological order.

With this revision, Don feels the report is complete.

CHECKLIST FOR PROGRESS REPORTS

1. Have I used the following documents to gather information: a proposal, laboratory or field reports, a journal of laboratory or field notes, earlier progress reports?

2. Have I used the tasks listed in the tasks/time schedule of my proposal to classify activities? If I have not submitted a proposal, have I classified work on my project into major tasks? Have I then listed activities under these tasks?

3. Have I included results when they will allow the reader to redirect the project or convince the reader that work is being accomplished effectively?

4. Have I briefly recalled the circumstances of the project in my introduction?

5. Have I included a preliminary assessment of results, if the assessment would provide useful information or convince the reader that work is proceeding satisfactorily? Have I placed the assessment effectively, in a section after the introduction or in the conclusion?

6. Have I arranged the body of my progress report effectively?

   A. Have I used a time organization for sequential or slightly overlapping tasks?

   B. Have I used a tasks organization for greatly overlapping or concurrent tasks?

   C. Have I used second-level headings within these organizational patterns to identify tasks or time?

   D. Have I grouped activities, then listed them in general chronological order?

7. Have I reported in detail on work in the period just past and work in the next period? Have I summarized work in the other preceding and succeeding periods? Do my work statements focus on actions done rather than on facts?

8. Have I informed the reader of the project's status in my conclusion and provided reasons for any delay?

9. Have I used headings and lists effectively to guide the reader and to highlight and segment activities?

10. Have I used a businesslike, direct tone and positive, definite language?

SAMPLE PROGRESS REPORTS

Progress report 1 is the second of two on Don's project for Merax Mining. Notice that work completed and described in detail in the first progress report is summarized in this report. Don has then expanded the summary of work in the third period, which was described under the heading "Work Projected" in his first report. Don has also used the same general format as in the first report, to meet reader expectations, but has briefly summarized work that report detailed.

PROGRESS REPORT 1

P.O. Box 2544
Denver, CO 80222
June 24, 1982
José Varga
Coordinator, Mining Operations

Merax Mining
343 Elm Street
Bozeman, MT 59715

Subject: Progress Report 2 — Preliminary Geological Exploration of the Galena Hill Area

Dear Mr. Varga:

On June 1, 1982, Simplex was contracted to perform a preliminary geological exploration of the Galena Hill area, Yukon Territory, Canada, for your company. Merax needed a preliminary survey of this area in order to determine its potential as a region for mining lead, zinc, or silver. Simplex has now completed the preliminary water and sediment surveys of the main creeks and will soon begin survey work on selected tributaries, in accordance with the following schedule:

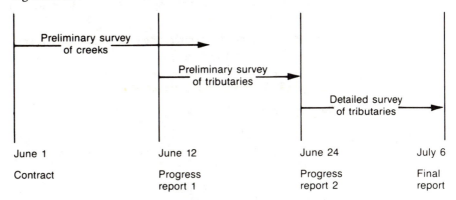

This progress report is the last to be submitted before work is completed.

PRELIMINARY ASSESSMENT OF RESULTS

Water and sediment surveys of Flat Creek showed an anomaly downstream from North Star Creek. Because no active mine exists in this tributary's location, it is the most promising area for an undiscovered ore body. The water survey also showed anomalies in Christal Creek downstream from Silver Gulch and in Duncan Creek downstream from Parent Creek. However, these anomalies may be due to active mines in the area of each tributary. Sediment surveys showed no significant anomalies.

WORK COMPLETED IN THE FIRST PERIOD

In the progress report of June 12, 1982, we reported the results of preliminary surveys of water and sediment in Christal Creek. We mapped the results of the survey and identified the anomaly downstream from Silver Gulch. We again show those results on Maps A and B, appended here.

WORK COMPLETED SINCE JUNE 12, 1982

Preliminary Survey of Creeks

1. We have completed the water survey of Flat and Duncan Creeks, which showed anomalies downstream from North Star and Parent Creeks, respectively. (See Map A for the location of these findings.)
2. We have also completed sediment surveys of Flat and Duncan Creeks, which showed a significant anomaly downstream from North Star Creek. (See Map B for the location of these findings.)

Preliminary Survey of Tributaries

1. We have completed preliminary water and sediment surveys, using a sample spacing of five per mile, in Silver Gulch, Parent Creek, and North Star Creek tributaries. All three will be scheduled for detailed surveys, as anomalies were present.
2. We have mapped and interpreted the results. (See Maps A and B for the location of these findings.)

WORK TO BE COMPLETED BY JULY 6, 1982

Detailed Survey of Tributaries

1. We will conduct a detailed survey of water and sediment, using situational spacing (i.e., spacing determined by the field worker's findings as the survey is conducted).
2. We will map and interpret the results.
3. We will draw conclusions from the results of all surveys.

If no serious problems are encountered, the final report will be delivered on July 6, 1982, as scheduled. If you have any questions about this report, please feel free to contact me through our main office.

Sincerely yours,
SIMPLEX GEOCHEMICAL

Don Murphy
Geologist

DM:sc
Enclosures (2): Maps
cc: Mr. Tom Kinsey

Sample Progress Reports 279

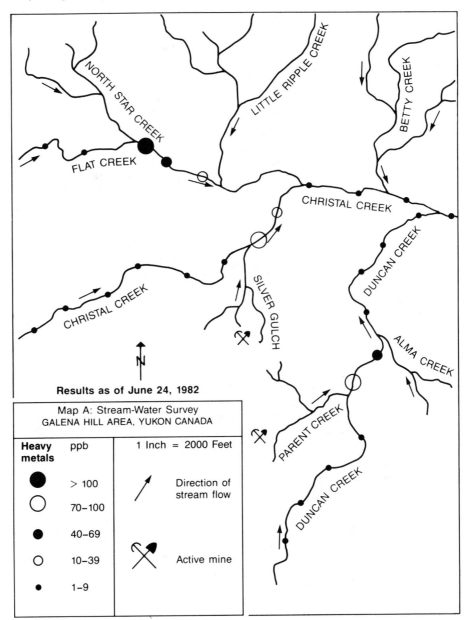

13 | Progress Reports

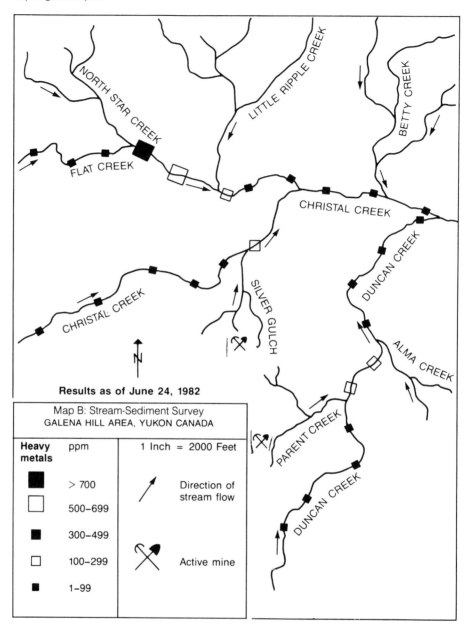

Sample Progress Reports   **281**

In our second sample, we show a progress report on a project that has greatly overlapping tasks. Therefore, organization by task is used. This project required only one progress report, an in-house memorandum.

PROGRESS REPORT 2
ATKIN FARM IMPLEMENT

To: Clarence Nelson
           Manufacturing Services Department
From: Alvin Caine *AC*
           Engineering Department
Date: November 1, 1983
Subject: Progress Report on the Presentation of Cemented-Carbide Cutting
           Tools for the Perishable-Tool Section

On October 15, 1983, you asked me to prepare a presentation and report on cemented-carbide cutting tools. The presentation will be given to the supervisors in the Perishable-Tool Section, who are not familiar with these tools. The presentation will familiarize the supervisors with the nature of cemented-carbide tools and their limitations, just before the new tools are phased in. This progress report outlines the work done on three tasks (literature search, tool tests, vendor research) and the work to be done on these tasks in order to complete the project by the assigned date of November 15th.

LITERATURE SEARCH

A survey of trade and professional journals was conducted to find published materials that might be used in the presentation.

Work Completed

All promising journals to which we have access have been surveyed. I discovered no useful information for the presentation.

TOOL TESTS

Tests comparing machining results in two categories are being conducted:

1. Plain carbide tools versus cemented-carbide tools
2. Different types and brand names of cemented-carbide tools

Work Completed

Tests comparing carbide and cemented-carbide tools are complete and showed that cutting speeds for cemented-carbide tools are as much as 200 surface feet per minute higher than speeds for plain carbide tools. This higher speed is a significant advantage.

Comparisons of types of cemented-carbide tools are also complete and showed that titanium-carbide coatings may extend tool life by 8 percent over uncoated cemented-carbide tools.

Work to Be Completed

Tests comparing brands of tools have not been conducted because no experienced machinists were available. However, these tests will be completed by the end of the week.

VENDOR RESEARCH

Valenite, Kennemetal, and Carbolog, the three tooling companies in our area, are being queried about characteristics of their cemented-carbide cutting tools.

Work Completed

Letters have been sent to all three vendors requesting machining data, tooling samples, and product information. Responses should arrive by the end of the week.

Work to Be Completed

Analysis of vendor information will be completed within five days of receiving responses. On November 10th, I will meet with John Thompson, a specialist on machine-tool materials at State University. He may provide valuable information for the presentation.

The project is slightly behind schedule because of the unavailability of machinists. However, I should have no difficulty completing the work, submitting the report, and conducting the presentation by the November 15th deadline. Preliminary results indicate that cemented-carbide tools with titanium-carbide coatings may be a significant improvement over plain carbide tools.

If you have questions or comments, please call me at extension 1313.

EXERCISES

Topics for Discussion

1. List the activities you perform when writing a paper. (You may specify the type of paper.) Classify those activities into major tasks. Then order these tasks and list individual activities under each.

Exercises **283**

2. Discuss whether progress reports on the following projects should be organized by time or by task. What would you use as first-level headings? Would second-level headings be useful? If so, what would they be? What specific activities would be listed under these headings?

Painting a house

Organizing a majors' club trip

Preparing for a senior-level comprehensive examination

Obtaining a job

Topics for Further Practice

1. Assume you work for Ellison Research, Inc., and your supervisor, Lisa Donne, has assigned you the task of taking a technical-writing course. Because writing is so important at Ellison and a major part of your work, Ms. Donne hopes the course will help you; she is also considering the course for other members of the staff. Midterm has just passed. Write a progress report on your work in this course for Ms. Donne. Include enough results to allow her to decide if the course has assisted you and if it would benefit others.

2. Some experiments you are conducting in a laboratory may take several days to complete. Select the midpoint of one of these experiments and write a progress report for your laboratory supervisor. Include sufficient results to convince him or her that work is progressing satisfactorily.

3. If you have begun a major project for your technical-writing course, prepare a progress report on that project for your instructor. Your purposes are to inform your instructor of its status and persuade him or her that work is being satisfactorily accomplished or that the project requires redirection.

4. Work toward a college degree can be considered a long-term project. Write a progress report to your academic adviser on the status of this project. Assume this report is not the first you have submitted: You have written similar reports at the close of each previous year. Include sufficient results to persuade the reader that work is progressing satisfactorily and to allow him or her to change the direction of your course of study, if necessary.

5. You have taken a job with a firm of your choice. Some weeks ago, you discerned that a change in procedure could benefit the firm. Your proposal to investigate the change was accepted. Now you must write a progress report to your supervisor on the work you have accomplished to date in your investigation and the work remaining. Write the progress report. (You may specify the firm, your position in it, and the change you are investigating.)

14

Final
Reports

THE NATURE OF FINAL REPORTS

You will often be required to submit a written report after completing all technical work on a project. This final report communicates and makes a record of information you have gathered, conclusions you have drawn, and any recommendations you may make.

Final reports vary in format, depending on your communication situation. Short in-house reports are often written as memorandums, while brief out-of-house reports are often submitted as letters. In our samples, we include reports in these forms. Longer reports are usually done in full report form, with the prefatory and supplemental material your audience requires. (See Chapter 17, Report Supplements, for descriptions of this material.) In all cases, however, final reports close your project.

In our illustration, we follow Sam Rose through the process of writing a final report. Sam is a chemist with Chemtech Research and Consulting, which has been hired by American Oil Institute to determine the most economical method for purifying hydrocarbon compounds. This institute requires 100-gm samples of ultrapure hydrocarbons (99.95% level of purity) to conduct thermodynamic tests. Since compounds at this level of purity are not commercially available, the institute must purify the compounds themselves.

Sam has compared three purification methods and determined the optimal one. He must now write a final report.

SAM'S FINAL REPORT

Prewriting

Analyzing the Communication Context

Analyzing the Audience

Although your final report may be directed to one person (e.g., your immediate supervisor) or one group (e.g., American Oil Institute), your audience will frequently be broader than this designation suggests. For example, a final report to your supervisor may also be sent to upper-level management and included in your company's files. A report to American Oil Institute may travel through all levels of the company hierarchy, from upper-level management to technicians who would carry out the purification process. Therefore, you must carefully analyze your audience, including both primary and secondary readers. Checking your company's files for records of past contact with a client and contacting other members of your organization who may have worked with this client can assist with this task.

Sam learns from his supervisor that his report will have two primary readers at American Oil: Charles Carter and Sheryl McDaniels. Charles Carter, manager of property accounting, controls expenditures for new projects. His educational background is in business, but he has been with American Oil Institute for nine years. Therefore, he will understand some of the technical data Sam must convey.

Sheryl McDaniels is director of the research division that will produce the ultrapure hydrocarbons. As a petroleum engineer, she may be familiar with the different purification methods, but Sam is not certain of this fact. She will, however, understand Sam's discussion of these methods and will be able to critique Sam's procedure for evaluating each.

Sam's secondary readers include the president, vice president, and board of directors for American Oil, since they will ratify the recommendations Charles Carter and Sheryl McDaniels make. Technicians in Sheryl's lab, who will purify the hydrocarbons, are also secondary readers. Upper management of American Oil will not be familiar with Sam's technical data, while the technicians will understand it even though they may be unfamiliar with the particular methods compared.

Analyzing Purpose and Use

Your primary purpose in writing a final report and your readers' uses of your report will vary, depending on your communication situation. At times, you may simply convey data to your readers. In this case, your primary purpose is to inform, and your readers will use your report to gain knowledge. At other times, you may draw conclusions on the basis of those data and recommend a course of action. In this case, your primary purpose is to persuade. Your readers will then use your report to make a decision for or against the recommended action.

Both these primary purposes depend on secondary purposes. If you are primarily informing readers, you must secondarily persuade them your data are valid, by the facts you provide and the methods you have used to obtain these facts. If you are primarily persuading readers, you must secondarily inform them of the data supporting your conclusions and recommendations. This information will serve to convince readers your recommendations are sound.

Sam's primary purpose is to persuade: He is recommending the most cost-efficient method of purification. His readers at American Oil Institute will use his report to choose a purification method. Sam's secondary purpose is to inform his readers about the cost data he has gathered on the three methods. The data he presents, the supporting information he provides, and the way he arranges that information will help him accomplish his persuasive purpose.

Gathering Information

You will gather most of the information for your final report from documents kept as part of your technical inquiry: note cards, notebook entries, daily work reports. However, if you have submitted a proposal or progress reports on the project, these documents may also provide material for your final report. For example, the proposal may have included a description of the problem, a project description, or a methodology section. These materials could be adapted for the final report. Progress reports will have provided a description of work done and results, which could furnish material for your discussion.

Sam has not submitted a proposal or progress reports to American Oil Institute because they hired Chemtech directly for a project of short duration. Therefore, his laboratory notes, where he has described his testing methods and results, provide the information for his final report.

Selecting and Arranging Content

Selection

When selecting content for your final report, you must choose only the data your readers need. Chapter 4, Selecting and Arranging Content, will assist you with this task.

To select content for his report, Sam lists the material his readers require, as you see in Figure 14.1.

Arrangement (Structure)

Your first steps in structuring your report are to classify the items on your list of content, prioritize your readers' needs, and segment your report to satisfy these needs. Sam feels the items on his list of content are well classified, so he prioritizes his readers' needs and segments his report.

Sheryl McDaniels is the primary decision maker for Sam's report because she will advise Charles Carter on technical questions concerning the purification methods. Therefore, Sam decides the information she requires should form the main text of his report. Charles Carter is the second decision maker, while upper management is third: Mr. Carter will advise upper management on whether or not to fund this new project, after which they will ratify his decision. Since these readers are not techni-

FIGURE 14.1
Sam's list of content

Primary Readers

Charles Carter
1. The three methods compared and the standards of comparison
2. Cost figures for each method
3. The recommended method

Sheryl McDaniels
1. Technical description of the methods
2. Evauation procedure used to weigh results
3. Standards for judging the methods and results
4. The recommended method

Secondary Readers

Upper Management
1. The problem giving rise to the comparison
2. Cost figures for the recommended method

Technicians
1. Technical process description of the recommended method

---

cally versed, Sam will satisfy their needs with an introduction, a conclusion which includes recommendations, and an executive summary sent separately from the final report. The technicians will act only if the project is funded. Thus, Sam will satisfy their need for technical material with an appendix.

Your next step is to consider whether standard reporting forms (general sequences for information in reports) will help you structure your report. (In Chapter 4, Selecting and Arranging Content, we give the procedure to follow when no standard reporting form suits your material.) These standard reporting forms are given below:

1. Problem/solution report (discusses the solution(s) to a problem)
    A. Introduction
        (1) Discussion of the problem
        (2) Brief description of the proposed solution(s)
    B. Body
        (1) Aspects of the solution(s)
        (2) Alternative solutions (optional)
    C. Conclusion
        (1) Conclusions
        (2) Recommendations
2. Design report (presents a design)
    A. Introduction
        (1) Discussion of the circumstances necessitating the design
        (2) Description of the design (if brief)
        (3) Design methods (if brief)

B. Body
 (1) Design methods (if lengthy)
 (2) Description of the design (if lengthy)
 (3) Advantages/applications of the design
 (4) Alternative designs (optional)
C. Conclusion
 (1) Conclusions
 (2) Recommendations

3. Laboratory report (presents research results)
 A. Introduction
 (1) Discussion of the problem being researched
 (2) Research methods (if brief)
 B. Body
 (1) Research methods (if lengthy)
 (2) Equipment and materials
 (3) Results
 (4) Discussion of results
 C. Conclusion
 (1) Conclusions
 (2) Recommendations (often concern future research on the problem)

4. Feasibility study (discusses the practicality of a subject)
 A. Introduction
 (1) Discussion of the subject being studied
 (2) Methodology (optional)
 B. Body
 (1) Standards for judging practicality
 (2) Alternatives (optional)
 C. Conclusion
 (1) Conclusions
 (2) Recommendations

5. Justification report (proves a subject is preferable)
 A. Introduction
 (1) Discussion of the subject requiring justification
 (2) Methodology (optional)
 B. Body
 (1) Standards for justification
 (2) Alternatives (optional)
 C. Conclusion
 (1) Conclusions
 (2) Recommendations

6. Comparison report (compares two or more subjects)
 A. Introduction
 (1) Discussion of the circumstances giving rise to the comparison
 (2) Methodology (optional)

B. Body
- (1) Comparison by standard:
  - Standard 1
    - Subject A
    - Subject B
  - Standard 2
    - Subject A
    - Subject B
    - (and so on)

or

- (2) Comparison by subject:
  - Subject A
    - Standard 1
    - Standard 2
  - Subject B
    - Standard 1
    - Standard 2
    - (and so on)

Choice of a pattern for the body of a comparison report depends on the reader effect desired. Comparison by standard allows for a tighter, point-by-point comparison, while comparison by subject collects all the data for each subject in one place. Comparison by standard may be most useful when the results of the comparison are not clearly favorable to one item or another. Comparison by subject may be most useful when your data are clear-cut.

C. Conclusion
- (1) Conclusions
- (2) Recommendations

Use of a standard reporting form depends on your communication situation. For example, Sam has decided that the comparison Sheryl McDaniels requires should form the body of his report. Therefore, he will use the standard reporting form for a comparison. The form then suggests a general order for his material. However, this order can be varied to suit a particular content: Optional sections can be omitted; sections can also be added or rearranged.

The next step in structuring your report involves overall order: You can write a deductive or an inductive report, as described in Chapter 4, Selecting and Arranging Content. Because Sam has obtained highly favorable cost figures for the purification method he will recommend, he feels his audience will agree with his conclusions and recommendation. Therefore, he will write a deductive report, placing his conclusions and recommendation first. Because they are long, he will place them in a section preceding the introduction.

The next step in structuring your report involves the order of sections in the body. Listing these sections in a logical order, as you see in Figure 14.2, will assist with this task.

FIGURE 14.2
Sam's list of sections

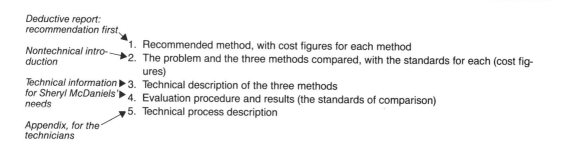

The remaining steps in structuring your report involve the order of information within sections. A checklist of possible details to include and outlines of your particular material will assist you in structuring the report's introduction, body, and conclusion.

The Introduction. The introduction to a final report provides readers with an overview necessary for understanding the report. This section is important because many readers will skim it briefly, to orient themselves to the report before turning to the conclusion or the body.

In Figure 14.3, we present possible details to include in any introduction, with a suggested order, then discuss Sam's selection.

FIGURE 14.3
Checklist of possible details to include in the introduction to a final report

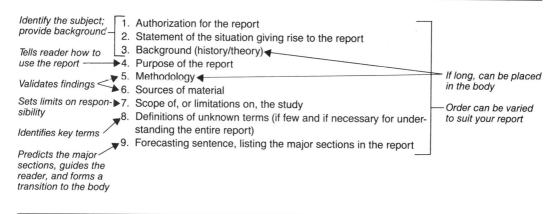

Sam outlines his introduction as seen in Figure 14.4, selecting only the details his readers require.

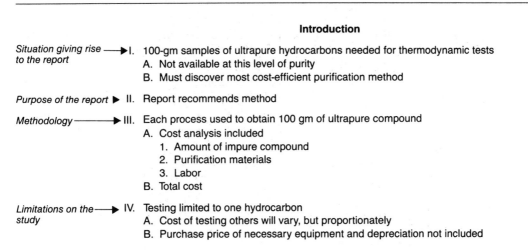

FIGURE 14.4
Sam's outline for his introduction

Because the tests he will discuss in the body are the source of Sam's material, he omits that section from his introduction. He also omits definitions, since none are necessary for understanding the entire report. Therefore, he will define in his text any words his readers might not know.

The Body. The body of the report conveys your information in detail. Although many audiences may not read the body in detail, they may skim it for important points. Therefore, the body must be well structured.

Your major decision in structuring this section involves ordering the points you will present. Chapter 8, Descriptions of Items, will assist with a design report because this report involves physical description. Chapter 9, Descriptions of Processes, will assist with a laboratory report because research methods are usually presented as a description of steps performed, and results are presented and discussed in the order obtained. In this chapter, we discuss the remaining reports.

In the body of problem/solution, feasibility, justification, and comparison reports, you are ordering lists. In a problem/solution report, you order a list of aspects or solutions; in the remaining reports, you order lists of standards. Therefore, principles of emphasis and patterns of development will assist you with structure. (See Chapter 4, Selecting and Arranging Content, for a discussion of these principles and patterns.)

14 | Final Reports

Using principles of emphasis, you may decide to place your most important aspect or standard first (in the position of most emphasis) or last (in the position second in stress). Your choice depends on whether you want your audience to receive your most vital information immediately or whether you feel you ought to build up to these data. You can then use the middle position for less important or negative aspects or standards. Using patterns of development, you may decide to proceed from most to least important or least to most important when arranging your aspects or standards, depending again on audience reaction to your material.

Sam decides to order his standards (costs of impure compound, labor, purification materials) according to principles of emphasis because he wishes to stress two favorable costs for his recommended method. He places the most significant cost (the impure compound) first because it will best distinguish the three methods he has tested and best support his recommendation. He places materials last because his rec-

FIGURE 14.5
Sam's outline for the body of his report

Body

I. Principles of the methods

II. Evaluation procedure
    A. Trace obtained from 0.005-gm sample as standard
    B. Traces from three methods compared for ultrapurity

III. Comparison of methods
    A. Purification by zone refining
        1. Methodology
        2. Results
            a. Impure compound: $450
            b. 10.3 lab hours: $257.50
            c. No materials
            d. Total: $707.50
    B: Purification by liquid chromatography
        1. Methodology
        2. Results
            a. Impure compound: $600
            b. 88 lab hours: $2,200
            c. Materials: $49
            d. Total: $2,849
    C. Purification by recrystallization
        1. Methodology
        2. Results
            a. Impure compound: $1,500
            b. 56 lab hours: $1,400
            c. Materials: $67.50
            d. Total: $2,967.50

ommended method has no materials costs. He de-emphasizes labor costs by placing them in the middle.

If you are writing a comparison, you have two further decisions to make when structuring the body of your report: whether you will compare by standard or by subject; and how you will order your subjects. Sam decides to compare by subject because he wants to present the three figures for the three methods together, giving his readers the whole picture at once. Comparison by subject will then highlight his recommended method, which is superior in all three areas. Because of this superiority, he also decides to place this method first, feeling his audience will not resist his information. He will then proceed from most to least important in ordering his subjects.

Figure 14.5 shows Sam's outline for the body of his report. Notice that background information and methodology have been included here because these discussions are too lengthy for introductory material and too technical to serve Charles Carter's needs.

The Conclusion. The conclusion contains the implications or the significance of your information: the meaning of your results and the action that ought to be taken on the basis of this meaning. Readers rely heavily on this section because it gives your assessment of the situation. Therefore, the conclusion is extremely important.

Ordering your conclusion section usually involves discussing conclusions drawn from your data in the sequence you have presented these data in the body and recommending an action on the basis of your conclusions. This recommendation can be placed before your conclusions, if your reader is likely to agree with it, or after, if your reader requires explanation to agree.

Sam does place his favorable recommendation before his discussion of conclusions. We give the outline for his conclusion in Figure 14.6.

FIGURE 14.6
Sam's outline for his conclusion

Conclusion

I. Zone refining is recommended
   (Include list of total costs)

II. Discussion
   A. Zone refining loses 33% of the impure compound, requires no materials, is nearly automatic
   B. Liquid chromatography loses 50% of the impure compound, involves high materials costs and prohibitive work hours
   C. Recrystallization loses 80% of the impure compound, has highest materials costs, and requires many work hours

Arrangement (Format)

Your major formatting decisions in a final report involve the use of headings and a numbering system. You should use informative headings for major sections and subsections of the report, to identify the precise information in each section, segment your material, and indicate its levels. These headings will also aid readers who wish to skim your report for important facts. You can also use a numbering system with your headings as further reader guides. Chapter 16, Format, will assist you with devising and formatting headings and with using a numbering system.

Sam decides he will use two levels of headings, to signal the major divisions of his report and to highlight each purification method in the body of his report. Because his introduction and conclusion are brief, he does not feel readers require second-level headings in these sections as guides to the material. He decides not to use a numbering system because he has only two levels of information. Therefore, clearly differentiating the levels of his headings by position and appearance will adequately segment his material.

Writing

The following is a first draft of Sam's final report. We have annotated important points.

RECOMMENDED METHOD FOR ULTRAPURIFICATION

Zone refining is recommended as the most cost-efficient method of purification. The following figures support this recommendation:

Total cost to ultrapurify 100 gm of 1,8-dimethylnaphthalene

List highlights cost figures. The recommended method is placed first, as most important.

Zone refining	$ 707.50
Liquid chromatography	2,849.00
Recrystallization	2,967.50

The discussion of each conclusion is nontechnical, to serve those readers' needs.

Zone refining is cost efficient because only 33% of the impure compound, which is extremely expensive, is lost in the purification process. The process is nearly automatic, reducing the work hours required to 10 minutes twice a day. No materials costs are involved in this process.

Liquid chromatography is very expensive. The number of work hours required to purify the compound makes the process especially cost prohibitive. Materials costs also contribute to the overall expense of this method.

Recrystallization is also expensive because 80% of the impure compound is lost in the purification process. A

The results are assessed and the implications are given, for nontechnical readers.

Sam's Final Report

large number of work hours are required. Materials costs
are the highest of the three methods.

Tabulations of these results are presented in Appendix ◄ *Mention of appendi-*
A. Appendix B contains a technical process description of   *ces informs readers of*
zone refining.   *additional data.*

THE PROBLEM WITH ULTRAPURIFICATION

Authorization ———► Chemtech was asked by the American Oil Institute (AOI)
to test methods for ultrapurifying hydrocarbon compounds.
Thermodynamic studies require 100 gm of ultrapure com- ◄ *Situation giving rise to*
pound (99.95% pure as measured by nuclear magnetic reso-   *the report*
nance). Because compounds at this level of purity are not
commercially available, AOI must develop purification
measures. This report recommends the most cost-efficient ◄ *Purpose of the report*
method.

Zone refining, liquid chromatography, and recrystalli- ◄ *Brief discussion of*
zation were used to obtain 100 gm of ultrapure compound.   *methodology, for non-*
The amount of impure compound needed to produce the   *technical readers*
final product, the amount of materials used, and the total
number of work hours needed to purify the compound
were determined. Cost figures for these standards and a
total cost figure were then derived.

The project was limited to evaluating cost efficiency ◄ *Limitations on the*
for purifying one compound, 1,8-dimethylnaphthalene.   *study*
However, the cost of ultrapurifying other compounds will
vary proportionately for the three methods. Initial purchase
price and depreciation of the equipment necessary for each
process were not included in the cost analysis because AOI
has this information on file.

This report presents principles of the methods, the ◄ *Forecasting sentence*
evaluation procedure, and a comparison of the methods, to
arrive at the most cost-efficient process.

PRINCIPLES OF THE METHODS

Zone refining, liquid chromatography, and recrystallization ◄ *Discussion of princi-*
are the only methods that will produce an ultrapure prod-   *ples. Provides techni-*
uct. Zone refining removes impurities on the principle of   *cal background for*
the difference in solubility of the impurity and the com-   *understanding the*
pound in the liquid and solid phases of the mixture. The   *methods. Serves*
zone refiner is an apparatus that slowly lifts a glass tube   *Sheryl McDaniels'*
containing the impure compound past a series of heaters,   *needs.*
then quickly lowers the tube to its starting point. Each of

the heaters melts a band of the solid. These liquid zones move down in the tube relative to the upward motion of the tube in the apparatus. Impurities soluble in the liquid phase are moved into the liquid zones at the bottom of the tube, while impurities more soluble in the solid phase are moved to the top of the tube. The tube is cycled until all impurities are at the top and bottom of the tube. The middle portion then contains the ultrapure product (Gutsche, 1975: p. 430).

Liquid chromatography uses the difference in attraction of the compound and the impurities to an organic material. The impure compound is dissolved in a solvent and injected into a chromatograph. The impure compound is carried in the solvent stream down a column of organic material. The compound and the impurities compete to bind to the limited amount of organic material. Either the compound or the impurity — whichever has the greatest attraction to the organic material — binds tightly. The other is washed from the column and out of the chromatograph in the solvent flow. The solvent containing the compound is collected and evaporated to obtain the ultrapure compound (Gutsche, 1975: p. 365).

In recrystallization, impurities are removed from the compound on the principle of the difference between the solubility of the impurity and the compound. A known amount of the compound is added to the solvent, and the solution is heated until all the solid dissolves. The solution is allowed to cool. The compound forms crystals, but the impurities remain dissolved in the solvent. The solvent and the impurities are separated from the ultrapure compound by filtration (Ault, 1979: p. 34).

EVALUATION PROCEDURE

A 0.005-gm sample of ultrapure 1,8-dimethylnaphthalene was added to a sample tube containing n-hexane solvent. The sample tube was placed in the nuclear magnetic resonance spectrometer, and an ultrapure trace was obtained. This trace was then used as the standard to which sample traces resulting from each method were compared. If the sample trace was identical to the standard, the sample was also 99.95% pure. If the sample trace had irregularities that did not appear in the standard, the sample was not 99.95% pure.

◄ *Description of evaluation procedure for results. Validates findings.*

Sam's Final Report

COMPARISON OF METHODS

Purification by Zone Refining

A 150-gm sample of the impure 1,8-dimethylnaphthalene was poured into the glass tube suspended in the zone refiner. Each of the 10 heaters of the zone refiner, which act to produce molten zones, were turned on, and the tube was allowed to cycle for 100 passes. Each pass took 2 hours. During the cycling period, the apparatus was checked twice a day to ensure proper operation.

Methodology conveys technical material to Sheryl McDaniels. Validates findings.

At the end of the cycling period, the glass tube was removed. The area of the tube containing the ultrapure compound was determined by shining ultraviolet light on the tube. The top and the bottom portions of the tube, which contained the impurities, appeared grey, while the middle, pure, region appeared white. The area containing the pure compound was marked on the outside of the tube, and the tube was cut crosswise with a glass saw at the marks. The pure compound in the middle section of the tube was removed by tapping the tube gently and allowing the compound to pour out into a container. The final product was determined to be ultrapure and weighed. No additional runs were necessary, as the purified sample weighed slightly more than 100 grams.

A cost analysis of zone refining is given below:

Informal table highlights cost figures.

Impure compound (150 gm)	$450.00
Labor (10.3 hours)	257.50
Materials	—
Total	$707.50

Purification by Liquid Chromatography

A mixture of equal parts of methanol and H_2O was cycled through a chromatograph. Five grams of 1,8-dimethylnaphthalene were dissolved in 0.01 l of the solvent mixture and injected into the chromatograph. The compound was allowed to cycle through the chromatograph for 2 hours; then the solvent containing the compound was collected.

The solvent was removed from the compound by rotary evaporation, a method used to convert the solvent to a vapor by heating a rotating flask under a vacuum. The vapor was pulled from the flask by the vacuum, so that only the compound in solid form remained. The pure compound was removed from the rotary evaporation flask by scraping with a spatula.

A small amount (0.05 l) of tetrahydrofuran solvent was injected into the liquid chromatograph to remove the impurities on the column. The tetrahydrofuran containing the impurities was collected from the chromatograph and discarded. The compound was determined to be ultrapure and weighed. The process was repeated until 100 gm of ultrapure compound had been produced.

A cost analysis of liquid chromatography is given below:

Impure compound (200 gm)	$ 600.00
Labor (88 hours)	2,200.00
Materials	
2 l tetrahydrofuran ($12.00/l)	
5 l methanol ($5.00/l)	49.00
Total	$2,849.00

Purification by Recrystallization

Fifty grams of the impure 1,8-dimethylnaphthalene were added to 0.5 l of acetone in a one-liter beaker. The mixture was heated on a hot plate until all the compound dissolved. The solution was allowed to cool so the compound could crystallize. The solvent and the dissolved impurities were removed from the crystalline compound by filtration through a Buchner funnel. The compound was allowed to dry for an hour in a desiccating vessel.

This process was repeated using n-hexane and toluene as solvents. The final product was determined to be ultrapure and weighed. The process was repeated until 100 gm of ultrapure compound had been produced.

A cost analysis of recrystallization is given below:

Impure compound (500 gm)	$1,500.00
Labor (56 hours)	1,400.00
Materials	
5 l acetone ($5.00/l)	
5 l n-hexane ($3.50/l)	
5 l toluene ($5.00/l)	67.50
Total	$2,967.50

Rewriting

Amount and kind of detail, appropriate emphasis, and logical progression of the material are your most important criteria when revising a final report. Since final reports are kept on file as part of a company's records, these reports must be understandable after the specific situation necessitating the report has passed. Therefore, final reports must be complete. In addition, these reports frequently involve conclusions and recommendations and may include standards on which conclusions are based. The details in these sections must receive appropriate emphasis and proceed logically, so that readers are led to agree with the recommended action.

When Sam rereads his report for amount and kind of detail, he realizes he has not given the manufacturer and model numbers of the equipment used in zone refining and liquid chromatography. This information is necessary because different manufacturers' brands and models may result in different cost figures. Sam adds the information to the end of his section on the evaluation procedure, revising the heading as follows:

Evaluation Procedure and Equipment
The equipment used in zone refining and liquid chromatography is listed below:
1. Princeton Organics Zone Refiner, Model 154
2. Waters Liquid Chromatograph, Prep LC Model 1200

Since recrystallization requires only standard laboratory glassware, Sam does not include data for this process.

When Sam rereads for appropriate emphasis and logical progression, he sees that he has presented the three cost standards in his introduction using a different order from the order he used in the conclusion and in the body. In his introduction, he placed materials costs second and labor third, while in the conclusion and the body he reversed this order. Thus, the three sections are not parallel. He revises his introduction to follow the order of the conclusion and the body, because he wants to emphasize materials costs as the more favorable detail.

Now that Sam has revised his report, he writes his executive summary.

SAM'S EXECUTIVE SUMMARY

An executive summary is a highly condensed version of a final report, written after the final report is finished. An effective executive summary is brief, but complete enough to make reading the report optional. In addition, an executive summary is usually directed toward readers who are not technically oriented.

An executive summary may be sent separately from the report or may accompany it. If the summary is sent separately from the report, a letter or memorandum of transmittal would be attached to the summary.

300   14 | Final Reports

FIGURE 14.7
Checklist of possible details to include in an executive summary

Orients the reader to ⟶ I. Introduction
the situation
        A. Description of the problem
        B. Purpose of the report
        C. Scope of the project
        D. Person to whom the report was sent (if the summary is sent separately)

Briefly identifies ⟶ II. Summary of the Report's Contents
important facts
        A. What was done to solve the problem (methods)
        B. Results, conclusions, and recommendations

FIGURE 14.8
Sam's executive summary

Executive Summary

Statement of the ⟶ American Oil Institute's research division needs 100-gram samples
problem of hydrocarbon compounds at 99.95% purity, in order to conduct
thermodynamic tests. Because compounds at this level of purity are
not commerically available, AOI must develop purification measures.

      Chemtech has completed a cost analysis of three methods to
ultrapurify one compound, 1,8-dimethylnaphthalene. The cost of ◀— *Scope of the project*
ultrapurifying other compounds will vary proportionately for the three
methods. We have sent a report recommending the most cost- ◀— *Purpose of the report*
Person to whom the efficient method to Mr. Charles Carter, manager of AOI's Property
report was sent Accounting Department.

      We evaluated the three processes (zone refining, liquid chro- ◀— *What was done to*
matography, and recrystallization) according to three criteria: the   *solve the problem*
cost of the amount of impure compound needed to produce 100
grams of ultrapure compound, the cost of labor, and the cost of ma-
terials. We found that zone refining is by far the most cost-efficient ◀— *Results, conclusions,*
method because only 33% of the impure compound is lost in the   *and recommenda-*
purification process. Also, the process is nearly automatic and   *tions.*
involves no materials costs. Liquid chromatography is very expensive
because of the large number of work hours involved and high materials costs. Recrystallization is also expensive because it loses 80%
of the impure compound and has the highest materials costs of the
three methods.

      The total cost of zone refining was $707.50 compared to
$2,849.00 for liquid chromatography and $2,967.50 for recrystallization. Therefore, we recommend that AOI use zone refining to prepare ultrapure samples for thermodynamic testing.

An executive summary differs from an abstract in being longer (up to 1000 words) and more complete. In addition, the order of the original document need not be followed, and personal pronouns and paragraphing can be used.

Figure 14.7 shows a checklist of possible details to include in an executive summary, with annotations on the purpose of each section.

To write an executive summary, condense the material in the introduction to your report, then briefly describe the major facts from the body and conclusion, using the first-level headings in the report as guides to important information in the summary. Remember, however, that the executive summary is written for the lay reader. Therefore, omit highly technical sections or express them in nontechnical language. In addition, visual aids are not included in an executive summary.

We show Sam's executive summary in Figure 14.8, with annotations. Notice that he has omitted the technical information in the body of his report from the summary, because the audience of his summary does not require these details.

When Sam has written his executive summary, he plans the necessary prefatory and supplemental material for his report, following the guidelines we present in Chapter 17, Report Supplements. He then has his report typed, edits it, and sends it to AOI.

CHECKLIST FOR FINAL REPORTS

1. Should my report be in memorandum, letter, or full report form?

2. Have I discovered all the information I can about the primary and secondary readers of my report?

3. Is my report primarily informative or primarily persuasive?

4. Have I used my technical notes, company files, and company personnel to gather material for my report? Were a proposal and progress reports submitted, which should be checked for information?

5. Have I listed the content all my readers require? Have I classified items on this list? Have I prioritized readers' needs, to segment the report and decide where to place material: in the body, the introduction, the conclusion, the appendix, or the executive summary?

6. Have I used a standard reporting form, if possible, to obtain a general idea of the overall structure for my report? Have I varied this reporting form to suit my material and my audience?

7. Have I placed my conclusions and/or recommendations first or last in the report? Is this placement effective? If placed first, should the conclusions and/or recommendations be in a separate section or in the introduction?

8. Have I selected appropriate material for my introduction?

9. Have I ordered the body of my report effectively? If I am writing a problem/ solution, justification, feasibility, or comparison report, have I arranged my

302   14 | Final Reports

aspects or standards in an appropriate sequence, following principles of emphasis or the pattern of most to least or least to most important?

10. Have I presented my conclusions in the order I have discussed them in the body? Is the placement of my recommendation effective, before or after my conclusions?

11. Have I used headings and/or a numbering system to guide my readers through my material?

12. Is my executive summary brief, yet complete? Is it written for the nontechnically oriented reader?

SAMPLE FINAL REPORTS

We annotate final report 1, a laboratory report. Because the writer and audience both worked in the same organization and because the report is very brief, it is written in memorandum form rather than letter or full report form.

The writer was asked by her supervisor to determine whether the activity of immobilized yeast cells is affected by carbon dioxide pressures up to 80 psig. Laboratory testing provided the answer to the question, and the report summarizes what happened in the tests.

The supervisor used the information to decide a research question: Would a high-pressure batch reactor be feasible in fermenting corn to ethanol? The fermentation process is driven by the activity of the yeast cells. If the cells' activity was not affected by pressure, a high-pressure batch reactor might be feasible. The laboratory report enabled the supervisor to decide if further research was warranted.

FINAL REPORT 1
ETHANOL PRODUCTION INC.
INTEROFFICE COMMUNICATION

To: Joan Shriver, Section Supervisor
From: Jacqueline Paulson, Laboratory Technician
Date: 5/1/83
Subject: Lab tests on the effects of high-pressure CO_2 on immobilized yeast cells

Introduction

Authorization: helps identify the project ▶ On April 15, 1983, you asked me to perform lab tests to determine the effects of CO_2 at pressures from 0 to 80 psig on the activity of immobilized yeast cells. Model-scale test- ◀ *Scope of the project*
ing was performed in our microbiology lab. The purpose of
Purpose of the report ▶ this report is to present the test results and the conclusion that CO_2 pressure up to 80 psig has no effect on the yeast ◀ *The conclusion*
cells' activity.

Sample Final Reports

Equipment

Only the most important pieces of equipment are listed, since others are obvious. The list highlights this equipment. ▶

100-ml high-pressure reactor
3-gal constant temperature bath
Pump and extrusion nozzle

Shaker bath
Overhead stirrer

Pressure gauge
(100 psig)

Methods

Preparation of immobilized yeast cells was done in three steps. The first step involved dissolving 1 gm of NaCl in 120 ml of H_2O, adjusting the pH to 4.5 with dilute HCl, and heating to 40°C. The second step involved adding 2 gm Na-alginate to the yeast mixture, then stirring the yeast mixture with an overhead stirrer until thoroughly mixed. The third step involved slowly pumping the yeast mixture through extrusion nozzles to form small drops that fell into a solution of 0.10 M $CaCl_2$. The drops solidified into small round beads or cells. The cells were refrigerated in H_2O for 20 hours before use.

◀ *Since the description of methods is a process description, it is organized chronologically. It is separated from the introduction because the section is long.*

The model testing was done at 36°C, because the immobolized yeast cells are highly active at this temperature. The temperature was kept constant so that any pressure change in the reactor would be due to CO_2 gas being produced.

Thirty milliliters of yeast cells and 60 ml of 10 percent by weight glucose solution were introduced through the reactor's valve, which was then closed. The immobilized yeast cells and glucose immediately began to react, forming ethanol and CO_2 gas. The CO_2 pressure was measured by a 100-psig pressure gauge located on top of the reactor. The CO_2 pressure was recorded every 5 minutes. The testing was stopped when the CO_2 pressure reached 80 psig.

Results and Conclusions

The pressure readings were tabulated and then graphed. ◀—— *Results*
(See Figure 1 on page 3.) The graph indicates a linear relationship between CO_2 pressure and reaction time. This lin- ◀ *Discussion of results*
ear relationship shows that the activity of immobilized yeast cells is independent of CO_2 pressures ranging from 0 ◀ *Interpretation of results*
to 80 psig. If the cells had been affected adversely by the increasing pressure, their reduced activity would have slowed the fermentation process. The production of CO_2 and the slope of the graph would then have decreased with

time, producing a curve. Thus, the test results indicate that CO_2 pressures from 0 to 80 psig have no effect on activity. ◀ Conclusion from the results

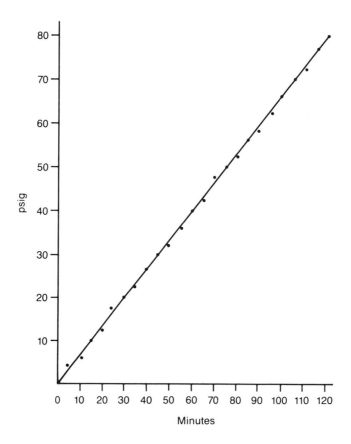

Figure 1: Graph of CO_2 pressure in reactor vessel versus time

Our second final report is a study of the feasibility of moving 50 mature white fir trees. The report was requested by the mayor and city council of Davis, Indiana, who wished to move the trees. The writer, a landscape architect with a local firm, submitted his report in letter form because the report was a brief out-of-house document.

Sample Final Reports | **305**

FINAL REPORT 2

150 Main Street
Davis, IN 60514

July 12, 1983

Mr. Thomas Cook, Mayor
1462 City Center, #1
Davis, IN 60514

Dear Mr. Cook:

I submit this report on the feasibility of moving 50 mature white fir trees now located along University Avenue near the Davis County Fairgrounds. Last week I met with several local tree surgeons at the site. We considered the option of moving the trees about forty feet east so they would become the new second row of trees. (See the attached diagram of this location.) I have evaluated the survival rate of the trees, the aesthetics, and the costs.

Survival Rate

The tree survival rate is the most important consideration of the project. If the survival rate is low, the cost of moving may not be justified. The size and type of tree, the time of year, and the conditions of moving must all be considered.

If moved in spring under normal conditions, mature white firs have an 80% chance of survival. Considering the value of these trees, an 80% survival rate would be worth the expense involved in transplanting.

Aesthetics

The city's current road-building plan calls for widening University Avenue and cutting down the row of fir trees closest to the road. Cutting these trees would not be aesthetically pleasing, because the West Davis Industrial Park would be exposed. On the other hand, having two rows of beautiful towering firs instead of one would be aesthetically appealing and provide a better screen than a fence. The transplanted trees would also define the straight-ahead view of the historic buildings on the Davis County Fairgrounds.

Cost

The costs involved are for transplanting the trees, not for moving per se: Since they will be dug up and moved only a few feet, no traveling expenses, such as road permits, escorts, or gas, are incurred.

All local tree movers charge by the trunk diameter. Transplanting a 20-in. tree costs $165. Since most of the white fir trees are 20 in. in diameter, the total would be $8,250.

The 50 white fir trees along University Avenue could feasibly be moved so the road can be widened. Trees of this maturity have an 80% survival rate.

Aesthetically, the transplanted trees would enclose and define the travel corridor. Since the trees do not have to be moved a long distance, the cost of transplanting would be $8,250. Therefore, I recommend that the city of Davis transplant the 50 white fir trees.

I appreciate the opportunity of doing this work for the city. If you have any questions concerning this report, please do not hesitate to call.

Sincerely yours,
MASTERS' LANDSCAPING SERVICE

Andy Masters
Andy Masters
Landscape Architect

AM:ca
Encl: diagram

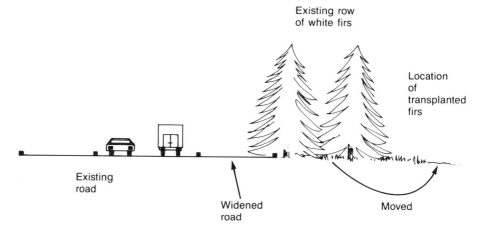

Location of transplanted trees

EXERCISES

Topics for Discussion

1. Choose a final report you have written and examine the sections included. Then answer the following questions:

    A. Does the report follow a standard reporting form? Which one?

    B. Is the report deductive or inductive? Why?

Exercises   **307**

    C. What details are included and omitted in the introduction?

    D. What information is included in the body? Is the information in an effective order?

    E. Are both conclusions and recommendations given in the conclusion? Is the order effective?

Topics for Further Practice

1. The subjects in the following list might be used for final reports. For each subject, determine possible aspects of the solution or standards of justification, feasibility, or comparison. Then discuss several ways of ordering these aspects or standards, giving the audience effect of each order.

    A. Justification of including a new course in your major

    B. Comparison of companies as possibilities for a field trip

    C. Solution(s) to the problem of bicycle congestion on your campus

    D. Feasibility of holding a professional meeting at your school

    E. Comparison of possible majors as careers

    F. Solution(s) to the problem of overenrolled, high-demand courses in your major

    G. Justification of a procedural change at your school (specify the change)

    H. Feasibility of a five-year degree program in your major

2. Designate several possible audiences for the reports in Question 1. For each audience, decide whether you would write a deductive or an inductive report, giving the reasons for your decision.

3. Choose one of the reports and audiences from Question 2. List the details you would include in the introduction to the report, giving reasons for selecting or omitting certain details. Then list the conclusions you might reach and the recommendation(s) you might make.

4. Write the final report for Question 3.

PART IV

PREPARING THE DOCUMENT: VISUAL AIDS, FORMAT, REPORT SUPPLEMENTS

15

Visual Aids

THE NATURE OF VISUAL AIDS

Visual aids are nonverbal forms of communication that help convey information to your readers. These visual aids are either symbolic or pictorial, depending on the information they present. Symbolic visual aids, such as tables, graphs, and charts, represent quantities. Pictorial visual aids, such as diagrams, photographs, and maps, represent the shape or physical appearance of objects. Visual aids are also classified as either tables or figures. Tables present data in columns and rows; figures include all other types of visual aids.

In this chapter, we first survey the general purposes of visual aids and the common types you might use. We then discuss appropriate visual aids and their integration with your text.

PURPOSES OF VISUAL AIDS

Visual aids will not substitute for a well-written text. However, they can complement and supplement your writing by performing several functions:

1. Shortening and organizing the text
2. Clarifying information
3. Indicating relationships
4. Emphasizing points
5. Summarizing material

311

Shortening and Organizing the Text

You can eliminate many words by using visual aids. Consider the following paragraph:

> The Carson Corporation offers four engines of varying specifications. The model S-9 is a 150-cc, two-stroke engine that develops 16 hp at 7000 rpm. The S-11 is a 175-cc, two-stroke engine that develops 17.5 hp at 7000 rpm. The R-10 is a four-stroke engine with 250 cc, and it develops 19 hp at 6000 rpm. The R-12 is the largest engine available. It is a 350-cc, four-stroke engine that develops 32 hp at 6000 rpm.

Presenting these data in the table shown in Figure 15.1 significantly shortens the amount of text needed and organizes the information for the reader.

FIGURE 15.1
Visual aid (table) shortening and organizing the text

Table 3: Carson Corporation's engines

Model	Displacement	Stroke	Horsepower	RPM
S-9	150 cc	2	16	7000
S-11	175 cc	2	17.5	7000
R-10	250 cc	4	19	6000
R-12	350 cc	4	32	6000

Clarifying Information

Visual aids allow you to express information in an alternative form to written text. Thus, visual aids clarify difficult material for the reader. Consider the following paragraph describing a euglena, a single-celled protozoan that lives in ponds:

> A euglena has a carrot-shaped body, thick at one end and tapering to a blunt point at the other. A whiplike appendage protrudes from the thick end, while short, hair-like structures protrude from around the entire body. Several dark dots, of varying sizes, may also be found on the body.

Because the euglena is difficult to picture in words, the paragraph does not give an accurate idea of the creature's appearance. The diagram of the euglena shown in Figure 15.2 more clearly represents this appearance.

FIGURE 15.2
Visual aid (diagram) clarifying information

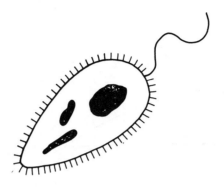

Indicating Relationships

Visual aids allow you to indicate relationships among data. These relationships might be obscured if presented in written text. For example, consider the following paragraph:

> The two species of eucalyptus, *E. nova-anglica* and *E. viminalis,* vary greatly in amount of growth with altitude. *E. nova-anglica* produces approximately 19 cords per acre up to 2400 ft. This species then produces approximately 18 cords at 3000 ft, 17 cords at 4000 ft, and 13 cords at 5000 ft. *E. viminalis,* however, shows a much more significant drop in growth with altitude. *E. viminalis* produces approximately 24¼ cords up to 500 ft, 24 cords at 1000 ft, 19 cords at 2000 ft, 15 cords at 3000 ft, 11 cords at 4000 ft and 9 cords at 5000 ft.

The fact that the reader must proceed linearly through this paragraph obscures the comparative nature of the figures given. The graph in Figure 15.3 on page 314, on the other hand, gives a physical picture of the trend described and allows for immediate comparison of the data.

Emphasizing Points

You can use visual aids to call attention to material you wish to emphasize. For example, consider the following sentence:

> Persons 20 years of age and under now account for over half the buying public for our store.

You might include a visual aid such as the bar graph in Figure 15.4 on page 315 with this sentence to dramatize this claim. Because the reader receives a visual picture of the percentage of buying public 20 years of age and under, this percentage is given greater emphasis.

FIGURE 15.3
Visual aid (graph) indicating relationships

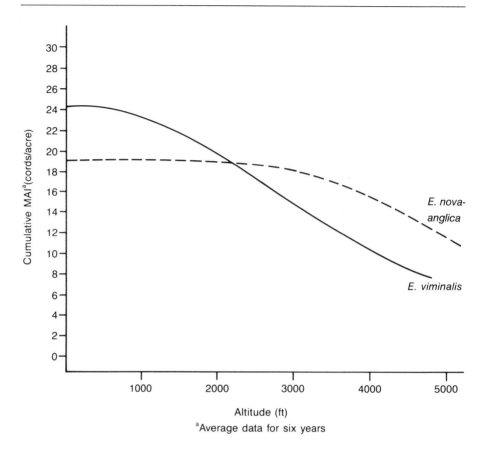

Figure 1:
Variation in Amount of Growth (MAI) With Altitude for Two Species of Eucalyptus

Summarizing Material

You can use visual aids to recapitulate information you have presented in words. For example, consider the following comparison:

> A pre-engineered metal building's initial cost is $25,300 ($6.22/sq ft). Because this type of building is built in modules, construction time is only 12 days. The building is virtually maintenance free because the finished exterior is metal; thus, no painting is required.
> A wood-framed structure initially costs $32,800 ($8.20/sq ft). Thirty working days are required to complete this type of building. Initial painting or

FIGURE 15.4
Visual aid (bar graph) emphasizing points

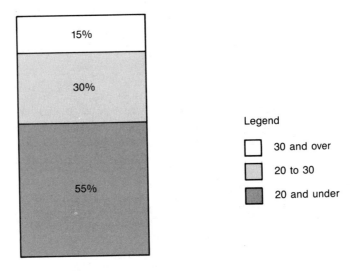

Figure 6:
Percentages of Age Groups in Brown's Buying Public

staining of the exterior is required. A moderate amount of maintenance is necessary in the form of painting or staining the exterior to keep it from decaying.

A concrete-masonry building costs $44,900 ($11.23/sq ft). A construction time of 60 days is required to finish this structure, with exterior painting necessary. This exterior painting is the only maintenance requirement.

A table such as the one shown in Figure 15.5 collects these data and allows the reader to review them quickly.

FIGURE 15.5
Visual aid (table) summarizing material

Table 1: Data on Buildings

Type	Initial Cost ($)	Construction Time (days)	Maintenance
Metal	25,300	12	None
Wood	32,800	30	Moderate
Masonry	44,900	60	Minimal

TYPES OF VISUAL AIDS

Symbolic Visual Aids

Symbolic visual aids include tables and those figures used to present quantities: graphs and charts.

Tables

Tables display data in columns and rows and should be used when you wish to stress exact amounts. Tables should also be used when you wish to present discrete, or individual, details rather than show comparisons or trends, because the arrangement of data stresses individual facts. However, comparisons or trends can be inferred from the data. Tables are usually composed of numbers, but the data can also be expressed in words.

Tables are often boxed by vertical and horizontal lines, but lines are usually omitted within the table. Instead, white space is used to separate columns and rows. Headings are placed at the top of each column and at the left of each row to identify the entries; for complex data, subheadings can also be used. If units of measurement (e.g., $, %, ft) apply to all the data in a column, the unit is placed in the heading to avoid repetition. In addition, information of secondary importance or information that cannot be included in the body of the table can be put in explanatory footnotes below the table and designated in the table by small superscripted letters. If the table is taken from a source, the source note is placed underneath the footnotes. In Figure 15.6, we show the general layout of a table.

FIGURE 15.6
General layout of a table

Table #: Title		
	Column Heading	
Row Heading[a]	Column Subheading[b]	Column Subheading
Row subheading		
Row subheading		
Row subheading		
Row subheading		

[a]Footnote
[b]Footnote
Source Note

When entering data in the columns, you should line up whole numbers by the right-hand digit. If decimal points appear, these points should be aligned.

Types of Visual Aids   **317**

If possible, a table should be completely contained on a single page. If the table is placed on a page with text, adequate white space (four to six spaces) should be used to highlight the table. When a table is too wide (has too many columns) to be shown on the page in the same direction as the text, the table can be shown "turned" on a page separate from the text. If the table must be continued on a second page, column and row headings should be repeated.

In Figures 15.7 below and 15.8 on page 318, we present two tables. The first table, which displays numerical data, indicates the percentage of neutrophils (protective cells) in the intestinal loops of piglets that have been injected with a virus. The second table, which displays data in words, gives information for troubleshooting well-drilling equipment.

FIGURE 15.7
Table displaying numerical data

Table 4: Percent neutrophils in different ligated loops
containing various strains of *E. coli*

Strain	Incubation time 2 Hours	4 Hours	6 Hours
TSB	0.8	11.5[a]	5.7[b]
123	0.3	27.6[a]	20.1
263	4.2	49.2	67.4[a]
431	0.5	26.5[a]	5.9[b]
987	4.3	18.9[a]	58.6[a]
1288	0.5	39.4[a]	64.2[a]
1459	0.3	18.1[a]	25.0

[a] Significant increase from previous incubation value.
[b] Significant decrease from previous incubation value.

Graphs

Graphs present numerical data in visual form. They are more useful than tables for showing comparisons and trends, because the pattern of the data is pictorially represented. However, graphs are generally less precise than tables, because exact individual amounts are not shown. Three common types of graphs are the bar graph, the line graph, and the pie graph.

Bar Graphs

Bar graphs represent numerical data by means of bars running horizontally or vertically on the page. (Vertical bar graphs are also called column graphs.) By placing two or more bars on a graph, you indicate comparisons and contrasts between the

FIGURE 15.8
Table displaying data in words

Table 4–1: Troubleshooting table

Trouble	Probable Cause	Remedy
No indication on gauge	Obstruction in hose	Remove obstruction
	Loss of instrument fluid (M-15)	Add instrument fluid
	Gauge internal mechanism damaged	Replace gauge
	Damper closed	Open and adjust damper
Gauge indication too high at zero	Gauge internal mechanism damaged	Replace gauge
	Gauge out of calibration	Calibrate gauge
	Too much hydraulic fluid (M-15)	Bleed off some instrument fluid
Gauge indication too low	Insufficient instrument fluid (M-15)	Add instrument fluid
	Gauge internal mechanism damaged	Replace gauge
	Air in system	Purge system
	Gauge out of calibration	Calibrate gauge

same entity at different times or different entities at the same time. For example, you might use a bar graph to show the number of students enrolled at your college in 1980, 1981, 1982, and 1983. You might also use a bar graph to show the number of men and women enrolled at your college in 1983.

Bar graphs have two axes, the x-axis and the y-axis, as shown in Figure 15.9. If the bars are run horizontally, the items to be compared are written along the y-axis and the units of measurement are written along the x-axis. If the bars are run vertically, the dependent variable is written along the y-axis and the independent variable (often time) is written along the x-axis. These axes must always be labeled so the data represented are clear. In addition, the label on the y-axis is traditionally written sideways.

FIGURE 15.9
The two axes of a bar graph

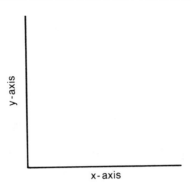

When constructing a bar graph, you should maintain adequate white space between the bars, so they are effectively highlighted: At least half the width of a bar is common. (You may join bars, but these graphs are more difficult to read.) You may also shade or mark the bars with lines, crossed lines (cross-hatching), or dots, to create interest and distinguish bars. The bars may be similarly shaded, or different shadings may be used. In Figures 15.10a and 15.10b, we show bars with and without shading to illustrate the effect.

FIGURE 15.10a
Bar graph without shading

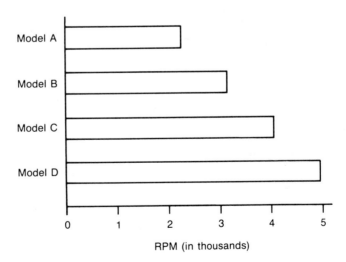

Figure 1:
Maximum RPM of Four Engine Models

FIGURE 15.10b
Bar graph with shading

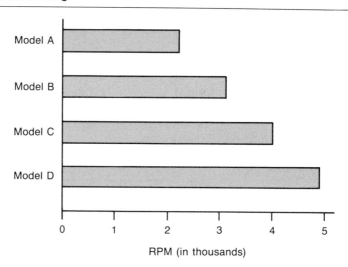

Figure 1:
Maximum RPM of Four Engine Models

The bars may also be arranged in any order you choose, depending on the information you wish to stress. For example, if the writer wished to emphasize model D's capacity in Figure 15.10, model D could be placed at the top of the graph.

Bar graphs should be entirely contained on a single page to make comparing the bars easy. The graphs should also be accurately scaled, so that comparisons are clearly

FIGURE 15.11a
Poorly scaled bar graph

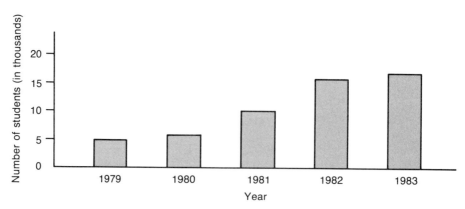

Figure 1:
Enrollment for a Five-Year Period

presented. Figure 15.11a shows a poorly scaled bar graph: Because the units are too large, the actual numbers represented are unclear. Figure 15.11b shows the same graph with an effective scale. In addition, if you wish to stress the numbers represented, you may write these numbers on or at the ends of the bars, as you see in Figure 15.11b.

FIGURE 15.11b
Effectively scaled bar graph

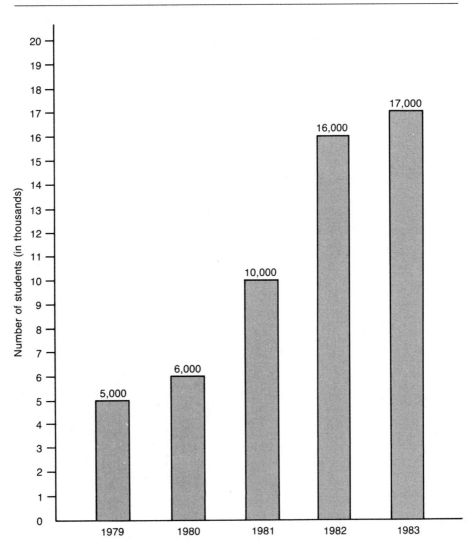

Figure 1:
Enrollment for a Five-Year Period

The graphs in Figures 15.10a and 15.10b and 15.11a and 15.11b are single-bar graphs because they picture only one entity for each unit measured (e.g., model A has 2000 RPM). The graphs are also partial bar graphs because each bar represents only part of the total (e.g., 5000 RPM). Three other types of bar graphs, the multiple bar graph, the segmented bar graph, and the 100% bar graph, allow you to picture several entities and compare entities as parts of a whole.

Multiple Bar Graphs. In multiple bar graphs, several entities may be plotted for each unit of measurement (usually time). The bars for each unit are differentiated by shading, which must be explained in a legend. Because the bars have a common baseline, the entities may be compared both within each unit and between units. In Figure 15.12, we show a multiple bar graph.

Multiple bar graphs may be used to picture up to three entities, after which the graph becomes difficult to read. Figure 15.13 shows a bar graph with too many bars.

Segmented Bar Graphs. In segmented bar graphs, one bar is subdivided to show the relative parts of a whole that each entity represents. The segments of the

FIGURE 15.12
Multiple bar graph

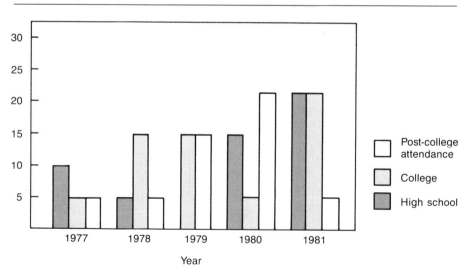

Figure 5:
Breakdown of New Employees by Educational Level, 1977–1981

FIGURE 15.13
Bar graph with too many bars

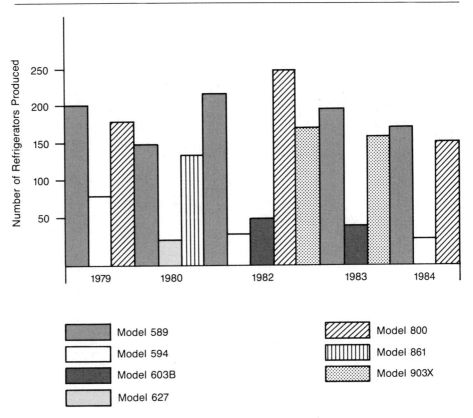

Figure 3:
Number of Refrigerators Produced for Seven Models Over a Five-Year Period

bar are then differentiated by shading, with the darkest shading beginning at the bottom, and a legend is provided.

In Figure 15.14, we show a segmented bar graph, using the same information given in Figure 15.12. Notice that the bars represent the total number of new employees for a given year. The bars are then subdivided to indicate educational level. However, because the segments of each bar do not have a common baseline, comparison of the components between bars is difficult. Instead, their relationship is stressed.

Because the total amount each component represents is difficult to discern, these amounts are often written on the bar graph, as in Figure 15.14.

FIGURE 15.14
Segmented bar graph

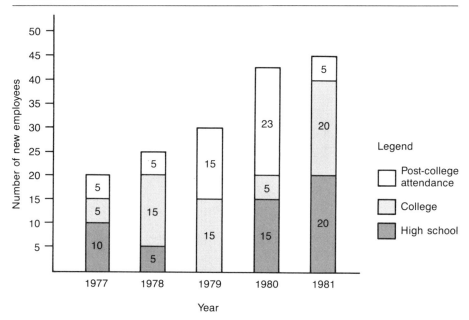

Figure 4:
Breakdown of New Employees by Educational Level, 1977–1981

The 100% Bar Graph. The 100% bar graph is a special form of segmented bar graph. In the 100% bar graph, a single bar representing 100 percent of an entity is subdivided into its components. These components are then shaded for differentiation, with the darkest shading beginning at the bottom, or at the left if a horizontal bar is used. Labeling may be done on the graph or in a legend. In addition, the percentage that each component represents is usually given because the 100% bar graph emphasizes percent.

Figures 15.15a and 15.15b illustrate vertical and horizontal 100% bar graphs for the same data.

Line Graphs

Line graphs represent numerical data by lines. These graphs are used when the data form a continuum: By drawing one or more lines to connect points on a scale, you indicate trends or rate of change. For example, you might use a line graph if you wished to stress a trend of increasing enrollment at your college from 1980 to 1983.

Like bar graphs, line graphs also have an x-axis and a y-axis. The dependent variable is written along the y-axis; the independent variable is written along the x-axis. These axes must be clearly labeled.

Types of Visual Aids **325**

FIGURE 15.15a
Vertical 100-percent bar graph

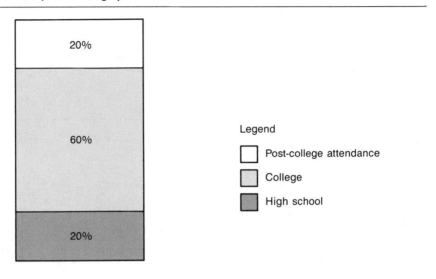

Figure 6:
Breakdown of New Employees by Educational Level, 1981

FIGURE 15.15b
Horizontal 100-percent bar graph

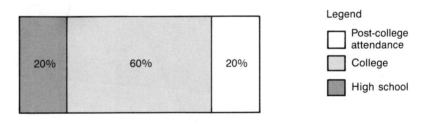

Figure 6:
Breakdown of New Employees by Educational Level, 1981

When constructing a line graph, you should select your scale carefully, so that the trend or rate of change is clearly discernible. If the increments along the y-axis are too small, the change between points will appear to be diminished. If the increments are too great, the change will be exaggerated. Figures 15.16a and 15.16b illustrate this difficulty.

As a general rule, the vertical scale of a line graph should be approximately three-quarters of the length of the horizontal axis.

FIGURE 15.16a
Line graph with diminished increments

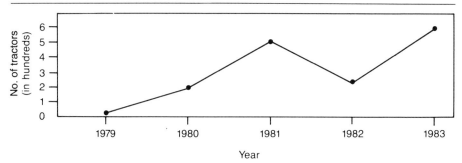

Figure 3:
Number of Tractors Sold for a Five-Year Period

FIGURE 15.16b
Line graph with exaggerated increments

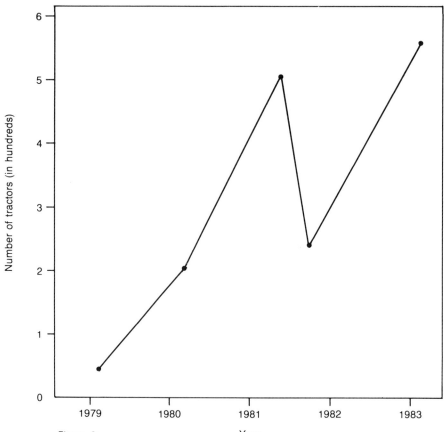

Figure 3:
Number of Tractors Sold for a Five-Year Period

The scale of a line graph does not have to start at zero when the data points begin at a significantly higher number. However, if the scale does not begin with a zero, this fact should be indicated by a jagged line or by a break in the axis. Failure to do so is called the fallacy of the suppressed zero, because the beginning point for the data is unclear.

If you are plotting one variable, your data points are marked, and a jagged line or a curve (called a faired curve) is drawn to connect the points: The jagged line tends to emphasize each point, while a faired curve stresses the trend. Figures 15.17a and 15.17b illustrate this difference.

FIGURE 15.17a
Line graph with jagged lines

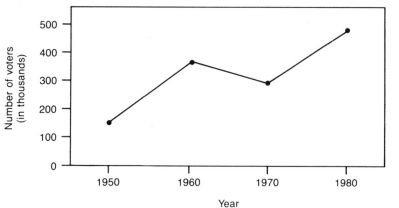

Figure 5:
Number of Voters in City Elections at Ten-Year Intervals

FIGURE 15.17b
Line graph with faired curve

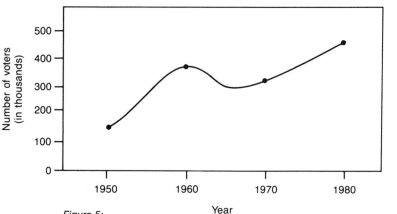

Figure 5:
Number of Voters in City Elections at Ten-Year Intervals

If you are plotting more than one variable, the lines or curves must be distinguished by using different colors or forms (e.g., solid, dashed, dash-dot-dash), and a legend must be included to identify these colors or forms. (However, remember that colors do not reproduce in photocopy.) Up to three variables may be plotted on a single line graph, after which the graph will appear crowded. For more than three variables, you should use separate graphs.

The data points on line graphs may be enclosed in various geometric shapes (e.g., circles, squares, triangles) to further highlight these points. In Figure 15.18, we show a multiple line graph with enclosed data points.

FIGURE 15.18
Multiple line graph with enclosed data points

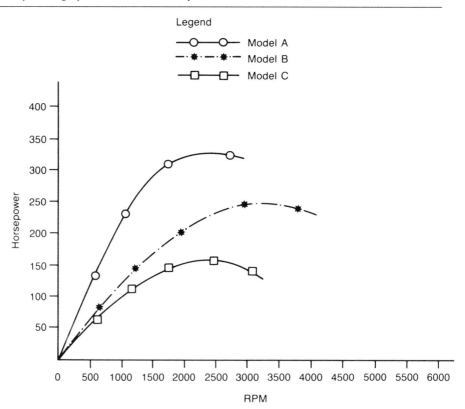

Figure 11:
Horsepower Developed by Three Engine Models

Pie Graphs

Pie graphs, like 100% bar graphs, indicate the components making up 100 percent of an entity. A circle, or "pie," represents the entire entity. Different-sized pieces of the pie then represent the percentage each component comprises of the whole, with each percent representing 3.6 degrees.

Because pie graphs illustrate relative parts of a whole, these graphs are most useful for dramatizing relationships. Pie graphs have been particularly popular for picturing financial data (e.g., the amount of money a utility company spends in a given year for each of its services) and are often used for the lay reader, because these graphs are easy to understand. However, pie graphs are not as accurate as tables for portraying exact amounts.

Pie graphs traditionally begin at 12 o'clock with the largest segment and proceed clockwise to the smallest segment, as shown in Figure 15.19.

FIGURE 15.19
Pie graph segments

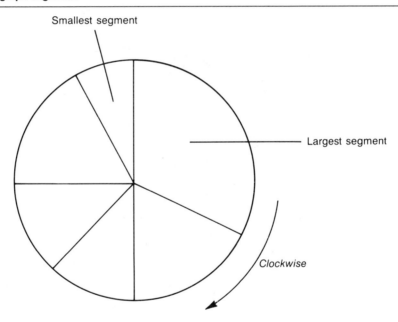

An exception occurs when several components too small to be represented separately are grouped in one segment under the designation "other": This segment always appears last.

If the segments are shaded, the shading usually proceeds clockwise from darker to lighter, beginning at 12 o'clock. Since the segments are labeled, a legend is not necessary.

Each segment should be designated with the component's name and the percentage of the whole the component represents. In addition, the absolute number the component represents may be given. These labels may be placed within the segments if space permits or just outside the circumference of the circle if space is limited. The labels should always be written so they can be read without turning the page.

A pie graph may effectively represent up to eight components, after which the graph becomes difficult to read. If you wish to represent more than eight components, consider grouping several components.

In Figure 15.20, we show a typical pie graph.

FIGURE 15.20
Pie graph

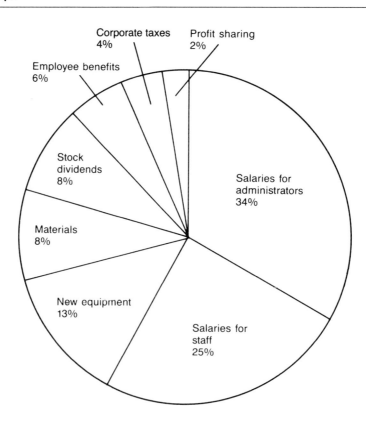

Figure 14:
Expenditures of Clarecorp, in Thousands of Dollars, for 1981 (Total Expenditures = $720,000)

Types of Visual Aids **331**

Charts

Charts visually represent nonnumerical data. As such, charts are useful for showing the overall relationship among components in a system. Two common types of charts are the flow chart and the organizational chart.

Flow Charts

Flow charts represent the direction, or "flow," of a process from beginning to end, thus providing an overview of the process. This process may involve actions and decisions (e.g., troubleshooting a system), or the process may be a sequence of actions or events taking place without the necessity of decision (e.g., the steps in proposal preparation; the circulation of Freon in a refrigeration system). Each action or event in the process is represented by a geometric figure, usually a square or rectangle, and the flow from action to action or event to event is represented by arrows drawn between figures. If the flow chart represents actions and decisions, the figure representing the decisions is usually a diamond. If the flow chart represents a sequence of events in an apparatus, the figure may be a simplified representation of the part where the event occurs.

In Figure 15.21 on page 332, we show a flow chart for a process involving actions and decisions. In Figure 15.22 on page 333, we show a flow chart for the sequence of events occuring in an apparatus.

Organizational Charts

Organizational charts represent the hierarchy of positions, relation of divisions, or distribution of functions in an organization. Positions are job roles (e.g., director of personnel); divisions are units within an organization (e.g., Office of Home Services); functions are the duties of the units within an organization (e.g., data processing). These topics should not be mixed in one chart. If you wish to picture several topics (e.g., a hierarchy of job roles and a hierarchy of duties), you should use separate charts.

The positions, divisions, or functions are arranged in boxes, with the highest position, division, or function (the one with the widest range of responsibility) at the top and others placed lower on the chart as the range of responsibilities narrows. Lines are used to show direct connections among positions, divisions, or functions.

In Figure 15.23 on page 334, we show a typical organizational chart representing positions.

If the washer will not operate, follow the steps outlined in this chart.

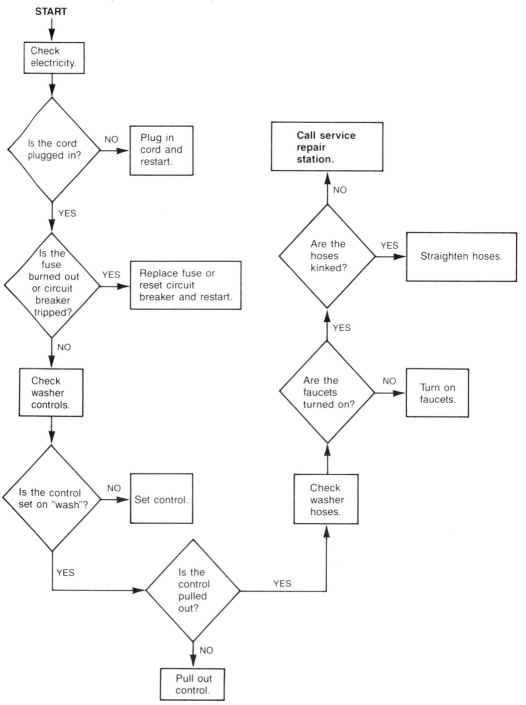

Figure 1: Flow Chart for Troubleshooting an Automatic Washer

Types of Visual Aids   **333**

◀ **FIGURE 15.21**
Flow chart for a process involving actions and decisions

FIGURE 15.22
Flow chart for the sequence of events occurring in an apparatus

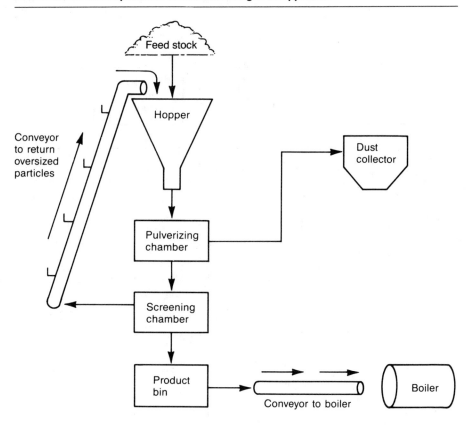

Figure 7:
Process for Preparing Boiler Fuel

FIGURE 15.23
Organizational chart showing positions

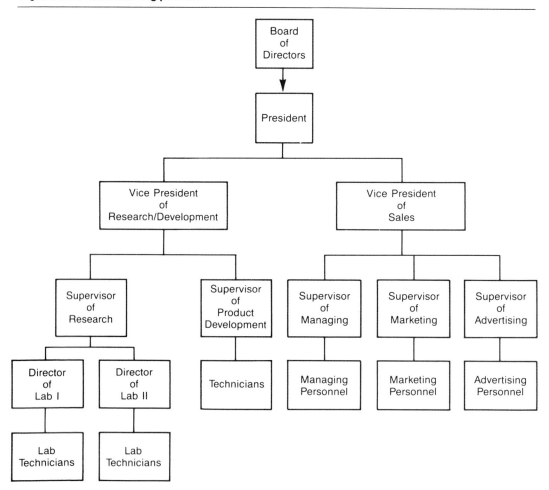

Figure 2:
Organizational Chart for Ecco Company

Pictorial Visual Aids

Pictorial visual aids include all figures representing the shape or appearance of objects or locations. Common pictorial visual aids include diagrams and drawings, photographs, maps, and samples.

Diagrams and Drawings

Schematic Diagrams

A schematic diagram is a simplified sketch of an object, representing only one dimension. Schematic diagrams are used when realism is not important, to show the general appearance of an object's outer surface. The parts of the object are usually represented by overall shape: e.g., as a circle, square, rectangle. If you wish to identify these parts, you may place labels on the diagram. However, if the object has many parts, you may place numbers on the diagram and list the parts in a legend.

In Figure 15.24, we show a schematic diagram of a pneumatic-hydraulic-angiogram injector, used to inject luminous fluid into a human body for an angiogram X-ray.

**FIGURE 15.24
Schematic diagram**

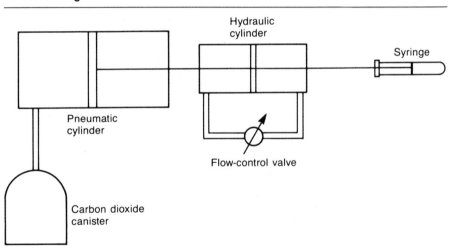

Figure 1:
Injector Arrangement

Shaded Sketches and Isometric Drawings

Shaded sketches and isometric drawings are used to represent three dimensions of objects. Sketches may be shaded to give the appearance of three dimensions, as shown in Figure 15.25.

FIGURE 15.25
Shaded sketch

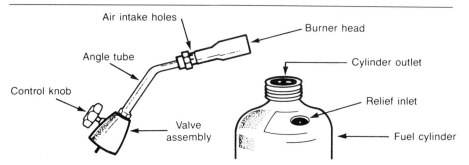

Figure 1:
Parts of a Propane Torch

An isometric drawing represents the three dimensions more realistically than a shaded sketch by positioning various parts of the subject. We show an example of an isometric drawing in Figure 15.26. Notice that the artist has also used shading to enhance the three-dimensional quality of the drawing.

FIGURE 15.26
Isometric drawing

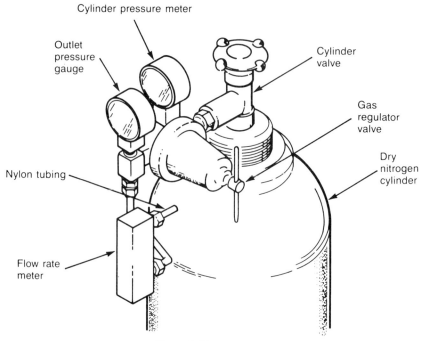

Figure 3-17:
Dry Nitrogen Purge Equipment

Types of Visual Aids **337**

Exploded-View Drawings

An exploded-view drawing shows the parts of an object pulled apart or exploded. These drawings also suggest how the parts are put together, thus indicating the relationship of parts in the assembled object. Exploded-view diagrams are often used in maintenance manuals, where directions for assembling and disassembling equipment are common. The parts may be identified in a legend separate from the drawing, as in the sample we show in Figure 15.27. These parts may also be identified on the drawing.

FIGURE 15.27
Exploded-view drawing

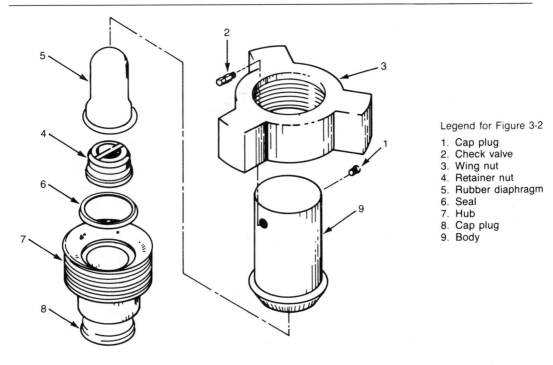

Legend for Figure 3-2
1. Cap plug
2. Check valve
3. Wing nut
4. Retainer nut
5. Rubber diaphragm
6. Seal
7. Hub
8. Cap plug
9. Body

Figure 3-2:
Gauge Protector, 2-inch

Cutaway Drawings

A cutaway drawing shows a portion of an object's outer surface broken away, allowing the reader to see the inside of the object. Thus, this drawing is used when the outer and inner parts of an object are important. In Figure 15.28, we show a cutaway drawing.

FIGURE 15.28
Cutaway drawing

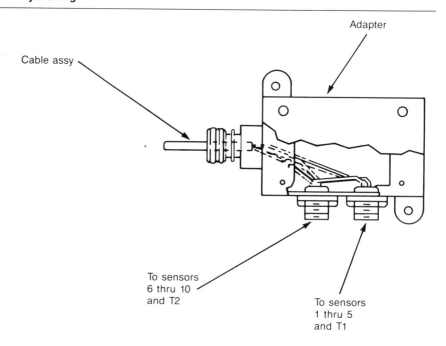

Figure 3-10:
Dual Adapter Cable Assembly

Maps

Maps represent the appearance of a geographic location. At times, elements that cannot be seen (e.g., state boundaries) are also included. Labels may be placed on the map to identify features, and a legend is included to identify symbols used for objects and the scale of the map, if important. In Figure 15.29, we show a map depicting the interstate highway system for Iowa and the midcontinent region.

Types of Visual Aids   **339**

FIGURE 15.29
Map

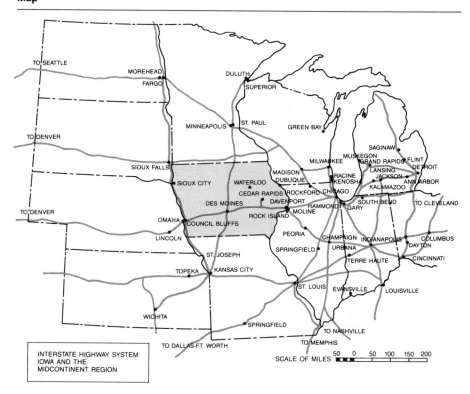

**INTERSTATE HIGHWAY SYSTEM
IOWA AND THE MIDCONTINENT REGION**

Source: *1977 Statistical Profile of Iowa,* compiled by the Iowa Development Commission, p. 18.

Photographs

Photographs should be used when a realistic representation of an object or location is important or when an object or location is so complex that other pictorial visual aids would not effectively picture it. For example, the cells in a cross-section of bread would be difficult to draw. A photograph would give a more useful idea of their appearance than a drawing.

Photographs may cause special problems. Photographs are expensive to duplicate, and because they show the entire object or location exactly as it appears, they do not allow the writer to select content or points of emphasis, as is possible with a drawing.

Since realism is the major advantage of a photograph, prints should be as glossy as possible, to enhance detail.

Samples

Samples are actual pieces of objects, such as swatches of textiles or pieces of metals. Samples allow the viewer to see what the components of a product or the product itself will look like. If small, these samples may be included in a report. If large, they may be sent under separate cover.

APPROPRIATE VISUAL AIDS

In order to complement and supplement your text effectively, your visual aids must be appropriate. That is, they must convey the information you wish to communicate, but no more, and must emphasize the points you want to stress.

Conveying Information

Visual aids may convey too much information to readers, making the task of assimilation more difficult. Consider the following example:

Gravel companies charge by the ton to deliver crushed rock. The approximate delivery cost per ton is listed below:

Delivery Cost of Crushed Rock (per ton)

1	2	3	4	5	6
$15.00	27.00	40.00	48.00	60.00	75.00

Since your driveway and walk will require 3 tons, cost of gravel will be approximately $40.00.

The reader of this document was a homeowner accepting bids for resurfacing her drive and walk. Notice that the only fact she required was the cost of 3 tons of crushed rock. Thus, the table showing all costs was unnecessary. A sentence stating the cost of 3 tons would have better served her needs.

Now consider the same facts used to support a different point:

Gravel companies charge by the ton to deliver crushed rock. The approximate delivery cost per ton is listed below:

Delivery Cost of Crushed Rock (per ton)

1	2	3	4	5	6
$15.00	27.00	40.00	48.00	60.00	75.00

Notice that the cost decreases significantly when larger amounts are ordered. Therefore, I recommend delivery of 6 tons per load rather than the 3 tons you initially requested.

The reader of this document was a builder who would order a number of truck-loads of crushed rock. He required all the data in the table so he would know costs of

Appropriate Visual Aids   **341**

various truckload sizes. Therefore, the table conveyed essential information and was correctly included in the text.

Emphasizing Points

Since different types of visual aids could be used to express the same data, you must choose the type that emphasizes the point you wish to make. For example, consider the following data for two area schools:

Enrollment Figures for Central University

Year	No. of Students
1979	7,000
1980	10,000
1981	9,500
1982	10,500
1983	11,000

Enrollment Figures for Bennett College

Year	No. of Students
1979	5,000
1980	7,500
1981	6,500
1982	6,000
1983	5,500

Three symbolic visual aids — a table, a bar graph, or a line graph — could be used to picture this information. We show these visual aids in Figures 15.30a-c and discuss their differing emphases.

FIGURE 15.30a
Table showing data

Table 1: Enrollment figures for two area schools

| Year | No. of Students | |
	Central University	*Bennett College*
1979	7,000	5,000
1980	10,000	7,500
1981	9,500	6,500
1982	10,500	6,000
1983	11,000	5,500

This table emphasizes the individual enrollments at each school for the given years. The table also stresses the actual enrollment figures. Therefore, the table might be used to support the following point:

Enrollment figures for two area colleges, years 1979–83, are given in Table 1.

— TABLE —

As the figures show, Central University enrolled 7,000 students in the year 1979, while Bennett College enrolled 5,000. By the year 1983, Central University's enrollment had increased to 11,000, while Bennett College's was 5,500.

FIGURE 15.30b
Bar graph showing data

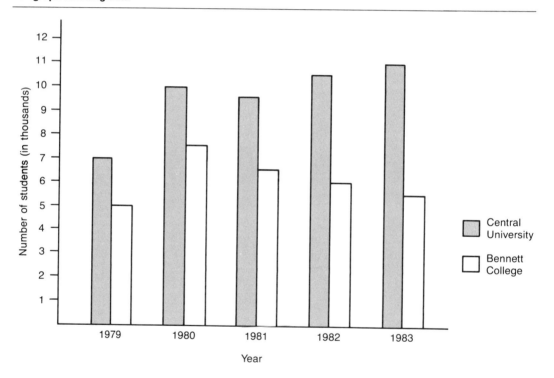

Figure 1:
Enrollment Figures for Two Area Schools

This bar graph emphasizes the comparison of enrollment figures within each year and as a whole. Therefore, the bar graph might be used to support the following point:

Appropriate Visual Aids

Enrollment figures for two area colleges, years 1979–83, are given in Figure 1.

— FIGURE —

Notice that Central University has had significantly higher enrollments than Bennett College for each year in the period.

FIGURE 15.30c
Line graph showing data

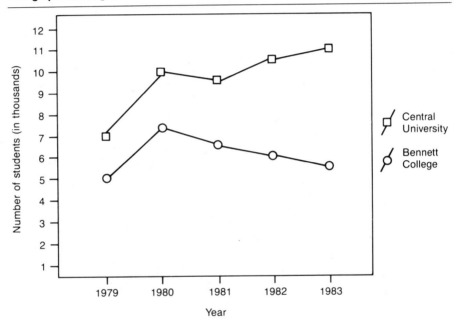

Figure 1:
Enrollment Figures for Two Area Schools

This line graph emphasizes the trend in enrollment figures at the two schools. Therefore, the line graph might be used to support the following point:

Enrollment figures for two area colleges, years 1979–83, are given in Figure 1.

— LINE GRAPH —

Notice that Central University has experienced enrollment growth since 1981, while Bennett College's enrollments have declined since 1980.

Different pictorial visual aids may also be chosen in order to create different emphases. For example, consider the diagram and photograph of a coating applicator for lightwave fibers, given in Figures 15.31a and 15.31b. The schematic diagram stresses the construction of the applicator because the diagram has been simplified: All extraneous information has been omitted. The photograph gives a better idea of the applicator's actual appearance.

FIGURE 15.31a
Diagram of the coating applicator

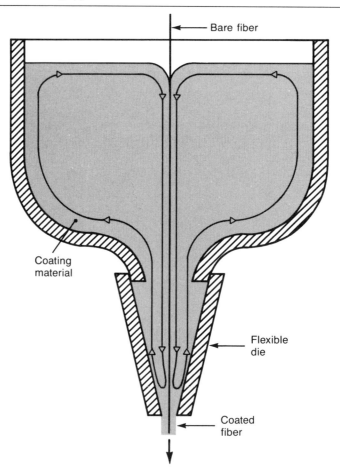

FIGURE 15.31b
Photograph of the coating applicator

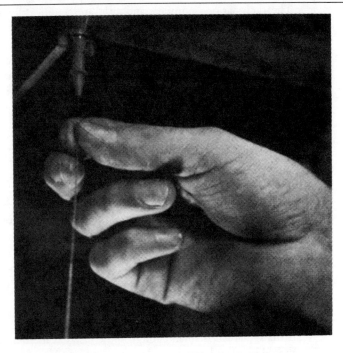

INTEGRATED VISUAL AIDS

In order for readers to utilize visual aids effectively, the aids must be integrated with the text. You integrate a visual aid by introducing and assessing it.

Introducing Visual Aids

Several operations must be performed to introduce a visual aid effectively. First, you must designate, number, and title the visual aid. Then, you must indicate the location of the visual aid and identify what the reader will find.

Designating, Numbering, and Titling

Every visual aid receives the appropriate designation (as a figure or table) and a number. Arabic numbers are traditionally used for figures, while either Arabic numbers or Roman numerals may be used for tables. These numbers are consecutive through the text, including figures and tables in the appendix, but figures and tables are numbered separately, e.g., Figure 1, Figure 2, Table 1.

The title should tell the reader as precisely as possible what the visual aid contains. Perhaps the perspective of the visual aid is important (e.g., *Cross-Section* of a

Diseased Heart; Condenser — *Top View*) or the type of visual aid (e.g., *Schematic* Diagram of a Solar Unit; *Flow Chart* of the Proposal Process).

The designation, number, and title are traditionally placed *below* figures and *above* tables in typed documents, although this placement varies in printed publications.

An exception to this rule of designating, numbering, and titling occurs for informal visual aids such as the table on page 340. Informal visual aids are simply incorporated in your text after an introductory sentence. Whether you use informal or formal visual aids depends on the nature of your document and your reader's reaction to the visual aids. If you are writing a very informal document or if formal designations, numbers, and titles are likely to overwhelm readers who are unused to them, you should simply insert the visual aids in your text.

Locating

Visual aids must be mentioned in your text if the reader is to use them effectively. This mention may take many forms: e.g., ''See Figure 2 for a cutaway drawing of a nuclear reactor.''; ''Data supporting this recommendation are given in Table 4.''; ''Figure 3 in the appendix shows an exploded diagram of the faucet assembly.''

The visual aid is then placed as close to its mention as possible. In a printed publication, this placement is determined during page makeup. But in a typed document, you have three options for placement:

1. Immediately following the mention
   If the visual aid is small enough to be inserted in your text, the visual aid should be placed directly below the sentence introducing it. You then continue your text after the visual aid.

2. On the next page
   If the visual aid is large enough to require a separate page, the visual aid should be placed on the page by itself immediately following the mention. No further text should be included on this page.

3. In the appendix
   If you have a series of visual aids that would significantly interrupt the text (e.g., a series of balance sheets showing the financial position of a company over several years) or an oversized visual aid (e.g., a foldout map), these visual aids should be placed in the appendix. In addition, visual aids depicting technical information in a report directed to lay readers ought to be appended. Technically proficient readers can then turn to these visual aids, while lay readers can concentrate on your text.

Identifying

Your introduction should identify what the reader will find in the visual aid: e.g., a cross-section of a diseased elm tree; the percentage of funds allocated to each department. Thus, you ensure that the reader who may be unfamiliar with the subject will know what the visual aid contains before viewing it.

Assessing Visual Aids

Your assessment of a visual aid directs reader attention to the points you wish to make. In this way, you ensure that the reader will see what you intend. For example, consider the table shown in Figure 15.32. The reader might notice several points about these data: e.g., precipitation steadily decreased in summer and steadily increased from fall through spring; January and February were the months with most precipitation; as temperature increased, precipitation decreased. Because a reader may not notice the point you wish to make, you must clarify this point in your assessment.

FIGURE 15.32
Table showing precipitation and temperature data

Table 5: Monthly precipitation and average temperature for Bear County, Kansas, 1983

Month	Precipitation (in.)	Temperature (°F)
Jan.	3.0	35°
Feb.	3.0	38°
March	2.9	45°
April	2.5	49°
May	2.0	55°
June	1.7	60°
July	1.2	68°
Aug.	0.9	72°
Sept.	1.4	65°
Oct.	2.2	45°
Nov.	2.4	40°
Dec.	2.9	38°

For example, an assessment of the table in Figure 15.32 might read as follows:

> Notice that precipitation in colder months was significantly higher than precipitation in warmer months, with the highest amount (3.0 in.) recorded in January and February and the lowest (0.9 in.) in August.

CHECKLIST FOR VISUAL AIDS

1. Have I used visual aids to shorten and organize my text, clarify information, indicate relationships, emphasize points, and summarize material?

2. Have I considered the various types of visual aids?

   A. Have I used tables to stress actual amounts or present individual details, either as numerical data or in words?

348   15 | Visual Aids

    B. Have I used graphs to present numerical data in visual form?

        (1) Have I used bar graphs to indicate comparisons and contrasts of the same entity at different times or different entities at the same time?

        (2) Have I used line graphs to indicate trends or rates of change?

        (3) Have I used 100% bar graphs or pie graphs to show the percentages making up a whole?

    C. Have I used charts to represent nonnumerical data in visual form?

        (1) Have I used flow charts to picture processes?

        (2) Have I used organizational charts to picture hierarchies of positions, divisions, or functions within institutions?

    D. Have I used diagrams and drawings to picture objects?

        (1) Have I used schematic diagrams to show the general appearance of objects' outer surfaces?

        (2) Have I used shaded sketches and isometric drawings to show three dimensions of objects?

    E. Have I used maps to represent the appearance of geographic locations?

    F. Have I used photographs to give a realistic view of objects or locations?

    G. Have I used samples to show the components of a product or the products themselves?

3. Have I created appropriate visual aids?

    A. Have I conveyed *only* the information my reader requires?

    B. Have I chosen visual aids that emphasize the points I wish to make?

4. Have I integrated my visual aids with my text?

    A. Have I introduced my visual aids by designating, numbering, and titling them? Have I located them effectively and identified what the reader will see?

    B. Have I assessed my visual aids by stressing the points I wish to make?

EXERCISES

Topics for Further Practice

1. Construct three different symbolic visual aids to represent the following data. Then introduce and assess each visual aid, stressing the point the visual aid is intended to convey.

    Three different offices of the same company, Illel, Inc., have supplied the following information on sales of athletic equipment.

    A. Chicago office: $142,000 in 1979; $183,500 in 1980; $168,500 in 1981.
    B. St. Louis office: $92,500 in 1979; $101,000 in 1980; $101,500 in 1981.
    C. New York office: $143,000 in 1979; $162,500 in 1980; $112,000 in 1981.

Exercises 349

2. Construct, introduce, and assess a pie chart showing the allocation of your college expenses for the current year — e.g., moneys spent on tuition, housing, books, and so on.

3. Construct, introduce, and assess a flow chart for a process in your field. The process may involve actions and decisions, or only actions or events.

4. Construct, introduce, and assess a flow chart that illustrates the function of an apparatus used in your field.

5. Construct, introduce, and assess an organizational chart for the hierarchy of positions in the administration of your school.

6. Construct, introduce, and assess four different pictorial visual aids of a single simple object. Be sure to stress appropriate points in your assessment.

7. Construct, introduce, and assess a map that represents your campus. Include a legend for data presented symbolically and a scale.

16

Format

THE IMPORTANCE OF FORMAT

Format, or the appearance of your writing on the page, greatly affects the reception of your document. An effective format aids your audience by making your document easier to read and by signaling important information. On the other hand, an ineffective format makes your reader's task more difficult, thus increasing the possibility that he or she will not understand your material. An ineffective format also reflects on you as a writer. Readers may believe you are careless or incompetent, which can affect their opinion of your technical work as well. Therefore, designing an effective format is an important skill.

In this chapter we discuss five elements of format: highlighting, white space, lists, headings, and numbering systems. We then consider the layout and standard parts of business letters and memorandums.

ELEMENTS OF FORMAT

Highlighting

Highlighting techniques are means of calling attention to words, phrases, or sentences you wish to stress. Since your readers' eyes are immediately drawn to your highlighted writing, you are able to achieve appropriate emphasis.

Many techniques exist for highlighting: underlining, listing, using white or empty space, punctuating (using dashes, ellipses or three dots, or parentheses), capitalizing, italicizing, using different type styles or colors. However, in technical writ-

350

Elements of Format 351

ing, underlining, listing, using white space, and punctuating are the most common techniques.

In the following example, a portion of a business letter, notice how these techniques are used to highlight the date, the event occurring on that date, and the location of the event:

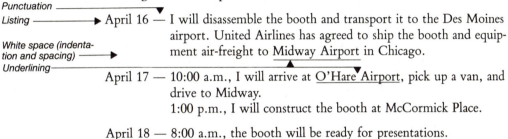

As we discussed last week over the phone, you are concerned about transporting your exhibit from Ames to the Chicago trade show.

I am glad to reply that all logistical plans are arranged and ready for implementing. Here is the plan:

April 16 — I will disassemble the booth and transport it to the Des Moines airport. United Airlines has agreed to ship the booth and equipment air-freight to Midway Airport in Chicago.

April 17 — 10:00 a.m., I will arrive at O'Hare Airport, pick up a van, and drive to Midway.
1:00 p.m., I will construct the booth at McCormick Place.

April 18 — 8:00 a.m., the booth will be ready for presentations.

White Space

White space refers to the empty areas you leave on the page. This space provides relief for the eye by breaking up long passages of prose just as silence provides relief for the ear by breaking up extended noise. You must consider both horizontal and vertical white space when formatting.

Horizontal white space is achieved by indentation. First, you must provide adequate margins around your text: 1 in. to 1¼ in. is standard for marginal space; 1½ in. should be used on the left-hand margin when your document is bound. Second, you may choose to indent your paragraphs five spaces. (An exception occurs when you use block form for letters or reports, where every line begins at the left margin.) Third, you may choose to indent your lists five spaces from the left and right margins. This spacing sets off and emphasizes the list and provides a break in the text.

Vertical white space breaks your text by separating parts as you read down the page. The amount of vertical white space you provide will vary, depending on whether you single space or double space your text. (In industry, reports are often single spaced. Letters and memorandums are always single spaced.) Table 16.1 provides some guidelines for incorporating vertical white space in documents.

In addition, visual aids are highlighted by surrounding them with white space. They should be separated from your text by four spaces in single-spaced reports and four to six spaces in double-spaced reports. This white space calls attention to the visual aids and ensures that they will be easy to read.

TABLE 16.1
Vertical white space

Location	Single-Spaced Documents	Double-Spaced Documents
Between major sections	1 in. white space or a new page (in lengthy reports)	1 in. white space or a new page (in lengthy reports)
Between units within sections	3 spaces	4 spaces
Between paragraphs	2 spaces	2 spaces
Between headings and text	2 spaces	2 spaces

Equations should also be set off from your text. The following format is usually employed:

$$F = kx \tag{1}$$

where

F = force needed to deflect the fiber,
k = a constant,
x = the offset of the fiber.

Notice that in this case the definition of the symbols in the equation is punctuated as a sentence, with commas after the values and a period at the end. An alternative style is to use no punctuation. In addition, all equations receive a number, written at the right margin. These numbers proceed consecutively throughout your document.

Lists

Lists are useful highlighting techniques when your material consists of a series of items. The list orders your details and segments them, thus facilitating reader retrieval of the information. Use of a list also allows you to emphasize certain details over others, since you can arrange the items in your list in any order you decide is effective. (Remember, however, similar items in a list are usually classified.)

Your list may appear as a numbered series down the page, in which case the list is usually indented five spaces from the left and right margins. You may also list your items in sentence form, if your list does not exceed three items and the items are simple. However, this technique does not highlight the list as sharply as do indentations.

To construct a list, place a colon after your introductory sentence, then number each item you present. (You may omit these numbers, if your listed items are short and simple.) If your list proceeds down the page, you should single space each item and double space between items.

Figures 16.1a and 16.1b show the proper layout of both types of lists.

Elements of Format **353**

FIGURE 16.1
(a) Layout of list down the page   **(b) Layout of list in sentence form**

(a)

Conduct your experiment in the following way: ◄─────────────

A colon precedes a list, after a complete introductory sentence.

1. or (1) Drain the column. ◄─────────────
2. or (2) Fill the column with hot H_2O.
3. or (3) Start the filter.

A period is required in a sentence list but is optional in a list of phrases.

(b)

Conduct your experiment in the following manner:

Optional numbers ► 1. or (1) drain the column; 2. or (2) fill the column with hot H_2O; 3. or (3) activate the filter.

Capitals are not necessary after a colon.

Commas may be used instead of semicolons, when your list is not introduced with numbers and does not contain interior commas.

Headings

Headings serve a number of purposes for your reader:

1. They provide reader guides, making the logical progression of your ideas clear and indicating levels of information.

2. They provide eye relief. Like white space, they break up passages of prose. (Use headings judiciously, however, to precede significant amounts of information. Too many short passages may interrupt the reader.)

3. They facilitate quick retrieval of information for the reader skimming or returning to the document.

Headings can perform the first function because they are related to the structure you have given your material: Headings signal the hierarchical divisions of your information. Therefore, headings are not all of equal weight. Consider an analogy to the outline form: Some topics are high-level generalizations (Roman numerals); other topics are categorized under these, in a descending hierarchical order, using Arabic numerals, and capital or lowercase letters. If you were to outline the headings of your document, those on the same level (e.g., all Roman numerals, all capital letters) would have the same weight. This fact means that, in formatting, they should also look the same, thus signaling the levels of information.

You have a choice of three methods for differentiating headings: position on the page, capitalization, and underlining. Position on the page refers to the placement you select for your headings. Figure 16.2 indicates possibilities for placement.

However, placement on the page does not sufficiently differentiate headings, because only four levels would be possible. In addition, placement does not effectively

FIGURE 16.2
Placement of headings

First in importance ——————————————————————► Center
Reader's eye is
drawn here

2nd in importance ——► Left Margin

3rd in importance ————————►Indent five spaces

4th in importance ————————► Indent five spaces as a paragraph lead. The text then continues on the same line
after the paragraph lead. Notice the period after the lead: It is *never* considered part of the
succeeding sentence.

highlight headings. You must combine position with capitalization and underlining to highlight and distinguish various levels of headings. Figure 16.3 presents two possibilities, carried down to six levels of headings. Remember, however, in short docu-

FIGURE 16.3
Two possibilities for formatting six levels of headings

First level	CENTER CAPITALIZE UNDERLINE

Second level	LEFT MARGIN CAPITALIZE UNDERLINE
Third level	LEFT MARGIN CAPITALIZE
Fourth level	INDENT FIVE SPACES CAPITALIZE
Fifth level	Indent five spaces
Sixth level	Indent five spaces as a paragraph lead. Paragraph leads are frequently underlined to highlight them.

or

First level	CENTER CAPITALIZE UNDERLINE

Second level	CENTER CAPITALIZE
Third level	LEFT MARGIN CAPITALIZE
Fourth level	Left Margin
Fifth level	Indent five spaces
Sixth level	Indent five spaces as a paragraph lead.

Elements of Format

355

ments you will rarely go below two or three levels of headings, because readers will not need further guidance to your material.

When considering format for headings, keep the following points in mind:

1. Format headings of equal weight (on the same level) identically.

2. Combine your three techniques (position on the page, capitalization, underlining) for the most effective format you can devise.

3. Avoid competition between techniques. For example, putting your first-level heading at the left margin and centering your second-level heading would make the second seem more important than the first. Similarly, underlining your second-level heading but not your first-level heading would put more emphasis on the less important heading.

Numbering Systems

Numbering systems can be used in conjunction with headings to help readers discern the levels of information in your document. Two common numbering systems are the outline form and the decimal form.

The outline form combines numerals and letters in alternation:

> I. First level
>   A. Second level
>   1. Third level
>   2.
>   a. Fourth level
>   b.
>   (1) Fifth level
>   (a) Sixth level
>   (b)
>   (2)
>   B. Second level
> II. First level

Notice that you cannot extend the outline form beyond fourth level without repeating numbers or letters. These must be differentiated from third-level numbers and fourth-level letters by using parentheses.

The outline form cannot be expanded beyond sixth level, since no further means of differentiation exist. Although you may not need to go beyond the second or third level in short documents, longer documents may require further segmentation.

The decimal form uses numbers and decimal points to differentiate levels:

> 1.0 First level
>   1.1 Second level
>   1.1.1 Third level

356   16 | Format

        1.1.2
            1.1.2.1 Fourth level
            1.1.2.2
                1.1.2.2.1 Fifth level
                1.1.2.2.2
                    1.1.2.2.2.1 Sixth level
                    1.1.2.2.2.2
    1.2 Second level
  2.0 First level

Notice that the decimal form, unlike the outline form, is infinitely expandable. More-over, readers can more quickly discern, simply from seeing the number of digits, the level of the section they are reading. However, the decimal form may appear unwieldy at lower levels of information because of the number of digits involved.

LAYOUT AND STANDARD PARTS OF LETTERS AND MEMORANDUMS

Letters

Layout

Several traditional styles are acceptable as layouts for business letters: block, modified block, modified block with paragraphs indented, and simplified. The following diagrams illustrate the placement of parts of the letter and the spacing for each of these styles.

Layout and Standard Parts of Letters and Memorandums   **357**

Block Style

(2-in. top margin, or center the letter in the middle of the page, if brief)

Heading ————————▶ 65 Main Street
                          Sherwood, NY 30617

Date ————————————▶ April 2, 1983

                          (4 spaces)

Inside address ————▶ Ms. Joan Kay
                          Vice President
                          Bissell Architects, Inc.
                          16 Oak Street
                          Brighton, NY 30514

                          (2 spaces)

Subject line (optional) ▶ Subject: Blueprints for Office Building

                          (2 spaces)

Salutation ——————▶ Dear Ms. Kay:
                          (2 spaces)

                          (1 space)

                          (2 spaces)

                          (2 spaces)

                          (2 spaces)

Complimentary close ▶ Sincerely yours,

Signed *signature* area ▶ *Dr. Henry Fuller*
                          (4 spaces)

Typed signature ———▶ Dr. Henry Fuller

                          (2 spaces)

Typist's initials ———▶ HF:fac
                          (2 spaces)

Enclosure notation ▶ Enclosure (2): Blueprints
                          (2 spaces)

Copies sent ————————▶ cc: Dr. Meredith Springer

358   16 | Format

Modified Block Style

(2-in. top margin, or center the letter in the middle of the page, if brief)

65 Main Street
Sherwood, NY 30617
April 2, 1983
(4 spaces)

Notice that the ⟶ Ms. Joan Kay, Vice President
addressee's title may
be placed after the   Bissell Architects, Inc.
name or on next line, 16 Oak Street
depending on length.   Brighton, NY 30514

(2 spaces)

Re (regarding) and ▶ RE: Blueprints for Office Building
subject are inter-
changeable.   *(2 spaces)*

Punctuation can be ▶ Dear Ms. Kay
omitted after the salu-
tation, but this is not *(1 space)*
standard.

(2 spaces)

(2 spaces)

(2 spaces)

If punctuation is omit- ─────────────────────⟶ Sincerely yours
ted after the saluta-
tion, punctuation is   *Dr. Henry Fuller*
also omitted after the
complimentary close.   *(4 spaces)*
                                                     Dr. Henry Fuller

                                                     (2–4 spaces)

HF:fac

(2 spaces)
Enclosures (2): Blueprints

(2 spaces)
cc: Dr. Meredith Springer

Layout and Standard Parts of Letters and Memorandums **359**

Modified Block Style with Paragraphs Indented

(2-in. top margin, or center the letter in the middle of the page, if brief)

469 Backen Boulevard
Columbia, SC 29216
April 2, 1982
(4 spaces)

Southeastern Medical Association
16 Fourth Street SW
Curry, SC 29208

(2 spaces)

An attention line can ▶ Attention: Ms. Davenport
be used to direct a
letter to a specific   *(2 spaces)*
person, if you are   Ladies and Gentlemen:
addressing a group.   *(2 spaces)*

(1 space)

(2 spaces)

(2 spaces)

Sincerely,
(1 space)
Your company's name ────────────────── ▶ EASTERN OFFICE SUPPLY
may be typed above
your signature, if you   *Gerald Cushman*
are not using a letter-
head.   *(4 spaces)*
                                              Gerald Cushman
Your title/job role   ▶ Promotion and Sales
should be included. ──────────────────
                                              (2–4 spaces)

GC:blw

(2 spaces)
Enclosure: brochure

"c" indicates that
photocopies rather   *(2 spaces)*
than carbon copies
have been circulated. ▶ c: Thomas Nickleby
                              Anne Bradford

360   16 | Format

Simplified Style

(2-in. top margin, or center the letter in the middle of the page, if brief)

2500 Second Street
Deering, TX 80513
June 13, 1982

(4 spaces)
William Unspoon, President
Big Bear Construction, Inc.
161 Atlantic Avenue
Houston, TX 82341

The salutation is ▶   *(2 spaces)*
omitted   *(1 space)*

(2 spaces)

(2 spaces)

The complimentary ▶   *Mark Able*
close is omitted

(4 spaces)
Mark Able

(2-4 spaces)
MA:dab
(2 spaces)
Encl (2): Budget Schedules
(2 spaces)
cc: Don Taylor

Although the simplified style is an acceptable form, it may seem informal in tone. Therefore, many businesses prefer the other three styles we have shown.

Layout and Standard Parts of Letters and Memorandums **361**

Standard Parts
Most business letters have seven standard parts.

The Heading
The heading consists of your address and the date the letter was sent. The address may be given on the letterhead or typed at the left margin, center, or ending at the right margin of your letter, depending on the letter style you have chosen. If you type your heading, *do not* include your name.

The Inside Address
The inside address includes the name of the person or group to whom you are sending the letter, the person's courtesy title (Dr., Ms.), business title (Personnel Manager), and address. The business title may be typed on the same line as the person's name, on the next line preceding the company's name, or on a line by itself. Choice depends on the length of the line.

The Salutation
The salutation is the opening greeting, followed by a colon or no mark of punctuation, if open punctuation is used. The salutation should reflect the first line of your inside address:

Ms. Brenda Hopps   Manager, Repair Department
. . .   . . .

Dear Ms. Hopps: Dear Manager:
("Ms." is often the standard form
of address for a woman, unless she
prefers another)

Barr Company   ABC Fraternity
. . .   . . .

Ladies and Gentlemen: Dear Members:
(When a mixed group is addressed)   or
                             Gentlemen:

The Body
The body of a business letter is traditionally single spaced, with double spacing between paragraphs. When your letter continues onto a second page, head that page with the addressee's name, the page number, and the date:

Ms. Brenda Hopps
Page 2
June 26, 1982

or

Ms. Brenda Hopps   2   June 26, 1982

This information allows the reader to reassemble the letter, if it should become disordered.

The Complimentary Close

Your complimentary close should suit the tone of your letter. Generally, one of the following is used: Sincerely or Sincerely yours; Yours truly; Cordially. Notice that only the first word of the complimentary close is capitalized. The complimentary close is followed by a comma, unless open punctuation is used. In that case, the comma is omitted.

The Signature Area

The signature area must contain your signature and typed name. The area may also include your company's name, if you have not used letterhead, and your business title.

Sincerely yours,
MSEG CONSULTANTS

Lawrence Peck

Lawrence Peck
President

If you are very familiar with your addressee, you may sign just your first name or nickname.

The Reference Section

Your initials, in capital letters as writer of the letter, and your typist's initials are always placed at the left margin.

If an item or items are included with the letter, an enclosure notation appears below these initials. The number of enclosures and their contents should be given, for identification purposes.

If copies of your letter are being sent to other people, their names should be listed in order of importance or in alphabetical order after the abbreviation cc (carbon copy) or c (photocopy).

Memorandums

Layout

The layout of memorandums varies greatly from company to company. (Many companies print their own forms to be used when writing memos.) However, all memorandum forms contain two parts: the heading and the body.

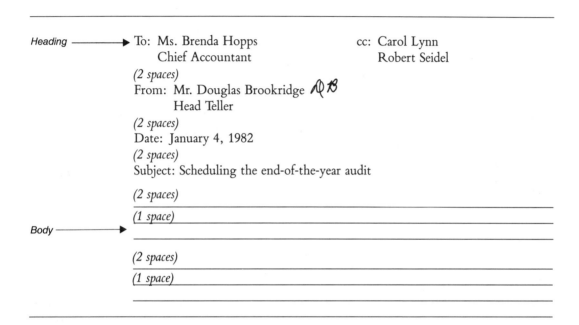

Standard Parts

The Heading

The heading is often aligned at the left margin or by the colons. (Other layouts are possible if the heading is preprinted.) This heading may include five items of information:

1. The person to whom the memo is sent

    Ms. Brenda Smith   Bob Smith
    Chief Accountant   or
        or   Bob
    Ms. Brenda Smith, Chief Accountant

If the courtesy title (Ms.) and the job role are unimportant or well known, these designations can be omitted, unless the memorandum will be filed. If so, this information should be included for future reference. In addition, if you are circulating this memo to a number of readers, all the names should be listed.

2. The person sending the memo
The form of this entry should parallel the form of the "To:" entry.

Mr. Douglas Brookridge	Douglas Brookridge
Head Teller	or
or	Doug
Mr. Douglas Brookridge, Head Teller	

You may initial your memo beside your name, to verify that you have authorized it.

3. The date the memo was sent

4. The subject of the memo
The subject line of a memo should be precise, in order to identify exactly what the memo is about. Consider the following subject lines.

Subject: Design Criteria
Subject: Design Criteria for the XR-50

Subject: Change in Procedures
Subject: Procedural Change for Billing System

In each case, the second subject line more precisely pinpoints the memo's contents.

5. The persons to whom copies have been sent

The Body

The body of a memo is traditionally single spaced, with double spacing between paragraphs. Identification of a second page is done as for a letter. Notice that the memo usually has no complimentary close or signature following the message, although memos may occasionally have a closing signature line. Instead, authorization is usually shown by your written initials beside your name in the heading.

CHECKLIST FOR FORMAT

Elements of Format

1. Have I used highlighting techniques to call attention to material I want to emphasize?

2. Have I used horizontal white space to provide eye relief and to highlight material? Have I used vertical white space to segment my text and to highlight material?

3. Have I used lists to highlight, order, segment, and emphasize material? Have I used an appropriate layout for my list: down the page or sentence form?

Exercises **365**

4. Have I used headings to guide my reader by indicating the logical progression and the levels of my information, to provide eye relief, and to facilitate quick retrieval of facts? Have I formatted these headings effectively, using position on the page, capitalization, and underlining? Are headings of equal weight formatted alike?

5. Have I used a numbering system to indicate levels of information?

Layout and Standard Parts of Letters and Memorandums

1. Have I followed the layout of the traditional business-letter style I have chosen?

2. Have I included the seven standard parts in my letter? Are they complete?

3. Does the heading of my memorandum include the traditional items of information?

4. Have I single spaced the body of my letter or memorandum and double spaced between paragraphs?

EXERCISES

Topics for Further Practice

1. Design the seven standard parts of a business letter for a letter to be sent to a firm, requesting information on a product they distribute. (You do not need to write the body of the letter.) Indicate the information in the body by drawing lines.

2. Design a memorandum heading for a memo to be sent to your technical writing teacher. The subject of the memo will be the topic of your final report.

17

Report
Supplements

THE NATURE OF REPORT SUPPLEMENTS

The body of your report gives most of the details you want to communicate to your readers. However, the body of a report may also be introduced by prefatory material and followed by supplemental material, which aid this communication. This material may include the following items:

Prefatory

1. Cover
2. Title page
3. Letter or memorandum of transmittal
4. Table of contents
5. List of illustrations
6. Nomenclature page or list of symbols
7. Abstract

Supplemental

1. Glossary
2. Footnotes (or endnotes) or references-cited page
3. Bibliography
4. Appendix or appendices

366

You will not need all these items in every report you write: Level of formality of the report and the information readers require will guide your choice. However, we discuss each item, so that you will understand how to compose them all.

In our discussion, we proceed from front to back of a report, according to the correct placement of each item.

PREFATORY MATERIAL

Cover

The cover immediately identifies your report, so that readers will not have to open it to discover its contents. The cover also protects your text so its appearance will create a favorable impression.

This cover traditionally gives two pieces of information: the title of the report and your name as author. Your title should be fully descriptive, featuring nouns for the subject of the report and giving the precise type of report. These nouns should be the major keywords for your report, to facilitate computer indexing for data bases. The following examples illustrate effective titles.

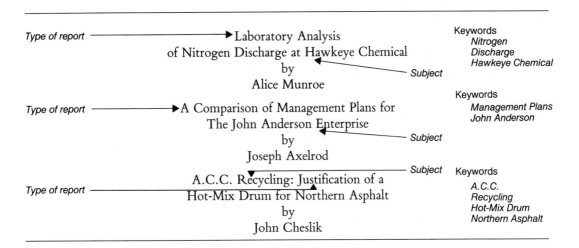

(Note that "A Report on. . ." does not convey sufficient information to identify the type of report.) The title may be expressed in one phrase (titles 1 and 2) or arranged as a title and subtitle (title 3). In addition, titles should be brief: ten words or less, if possible. For example, the following title contains words that could be omitted:

~~A~~ Structural Design ~~with~~ Reco~~mm~~ended Appl~~ica~~tions
of an Expanded Office Facility for the
Cue Insurance Agency

368   17 | Report Supplements

A more effective title would read as follows:

Type of report ————————→ Design of an Expanded Office Facility for the
Cue Insurance Agency ◄————————— *Subject*

Your cover should be sturdy cardboard, to protect your report. You then paste a *typed* label containing the title and your name on the cover: A hand-written label will appear sloppy. If your cover contains a window, you should type a cover page, placing the information so that it will show through the window.

Title Page

The title page contains four obligatory pieces of information: the title of the report, the person for whom the report has been prepared, the person preparing the report, the date.

1. Title
   This title is identical to the one appearing on your cover.

2. Person for whom the report is prepared
   Three items of information are usually listed: name, job role/title, affiliation:

   Sherwood Andrews   Sherwood Andrews, Project Manager
   Project Manager   or   Northern Asphalt
   Northern Asphalt

   If the report is an in-house document, the affiliation can be omitted.

3. Person preparing the report
   The same items of information are given for the person preparing the report as for the person for whom the report is prepared, in identical format:

   John Cheslik   John Cheslik, Project Analyst
   Project Analyst   or   Construction Methods and
   Construction Methods   Machines, Inc.
   and Machines, Inc.

4. Date
   The date is that on which the report is submitted.

In addition, title pages frequently contain other information: a descriptive abstract, the file number of the report, contract numbers under which the work was done, the name and signature of an authorizer other than the writer, a company emblem. Each of these should be given space on the title page appropriate to their importance and formatted effectively. Figure 17.1 shows the title page for a report, with annotations.

FIGURE 17.1
Title page for a report

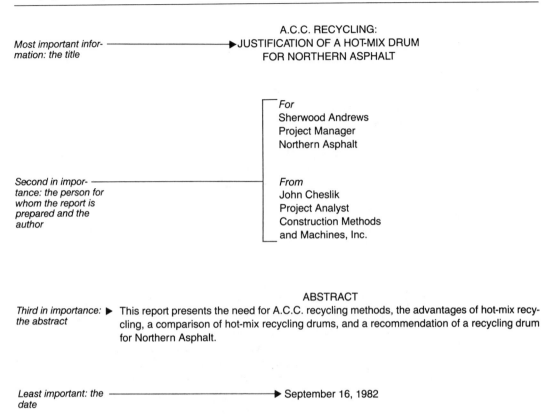

Letter or Memorandum of Transmittal

The letter or memorandum of transmittal delivers the report to its audience. Therefore, its primary purpose is to inform. However, it may also persuade the reader that the report's contents are necessary and valid. Thus, the transmittal document may play an important role in forming your reader's reaction to your report.

You have two choices for placing the transmittal document. You can bind it in the report immediately following the title page, if your letter or memorandum is persuasive and is intended to become part of a permanent record with the report. Or, you can attach the letter or memorandum to the cover, if the letter or memorandum is largely informative and is not intended to become part of a permanent record with the report. Your reader can then detach the transmittal document and send the report on its way through the company.

The letter or memorandum of transmittal should be in standard business-correspondence form. Figure 17.2 gives a checklist of possible details to include:

FIGURE 17.2
Checklist of possible details to include in a letter or memorandum of transmittal

I. Introduction
    A. Transmittal statement
    B. Authorization statement
    C. Subject of the report

II. Body
    A. Purpose of the report
    B. Points to which you wish to call reader attention
    C. Results, conclusions, and/or recommendations
    D. Acknowledgments

III. Conclusion
    A. Offer to answer questions
    B. Courteous close

The purpose of the introduction is to inform the reader that the report has arrived, as well as to identify its subject. The authorizer of the report may also be included, if readers require this information.

The purpose of the body is to call your reader's attention to any information you wish to stress. For example, you can place your major results, conclusions, or recommendations in the body, thus ensuring that the reader will note these points. You can also direct attention to significant findings in support of these points, thus persuading the reader that the contents of your report are valid.

The purpose of the conclusion is to end the letter or memorandum on a courteous note and to build good will with your audience.

You should also remember, however, that the letter or memorandum of transmittal is a brief document, not a summary of the entire report. In Figure 17.3, we have annotated a typical letter of transmittal.

Table of Contents

The table of contents, an abbreviated outline of your report, serves several purposes:

1. It makes the structure of your report visible, by indicating a hierarchy of ideas.

2. It forecasts the information contained in the report.

3. It allows your audience to access specific information easily without having to read the entire report.

FIGURE 17.3
Letter of transmittal

CONSTRUCTION METHODS & MACHINES, INC.
11th & High Street
Des Moines, Iowa 50301
Phone (515) 277-5322

July 15, 1982

Mr. Sherwood Andrews
Project Manager
Northern Asphalt
1131 Lakeview Street
Clear Lake, Iowa 50501

Dear Mr. Andrews:

Transmittal statement ▶ I am sending the report entitled "A.C.C. Recycling: Justification of a ◀ *Subject of the report*
Hot-Mix Drum for Northern Asphalt," as agreed in our contract. ◀ *Authorization state-*
ment

Contents of the report ▶ This report provides information on the need for A.C.C. recycling
and its purpose and the advantages of hot mix, and recommends a specific drum for *Calls reader's atten-*
Major conclusions ▶ your plant. You should note that the recommended drum is easier to *tion to an important*
motivate the reader to install and considerably less costly than other drums. The table on ◀ *point, to ensure it will*
accept the report page 4 supports these conclusions. *be noted*

I will be contacting you at the end of this week to answer any ques- ◀ *Offer to answer read-*
Courteous close ——▶ tions on the report or discuss our recommendation. Until then, thank *er's questions*
you for the opportunity to be of service.

Sincerely,

John Cheslik

John Cheslik
Senior Project Analyst

JC:blw

cc: Walter Blake
James Hill

To perform the first function, the sections in your table of contents must be effectively classified. To perform the next functions, your headings must be individualized and must reflect the contents of the sections they precede. In addition, headings of equal weight must be phrased in a parallel grammatical form, and your table of contents must be effectively formatted.

Classifying Sections

Sections of your report should be effectively classified, since too many or too few first-level headings will not make the hierarchy of your ideas visible to your readers.

Too Many First-Level Headings

The table of contents in Figure 17.4a contains too many first-level headings.

FIGURE 17.4a
Table of contents with too many first-level headings

1.0 Why Control the Mosquito?

2.0 Species Infesting the Area

3.0 Life Cycle

4.0 Daily Activities

5.0 Food Sources

6.0 Control Methods
    6.1 Selective Spraying
    6.2 Aerial Spraying
    6.3 Natural Predators
    6.4 Bush Clearing

7.0 Control Plan

Notice that sections 1.0 and 2.0 both contain introductory material, while sections 3.0, 4.0, and 5.0 all concern characteristics of the mosquito. Thus, these sections could be classified, as you see in Figure 17.4b.

FIGURE 17.4b
Reclassified table of contents

1.0 The Problem With the Mosquito
    1.1 Why Control the Mosquito?
    1.2 Species Infesting the Area

2.0 Characteristics of the Mosquito
    2.1 Life Cycle
    2.2 Daily Activities
    2.3 Food Sources

3.0 Control Methods
    3.1 Selective Spraying
    3.2 Aerial Spraying
    3.3 Natural Predators
    3.4 Bush Clearing

4.0 Control Plan

Prefatory Material **373**

Too Few First-Level Headings

The table of contents in Figure 17.5a contains too few first-level headings.

FIGURE 17.5a
Table of contents with too few first-level headings

I.   Introduction to Energy Conservation in Hotels
II. Methods Considered
   A. Old Hotels
      1. Conservation methods
      2. Systems changes
   B. New Hotels
      1. Layout
      2. Roof design
      3. Wall construction
      4. Building orientation
III. Recommendations for Energy Conservation

The first-level heading "Methods Considered" classifies the entire body of this report under one heading. However, the information in this section falls into two categories: data for old and new hotels. Since these topics are of equal weight, the structure of the information would be highlighted if these categories were used as first-level headings, as in Figure 17.5b.

FIGURE 17.5b
Reclassified table of contents

I.   Introduction to Energy Conservation in Hotels
II. Old Hotels
   A. Conservation methods
   B. Systems changes
III. New Hotels
   A. Layout
   B. Roof design
   C. Wall construction
   D. Building orientation
IV. Recommendations for Energy Conservation

Individualizing Headings

In order to identify information and provide easy access to it, the headings in your table of contents should be individualized. Consider the following partial table of contents:

1.0 Introduction
    1.1 Problem statement
    1.2 Purpose statement
    1.3 Limitations on the study

These headings could describe the contents of the opening section of any report. Therefore, the audience does not know the precise information in the introduction and cannot access this information easily without reading the report. In Figure 17.6, on the other hand, individualized headings have been used. Therefore, readers know the precise information included in each section and can turn with ease to the facts they need.

FIGURE 17.6
Table of contents with individualized headings

TABLE OF CONTENTS

LETTER OF TRANSMITTAL	ii
SUMMARY	iv
1.0 INTRODUCTION TO ASPHALT CEMENT CONCRETE RECYCLING	1
1.1 The Need for A.C.C.	1
1.2 A.C.C. Recycling Methods	1
1.3 Advantages of Hot-Mix Recycling	1
2.0 COMPARISON OF FOUR HOT-MIX RECYCLING DRUMS	3
2.1 Drum-In-Drum	3
2.2 Low Temperature Convection Heating	3
2.3 Dual-Zone	4
2.4 Dual-Zone With Forced Air	4
3.0 RECOMMENDATION WITH IMPLEMENTATION PROCEDURES	5
3.1 Recommendation: Dual-Zone Drum With Forced Air	5
3.2 Implementation Procedure	6
REFERENCES	7
APPENDICES	8
Appendix A: Details Of The Dual-Zone Drum With Forced Air	8
Appendix B: Road Construction With Recycled A.C.C.	10
Appendix C: Mixing and Testing Procedure With Recycled A.C.C.	11

Prefatory Material **375**

Notice several other points about the table of contents in Figure 17.6.

1. All prefatory and supplemental materials except the cover, title page, and table of contents are listed in the table of contents.

2. These prefatory and supplemental materials are *not* included in the numbering system of the report (i.e., the outline or decimal system). This numbering system applies only to the text of the report.

3. The appendices are listed separately with their titles. (If you have only one appendix, it would be listed as ''Appendix,'' with its title.)

4. Arabic pagination begins with the text of your report and proceeds through the appendices. The prefatory material receives lower-case Roman numerals, with the title page being considered i. However, this number never appears on the title page or the letter of transmittal. On the other pages, you may place this lower-case Roman numeral or Arabic number in the center of the page — top or bottom — or in the upper right-hand corner.

In addition, all the headings in your table of contents must appear in the report, worded exactly as they are in the table of contents. However, all the subheadings in your report *need not* appear in the table of contents. In particular, third-level headings and below are often omitted from the tables of contents of short reports. Since these third-level headings frequently appear close together in a report, the reader does not require their pages in the table of contents in order to find the information needed.

Reflecting the Information in the Section

Your headings should reflect the information they precede. Consider the ineffective example in Figure 17.7. The first-level heading ''Cooling Load'' does not adequately reflect the information in this section, much of which concerns heat gain. A more effective first-level heading would be ''Heat Gain and Cooling Load.''

FIGURE 17.7
Heading that does not reflect the information in the section

1.0 COOLING LOAD
    1.1 Outdoor and Indoor Conditions
    1.2 Heat Gain
        1.2.1 Through Exterior Walls and Roofs
        1.2.2 Within the Conditioned Space
        1.2.3 By Infiltration
    1.3 Summation of Heat Gain and Cooling Load

Expressing Headings in Parallel Grammatical Form

Headings equal in importance (i.e., on the same level) must be expressed in parallel grammatical form. Consider the headings in Figure 17.8a. Subheadings 2.1 and 2.2 are not parallel.

FIGURE 17.8a
Nonparallel headings

2.0 CONSIDERATIONS FOR AIR CONDITIONING
    2.1 Cleaning the Air for Comfort and Health
    2.2 Environmental Indices

FIGURE 17.8b
Headings expressed in parallel grammatical form

2.0 CONSIDERATIONS FOR AIR CONDITIONING
    2.1 Physiological Factors
    2.2 Environmental Indices

More effective subheadings are seen in Figure 17.8b.

Formatting the Table of Contents

Capitalization and position on the page should be used to format an effective table of contents. Consider the ineffective example in Figure 17.9a. No differentiation of levels exists in this table of contents through either capitalization or indentation. Therefore, the hierarchical relationship of the information is not immediately visible.

Figure 17.9b also shows an ineffectively formatted table of contents.

Although indentation has been used to differentiate levels, capitalization has not. Therefore, the differences in levels are not emphasized. However, the table of contents in Figure 17.6 on page 374 is effectively formatted.

FIGURE 17.9a
Ineffective format for a table of contents

TABLE OF CONTENTS

  I.   NEED FOR NEW FACILITY 1

  II. OBJECTIVE OF DESIGN STUDY 1

  III. CRITERIA FOR THE DESIGN 2
      MACHINERY 2
      LAYOUT 3
      PERSONNEL REQUIREMENTS 4

  IV. IMPLEMENTING THE DESIGN 5

Prefatory Material

FIGURE 17.9b
Ineffective format for a table of contents

Table of Contents

1.0 Problems with Present Controllers	1
2.0 Developing the Controller	3
2.1 Researching Needs	3
2.2 Acquiring Materials	3
2.3 Constructing and Correcting the Controller	4
2.4 Developing the Program	5
3.0 Testing the Controller	6
4.0 Future Design Possibilities	7

List of Illustrations

If you have used visual aids in your report, they are identified and their location is given in the list of illustrations. The title of this page may vary. If you have included both figures and tables in your report, the correct title of the page is List of Illustrations or Table of Illustrations. If you have used only figures, the correct title is List of Figures or Table of Figures. If you have used only tables, the correct title is List of Tables.

The placement of this list also varies. If you have a large number of visual aids, they should be listed on a page by themselves immediately after the table of contents. If you have only a few visual aids and space permits listing them on the table of contents page, you may place the list after your other entries. We show both arrangements in Figures 17.10a and 17.10b.

Notice several points about a list of illustrations:

1. If placed on a page by itself, the list appears as an item in the table of contents. If placed on the table-of-contents page, the list does *not* appear in the table of contents.

2. Traditionally, figures are listed first, then tables, separated by white space. At times, figures and tables may be listed together, following the order in which they occur in the report. However, this method does not distinguish types of visual aids as completely for your reader.

3. Figures and tables are numbered *consecutively* and *separately* throughout the text, *including* those figures or tables in the appendix.

FIGURE 17.10a
List of illustrations on a separate page

LIST OF ILLUSTRATIONS

FIGURES
1. THE SOLAR COLLECTOR 2
2. INTEGRATED SCHEMATIC 8
3. SUMMARY INFORMATION SHEET 17
4. PIPE-DIAMETER DETERMINATION 21
5. PUMP-SIZE DETERMINATION 22
6 STORAGE-SIZE DETERMINATION 23

TABLES
1. SOLAR-SYSTEM COMPONENTS 7
2. COST ESTIMATE 9
3. ENERGY REQUIREMENTS FOR HEATING WATER 15
4. SOLAR-SYSTEM COSTS 19
5. SYSTEM-OUTPUT PERCENTAGES 20

FIGURE 17.10b
List of illustrations on the table of contents page

TABLE OF CONTENTS

LETTER OF TRANSMITTAL ii
SUMMARY iv

1.0 AGRIBUSINESS TODAY 1

2.0 GOALS FOR LONGMAN FARMS 2

3.0 FARM ENTERPRISES 3
    3.1 Crop Enterprise 3
    3.2 Dairy Enterprise 4
    3.3 Swine Enterprise 5

4.0 LONGMAN FARMS OF THE FUTURE 7

    APPENDICES
    APPENDIX A: ESTIMATED DAIRY-ENTERPRISE BUDGET 8
    APPENDIX B: ESTIMATED SWINE-ENTERPRISE BUDGET 9
    APPENDIX C: PROJECTED WHOLE-FARM BUDGET 10

LIST OF ILLUSTRATIONS

Figure 1. Diagram of Longman Farms 3

Figure 2. Flow Diagram of Capital 6

Table I.   Expenditures for Longman Farms 11

Prefatory Material   **379**

4. Figures are numbered with Arabic numbers. Tables may be numbered with either Arabic numbers or Roman numerals.

5. The figure or table number, its title, and the page on which it appears are given, to aid the reader in locating the visual aid.

Nomenclature Page or List of Symbols

The nomenclature page or list of symbols gives the meaning of the symbols appearing in your report. If you have used a large number of symbols or equations and do not wish to define the symbols in your text, you may use a nomenclature page. Symbols are entered on this page as you see in Figure 17.11.

FIGURE 17.11
Symbols on nomenclature page

NOMENCLATURE

Btu   British thermal unit
pcf   pounds per cubic foot
        etc.

The Abstract

An abstract is a brief summary of your report. Two types of abstracts exist: informative or descriptive. If used, the descriptive abstract appears on the title page, as in Figure 17.1. The informative abstract is the last prefatory page before the text of your report. Occasionally, descriptive and informative abstracts will be included in the same report, in which case the informative abstract is called the summary, to differentiate it from the descriptive abstract.

In Figure 17.12, we give descriptive and informative abstracts for a report on highway-surface design. Notice the differences between the two.

The descriptive abstract lists the topics the report presents. This abstract gives no data or results; the reader must go to the report to learn those. Therefore, the descriptive abstract cannot stand alone as a substitute for the report. Instead, readers use this abstract to determine if they wish to read the document. In addition, the descriptive abstract is very brief, usually one to three sentences.

The informative abstract gives the important facts contained in the report. Therefore, this abstract can stand alone. In fact, since many readers will read it instead of the report, the informative abstract becomes an important document in itself.

The informative abstract is also short. It will usually be no more than 5 to 10 percent of the report's length but may be limited to 100–150 words regardless of the

FIGURE 17.12
Descriptive and informative abstracts

Descriptive Abstract

The need for asphalt-pavement design standards in Ames, Iowa; the analysis methods; determination guidelines; and recommended standards are discussed.

Informative Abstract

This report presents the need, analysis methods, determination guidelines, and recommendations for asphalt-pavement design standards in Ames, Iowa. Design standards are needed to ensure low road-maintenance costs and uniform road surfaces. For testing purposes, constituents were chosen from locally available materials and the mix was designed by the Marshall method, which uses density-voids analysis and stability-flow tests. Determination of an optimal mix was based on properties of specimens and accepted specifications for these properties to ensure maximum stability and unit weight, and a median range of unit air voids. Economy was also considered: Asphalt cement and limestone dust were conserved. The recommended mix is 4.8% asphalt cement, 5.7% limestone dust, 42.8% concrete sand, and 46.7% half-inch limestone. Field-quality monitoring and further pavement specification are also recommended.

---

length of the report. This abstract should *never* exceed one page. The informative abstract is often written as one continuous paragraph, although division into paragraphs is sometimes used. The abstract is double spaced if spacing permits, but single spaced if placed at the beginning of a document where space is limited.

Several important criteria govern writing informative abstracts:

1. The informative abstract summarizes the report. Therefore, no facts appear in the abstract that are not in the report.

2. The order of information in the report is retained.

3. The weight or relative importance of the information in the report is retained: Sections containing more important data (e.g., discussion sections, results, conclusions, recommendations) receive more space than sections containing less important data (e.g., introductions).

4. Exact data are given in the recommendation section. Thus, the reader knows the precise recommendation without having to turn to the report.

5. Personal opinion is avoided. All statements must be supported by facts.

6. The personal pronouns *I* or *we* are avoided. The informative abstract is always written from an objective point of view.

You use your table of contents to write abstracts. In Figure 17.13, we give the table of contents for the abstracts in Figure 17.12, then illustrate how the table of contents is employed to write the descriptive abstract and two types of informative abstracts, one with a predicting sentence first and one with major findings or recommendations first. We then discuss how to revise the informative abstract.

Prefatory Material **381**

FIGURE 17.13
Table of contents for abstracts in Figure 17.12

TABLE OF CONTENTS

TABLE OF FIGURES	iii
SUMMARY	iv
1.0 THE NEED FOR ASPHALT-PAVEMENT DESIGN STANDARDS IN AMES, IOWA	1
2.0 METHODS OF ANALYSIS	2
2.1 Choice of Constituents	2
2.2 Description of Design Technique	3
3.0 DETERMINATION OF OPTIMAL MIX	5
3.1 Design Criteria	5
3.2 Economic Considerations	5
4.0 RECOMMENDATIONS	7
4.1 Mix Proportions	7
4.2 Quality Monitoring in the Field	7
4.3 Further Pavement Specification	7
GLOSSARY	9
REFERENCES	10
APPENDICES	
Appendix A: Aggregate Grading	11
Appendix B: Properties Charts for Specimens	12

Using the Table of Contents to Write Abstracts

Informative Abstract with Predicting Sentence First

The informative abstract in Figure 17.12 is an abstract with a predicting sentence first. You should prepare this type of informative abstract when the most important information to your readers is an overview of your report. In this type, the first sentence, called the thesis statement, reflects the information in the first-level headings of the table of contents. Thus, this sentence predicts both the information given in the informative abstract and in the report, and sets out the order of that information. After reading this sentence, your audience will know whether they wish to continue further. (Notice that this thesis statement is rephrased as the descriptive abstract, which also reflects the information in the first-level headings of the table of contents.)

To write the predicting sentence, phrase each major heading as an item in a sentence list. Then write the rest of the informative abstract by summarizing the major facts in each section of the report, thus retaining the sequence of information in the report. You should give most space to your most important sections: In the informative abstract in Figure 17.12, sections 3.0 and 4.0 receive two sentences each because these sections contain data essential for the reader. Moreover, you should include any exact facts and figures you have obtained, as this writer has included the recommended mix.

Notice that the writer has also followed the remaining criteria governing informative abstracts. He has avoided personal opinion (e.g., ''This mix will solve all of Ames' pavement-design problems.''). He has also avoided using personal pronouns (e.g., ''For testing purposes, *I* chose. . .''). In addition, the informative abstract is short — between 100 and 150 words.

Informative Abstract with Major Findings
or Recommendations First

You should prepare an informative abstract with major findings or recommendations first when your readers require these rather than an overview of your report, as is frequently the case with scientific papers and laboratory reports. In this type of informative abstract, the thesis statement contains your findings or recommendations rather than a reflection of the major divisions of your table of contents. The partial abstract in Figure 17.14 illustrates this type of thesis statement.

FIGURE 17.14
Partial informative abstract with major findings or recommendations first

The recommended mix for asphalt pavement in Ames, Iowa, is 4.8% asphalt cement, 5.7% lime-stone dust, 42.8% concrete sand, and 46.7% half-inch limestone. Design standards are needed to ensure. . .

The remainder of the abstract then summarizes the facts in the major divisions of the table of contents, as in the informative abstract in Figure 17.12.

Revising Abstracts

Once you have written your informative abstract, you will frequently find it is too long. In Figure 17.15, we give a portion of the informative abstract in Figure 17.12 in an unrevised version, then discuss revision procedures.

FIGURE 17.15
Portion of an unrevised informative abstract

Design standards are needed in order to ensure that road-maintenance costs remain low and that road surfaces are uniform in construction. For the purposes of the tests conducted, constituents in the mix were chosen only from materials that were locally available. In addition, the mix was designed by the Marshall method. . .

After writing your abstract, revise for conciseness, according to the following criteria:

1. Omit all nonsubstantive words, i.e., those adding no information.
   Design standards are needed ~~in order~~ to ensure. . .
2. Omit transitions.
   ~~In addition~~, the mix was designed. . .
3. Omit obvious words.
   . . . uniform ~~in construction~~.
4. Reduce phrases to fewer words.
   . . . to ensure that road-maintenance costs remain low and that road surfaces are uniform in construction =
   . . . to ensure low road-maintenance costs and uniform road surfaces.

You should retain all words that add substance by conveying information, unless your informative abstract is still too long. In this case, you must omit information, cutting facts of lesser importance first. You will then be writing a combination of a descriptive and informative abstract, as you see in Figure 17.16.

FIGURE 17.16
Combination of descriptive and informative abstracts

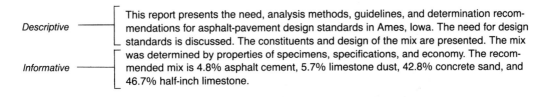

Notice that the writer has retained the data in his recommendation as the most important facts in his report.

SUPPLEMENTAL MATERIAL

Glossary

A glossary is a list of terms used in your report, with their definitions. You should use a glossary instead of other means of definition (e.g., placing definitions in the introduction, in the text, or in footnotes) when you have many terms to define and when they pertain to the entire report. In this case, the glossary allows you to collect these terms in one place for your reader.

We illustrate a typical glossary in Figure 17.17.

FIGURE 17.17
Glossary

GLOSSARY

Normalize Data:	To alter data by fitting them to a given range or subtracting out spurious effects
Periapsis:	The point in an orbit closest to the main body; in this report, the point where a spacecraft is closest to Venus
Phase Angle:	A number that indicates how much of the planet is sunlit in an image. When the planet is full (new), the phase angle is 0° (180°).
Pixel:	A picture element or dot
Standard Deviation:	A measure of how much a set of data varies from the mean value
Subsolar Point:	The point on a planet where the sun is directly overhead
Subspacecraft Point:	The point on a planet where the spacecraft is directly overhead
Wave Number:	A number representing the number of waves of a given length that could circle the planet at its equator

---

Notice that to shorten the glossary the writer has used phrases instead of sentences to define terms. If your reader requires more information than one phrase provides, you may expand your definitions, as you see in the definition of "phase angle."

Footnotes (or Endnotes) or References-Cited Page

The notes or the references-cited page documents with full bibliographic information any material you wish to cite. In Chapter 3, Gathering Information II: Library Research, we present guidelines for when to document information.

Because forms of documentation vary widely with discipline, you should consult a style manual in your field for the form used in your descipline. Following is the bibliographic information for several such manuals:

1. American Chemical Society. *Handbook for Authors of Papers in American Chemical Society Publications.* 3d ed. Washington, D.C.: American Chemical Society, 1978.

2. American Institute of Physics. *Style Manual for Guidance in the Preparation of Papers for Journals Published by the American Institute of Physics.* Rev. ed. New York: American Institute of Physics, 1978.

3. American Psychological Association. *Publication Manual.* 2d ed. Baltimore: American Psychological Association, 1974.

Supplemental Material 385

4. *The Chicago Manual of Style.* 13th ed. Chicago: Univ. of Chicago Press, 1982.
5. Council of Biology Editors. Committee on Form and Style. *CBE Style Manual.* 4th ed. Arlington, Va: American Institute of Biological Sciences, 1978.
6. Modern Language Association of America. *MLA Handbook for Writers of Research Papers, Theses, and Dissertations.* New York: Modern Language Association, 1977.
7. American Society of Agronomy. *Handbook and Style Manual for ASA, CSSA, and SSSA Publications.* Madison, Wisc: American Society of Agronomy, 1976.
8. Turabian, Kate L. *A Manual for Writers of Term Papers, Theses, and Dissertations.* 4th ed. Chicago: Univ. of Chicago Press, 1973.

Of these manuals, *The Chicago Manual of Style,* published by the University of Chicago Press, is the most comprehensive. However, the *CBE Style Manual* and the manual of the American Psychological Association are also widely used in their fields. Since these manuals are updated periodically, be certain yours is the latest edition.

We now discuss three types of documentation: references by order of citation in the text, author-year designation, and footnotes (or endnotes) with optional bibliography.

References by Order of Citation in the Text

With this system of documentation, a reference number (and sometimes a page number) appears in parentheses immediately following the material you wish to document:

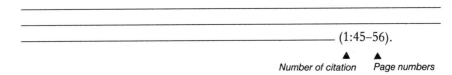

The first number refers your reader to a references-cited page at the end of your document where full bibliographic information is given. The references-cited page is usually not arranged in alphabetical order. Instead, the references are listed according to the order in which they appear in your document:

References Cited

1. Ringer, Jack. *Plant Physiology.* New York: Macmillan, 1970.
2. Bouker, Linda. *Plants and Their Properties.* San Francisco: Prince Publishers, 1975.

When you wish to refer your reader to a reference for a second or subsequent time, you use the same number of citation again, changing only the page numbers as necessary:

_____
_____
_____ (1:60–85).

This procedure is sometimes reversed: The references-cited page is sometimes put in alphabetical order first and then its reference numbers are cited in your text.

Author-Year Designation

When using the author-year system of documentation, you list the author's last name, the date of the reference, and any page numbers in parentheses immediately following the material you wish to document:

_____
_____
_____ (Blick, 1980, pp. 4–5).

This citation refers readers to a references-cited page at the end of your document, where references are listed alphabetically, with full bibliographic information for each:

References Cited

Blick, Jane. 1980. *Textiles and Weaving.* Boston: Brown Press.

Daws, Harold. 1975. *Flax.* Chicago: Rosen Press.

In subsequent citations, you repeat the parenthetical references, changing the page numbers as necessary:

_____
_____
_____ (Blick, 1980, pp. 10–14).

Footnotes (or Endnotes) with Optional Bibliography

In this system of documentation, you designate the material cited by a superscripted Arabic numeral placed after the material in the text. This numeral refers readers to the bottom of the page, where they will find the appropriate footnote, or to a separate endnote page located at the end of the document. (Use of a separate endnote page is most common.)

Sometimes a bibliography as well as footnotes will be included in the document. The bibliography, appearing immediately after the endnote page, lists all the references you have consulted, whereas the endnote page lists only the references you have cited in the document. Therefore, the bibliography provides readers with additional sources of information they may consult.

Bibliography

If you choose to include a list of all the references you have consulted, place this information on a bibliography page immediately after the endnote or references-cited page. List these references alphabetically, single spacing the entries and double spacing between entries.

Appendix or Appendices

The appendix is the last supplemental part of a report. You can use an appendix to present the following items:

1. Important material which, because of length, would interrupt the text (e.g., a series of visual aids, budgets, or balance sheets)
2. Material that will interest your readers but is not necessary for understanding the text
3. Technical material in a report also read by lay audiences
4. Certain large-scale items (e.g., fold-out maps). Bulky large-scale items, however, should be sent under separate cover.

Your appendix should be labeled with the word *Appendix* and its title. If you have several appendices, each receives a letter and an individual title. In addition, if the appendices consist of several pages, you may want to provide a separate cover for each appendix with its letter and title. These covers will help to segment your appendices for your readers.

CHECKLIST FOR REPORT SUPPLEMENTS

Prefatory Material

1. Does my cover identify my report by giving the title and name of the author? Does my title include the subject and type of report and the major key words? Is my title informative and brief?
2. Does my title page identify my report by including the title, the person for whom the report was prepared, the person preparing the report, and the date? Should I include any optional information?
3. Should my letter or memorandum of transmittal include both informative and persuasive details? Should the letter be attached to the cover or bound in the report as a permanent record?
4. Is my table of contents effective?
   A. Have I classified the sections effectively?
   B. Have I individualized my headings?

388   17 | Report Supplements

    C. Do my headings reflect the information in each section?

    D. Are they expressed in parallel grammatical form?

    E. Have I signaled levels of information by format?

5. Should I include a nomenclature page giving the meaning of the symbols used in my report?

6. Should I include a descriptive and an informative abstract?

    A. Is the descriptive abstract a brief, one- to three-sentence summary of the topics of the report?

    B. Is the informative abstract a brief summary (100–150 words) of the facts in the report?

      (1) Does my thesis sentence predict the topics of the report, or give my major findings or recommendations?

      (2) Have I included the important facts for each topic in my report?

      (3) Have I retained the order and the weight of the information in the report?

      (4) Have I avoided opinion in the abstract?

      (5) Have I avoided using personal pronouns?

Supplemental Material

1. Should I include a glossary to define a number of terms necessary for understanding the entire report?

2. Should I include a notes or references-cited page to document materials I wish to cite? Should my system of documentation be order of citation in the text, author-year designation, or footnote (or endnotes) with optional bibliography?

3. Should I include a bibliography to list alphabetically all the sources I consulted and to provide additional references for my reader?

4. Have I placed the following information in an appendix or appendices?

    A. Material that would interrupt the text

    B. Interesting material not necessary to understanding the report

    C. Technical material in a report read by lay readers

    D. Large-scale items

EXERCISES

Topics for Discussion

1. Discuss the strengths or weaknesses of the following titles:

    Modern Feed Facilities for Beef

Exercises   **389**

High-Level Waste-Storage Systems at Pinto with a Design for an Optimal System

The Reduction of the Danger of the Hazardous-Waste-Disposal System at Barkley Chemical

A Comparison Report of Superconductors in Fusion Reactors

A Solar Analysis of the Needs of the Livestock Barns at the S.U. Breeding Station

Windpower for Central Illinois Farms

Coagulants in Gels

Revise the titles you feel are ineffective: In each case, give the keywords that would be entered into a computer.

2. Using the following information, lay out an effective title page. Remember to consider placement of information according to importance.

Title: A Whole-Farm Management Plan for Arnold Family Farms, Inc.

Prepared for: Mr. James Arnold, President, Arnold Family Farms, Inc.

Prepared by: Williams Johns, Management Consultant, Johns' Farm Business, Inc.

Date: July 15, 1982

3. Identify the information in the text of the following letter of transmittal. Use our checklist of possible details to include (Figure 17.2) to assist you.

318 Pearson Avenue
Chicago, IL 50010
May 13, 1981

Mr. Philip Brown, President
Dettinger Company
Chicago Circle
Chicago, IL 50010

Dear Mr. Brown:
Enclosed is my recommendation report on the financial situation of the Dettinger Corporation.

This report cites the need for financial analysis, the current financial conditions, the problems associated with Dettinger's financial situation, and several recommendations for change. Please notice Tables 1 and 2. These tables clearly portray the severity of Dettinger's financial difficulties. As you can see, the deficit has steadily increased on the whole, as well as in relation to revenue.

I would like to extend a special thanks to Mike Terry of Dettinger Corporation for his cooperation and time. His assessments were highly beneficial in producing this report.

In conclusion, I believe Dettinger must act promptly to alleviate their financial difficulties. If you have any questions regarding this report, please feel free to contact me at 292-3806 at any time.

Sincerely,

4. The following letter of transmittal is not effective, although all the traditional parts are included in a logical order. Discuss specific revisions you would make in content and style.

125 North Hyland Street
Couler, IL 60010
May 15, 1981

John Fitz
598 Coover Street
Bloom, IL 60011

Dear Mr. Fitz:

Here is the report on my recent computer project.

This report tells how I went about my development, from researching through testing. The report explains why a microcomputer is an integral part of my design.

I would like to thank Dr. Jack Thompson for advising me during this project.

Any questions may be directed to me at the above address. I would be happy to answer any you may have.

Sincerely,

5. Read the following table of contents and informative abstract, then critique the abstract, pointing out specific weaknesses.

TABLE OF CONTENTS

TABLE OF ILLUSTRATIONS iii

SUMMARY iv

1.0 NEED FOR A NEW FACILITY FOR PARKRIDGE CENTER 1

2.0 REQUIREMENTS OF THE NEW FACILITY 2
 2.1 Space 2
 2.2 Parking 3

3.0 FEASIBILITY OF RELOCATING IN THE TAYLOR BUILDING . . . 4
 3.1 Available Space 4
 3.2 Parking Facilities 4
 3.3 Renovation Cost 5

4.0 RECOMMENDATION FOR RELOCATION 6

SUMMARY

This report discusses the feasibility of relocating Parkridge Center to the Taylor Building. Parkridge Center requires parking facilities for 20 cars, to accommodate clients. In addition, office space in the current facility is cramped. Room must be found for the two secretaries now employed by the center and the additional receptionist/secretary to be hired. Two large recreational rooms and two smaller rooms for classes of approximately 15 clients are necessary. The Taylor Building will precisely suit our needs at a cost of $1000 for minor renovations. I recommend that Parkridge Center relocate in the Taylor Building.

Exercises   **391**

6. Edit the following wordy abstract.

This report recommends that a substantial addition be built to Fourier Gym, to accommodate a large running track and two indoor tennis courts in addition to the two courts now located in the main gym. At the present time, in excess of 200 men and women use the recreational facilities at Fourier Gym on a daily basis. The noon hour is especially crowded; often 60 persons compete for running space on the small indoor track, while waiting lines for the tennis courts have sometimes been counted at 20 persons. Clearly an addition to Fourier Gym is needed to provide sufficient recreational space in order to satisfy the current demand. The addition would contain a quarter-mile track (a larger running track than that presently in use) and two more full-sized tennis courts. The addition itself would be oval in shape, in order to surround the track, with a domed roof. The addition, if approved, would be financed by means of student activity fees, which would then be offset by small admission charges. Finally, sufficient land is available to place the addition just south of Fourier Gym, which will allow for a direct connection between the two buildings by means of a short passage. Therefore, the addition ought to be built.

7. Determine the bibliographic form used in your field by consulting a style manual from your discipline. (Your reference librarian can assist you in locating this manual.) Give the bibliographic entry you would use for a book and a journal article.

Topics for Further Practice

1. Choose a report you have prepared for another class. Discuss the prefatory and supplemental material you included, giving your reasons for including this material. Discuss the material you omitted, giving reasons for the omission. Then critique the material included, indicating any revisions you would make.

2. Choose a report you have written that does not include any prefatory or supplemental material.
   A. Prepare a table of contents for that report.
   B. Write both a descriptive and informative abstract, based on that table of contents.
   C. Prepare a bibliography of sources you consulted in writing the report.
   D. Prepare all other necessary prefatory and supplemental material.

PART V

HANDBOOK OF STYLE: CAPITALIZATION, GRAMMAR, PUNCTUATION, SPELLING, USAGE

CAPITALIZATION

1. Capitalize proper nouns (names).
   William Booker
   Iowa State University
   Main Street
   Englishmen
   Staley's Feed Store

2. Capitalize words derived from proper nouns.
   Swedish
   Darwinism
   Americanize

3. Capitalize acronyms (designations made from the first letters of words in a name) and most abbreviations derived from capitalized words.
   NASA
   Btu (British thermal unit)
   F (Fahrenheit)

4. Capitalize a person's title when it precedes the name but not when the title follows the name, unless the title is part of the address in a business letter or the heading in a memorandum.
   President Cooper *but* John Cooper, president
   Dr. Mathews *but* Betty Mathews, doctor of internal medicine

   Do not capitalize a title used in apposition to a name.
   the doctor, Betty Mathews

   Do not capitalize a title (except President, when referring to a specific person) when the title is not followed by a name.
   The doctor *but* the President of the United States

5. Capitalize the full name of corporate bodies but not incomplete designations.
   Sharon City Council *but* the city council's decision

6. Capitalize titles of books, pamphlets, journals, and articles.
   Computer Engineering
   Living with the Sun
   The Journal of Psychiatry
   ''Coatings and Jackets''

7. Capitalize directional words *only* when they refer to geographical areas.
   the Midwest
   the South *but* ''We drove south.''
   the Pacific Northwest

8. Capitalize the names of days of the week and months, but not seasons.
   Friday
   September *but* fall, spring, etc.

395

396 Handbook of Style

9. Capitalize the names of disciplines only if derived from proper nouns or when used as part of the name of a department.

English
English Department
Biomedical Engineering Department *but* biomedical engineering
Computer Science Department *but* computer science

GRAMMAR

Agreement

1. Pronouns

A pronoun must agree in number (i.e., singular or plural) and gender with the word or words to which it refers.

The *committee* reviewed *its* decision.

Everyone performed *his or her* duty well.
(''His or her'' is used to avoid discrimination by sex.)

2. Subject–Verb

A. The verb of a sentence must agree in number with the subject of the sentence (i.e., either both singular or both plural).

       subject   *verb*
Wrong: A group of botanists *are attending* the conference.

       subject   *verb*
Correct: A group of botanists *is attending* the conference.

B. Collective nouns are considered singular when acting as a whole and plural when acting individually.

The group is cooperating well in the changeover.
The board of directors have split their vote.

C. The pronouns *anybody, anyone, each, everyone, someone* are usually considered singular.

Everyone is cooperating in the changeover.

D. Double subjects

(1) When two subjects are connected by either . . . or/neither . . . nor, the verb is singular if both subjects are singular.

Either answer 1 or answer 2 is correct.

(2) If one subject is singular and one is plural, the verb agrees with the closest subject.

Either answer 1 or answers 2–4 are correct.

(3) If both subjects are plural, the verb is plural.

Neither rulers nor protractors were sold.

(4) When two subjects are connected by both . . . and, the verb is plural.

Both answer 1 and answer 2 are correct.

Grammar **397**

Modifiers

1. Dangling modifiers
   A modifier is a word or group of words that acts as an adjective (a describing word). Because the modifier has no subject, the subject of the sentence becomes the subject of the modifier. If the subject of the sentence cannot logically be the subject of the modifier, the modifier "dangles."

   Wrong: After experimenting with several evaporators, it was determined that model no. 42A worked most efficiently.

   Correct: After experimenting with several evaporators, we determined that model no. 42A worked most efficiently.

2. Misplaced modifiers
   A modifier should be placed as close as possible to the noun or pronoun it describes. Otherwise, the modifier may seem to describe another word. This error is called a misplaced modifier.

   Wrong: The report on your desk *which you wanted to read* may be out-of-date.

   Correct: On your desk is the report *which you wanted to read.* It may be out-of-date.

Pronoun Reference

1. Case
   A. A pronoun may be in the nominative, objective, or possessive case:

      They broke *their* records, which pleased *them.*

   B. The nominative case is used when the pronoun is the subject of the sentence or follows a form of the verb *to be.*
      She will be in shortly.
      The person responsible is *she.*

   C. The objective case is used when the pronoun is the object of a verb or preposition.
      The laboratory honored *her* at its annual dinner.
      The laboratory could not function without *her.*

   D. The possessive case is used to indicate ownership.
      Its value is undetermined.
      The board, *whose* main function is advisory, will meet tomorrow.

398   Handbook of Style

2. Reference
   A. The word or words to which a pronoun refers must be clear.

   Wrong: The project manager disagreed with his employer on *his* standards.

   (The reference for the second ''his'' is unclear: project manager or employer.)

   Correct: The project manager disagreed with his employer on the project manager's standards.

   B. This, these, and it, as initial pronouns, cannot replace whole ideas.

   Wrong: The design will be completed by 1 July 1983. *This* will enable us to submit our report by 1 August 1983.

   Correct: The design will be completed by 1 July 1983. The completion will enable us to submit our report by 1 August 1983.

Sentences

1. Comma splice

   If two or more complete sentences are joined with commas, the error is called a comma splice.

   Wrong: The company delivered the chairs, they were broken.

   Correct: The company delivered the chairs. They were broken.

2. Incomplete sentence

   If a dependent clause (one which cannot stand alone) is punctuated as a sentence, the error is called an incomplete sentence.

   Wrong: When the company delivered the chairs. They were broken.

   Correct: When the company delivered the chairs, they were broken.

3. Run-on sentence

   If two or more complete sentences are run together with no punctuation, the error is called a run-on sentence.

   Wrong: The company delivered the chairs they were broken.

   Correct: The company delivered the chairs. They were broken.

PUNCTUATION

Apostrophes

1. An apostrophe signals an omitted letter or letters in a contraction (a combination of a pronoun and a verb or a verb and the word *not*).

I am = I'm	cannot = can't
it is = it's	do not = don't
she would = she'd	have not = haven't
they are = they're	is not = isn't
we will = we'll	would not = wouldn't
who would = who'd	

Punctuation **399**

2. An apostrophe indicates the plural of percentages, letters, and some abbreviations.
   The 1%'s were clearly indicated.
   The h's on the sign are blurred.
   Ph.D's are in demand.

3. An apostrophe indicates possession.

   A. An apostrophe plus s indicates possession after a singular noun or a plural noun that does not end in s.
      The ruler's graduation
      The children's knowledge

   B. An apostrophe indicates possession after a plural noun that does end in s.
      The heat exchangers' capacities
      The manometers' accuracy

   C. An apostrophe indicates joint possession.
      Iowa State University and Oklahoma State University's programs

      If each entity owns specific items, use two apostrophes:
      The project manager's and the engineers' directives

   D. An apostrophe plus s after the last word indicates possession in word groups.
      my brother-in-law's case

Brackets

Brackets are used within quotations to add material not in the original.
   "They [upper management] will make a decision tomorrow."

Colons

1. Colons are used after introductory sentences, followed by an explanation or a list.
   The equipment was outmoded: dirty, in poor repair, and old.

   We ordered the following components: two solar panels, one piping system, two pumps, one backup furnace.

   Colons are used *only* after *complete* introductory sentences, never after phrases:
   Wrong: The components ordered include: two solar panels, one piping system, two pumps, one backup furnace.
   (To correct this sentence, omit the colon.)

2. Colons can be used to separate two sentences when the second continues or completes the first.
   The failure was clearly unexpected: The dryer line simply stopped.

   The first word of a complete sentence following a colon may be capitalized or not capitalized. Maintain consistency throughout a document. (The first word of a *phrase* is not capitalized.)

3. Colons are used after the salutation of a business letter and between a title and subtitles.

   Dear Mr. Shorter:

   Replacement of TX Company's Automatic-Dryer Line: A Recommendation Report

Commas

1. Commas are used *before* coordinating conjunctions (and, but, or, nor, so, yet, for) joining two complete sentences.

   The components were back-ordered, so the delay was expected.

2. Commas are usually placed *after* long, introductory, dependent clauses.

   Since the components had to be back-ordered, the delay was expected.

3. Commas surround phrases that are not necessary to the meaning of the sentence — phrases that add details but could be removed without altering the sentence's meaning.

   Engineers, very experienced people, ought to understand this problem well.

   If a phrase is necessary to the meaning of the sentence, the phrase is *not* surrounded by commas.

   A chemist who is an expert in water analysis ought to be hired.

4. Commas separate three or more items in a series, if the series does not contain internal commas.

   The hand auger consists of a handle, a rod, and a cutting head.

   If the series contains internal commas (e.g., Boston, Mass.; Chicago, Ill.; and San Francisco, Calif.), the semicolon is correctly used.

   The comma before *and* is included to prevent confusion and to correctly distinguish items.

   The machine has a power switch, a motor, and source-and-gain controls.

5. Commas may be used to set off short introductory words or phrases.

   In fact, the law of supply and demand is not difficult to comprehend.

   However, students frequently find the law confusing.

   This usage is optional.

6. Commas set off an appositive (a word or words meaning the same thing as another word), quoted material, and parenthetical expressions.

   The law, a basic economic concept, must be understood.

   The speaker wrote, "I accept," before he read the terms offered.

   Our dryer line, seemingly, is in need of repair.

7. Commas are used in certain expressions:

   A. Addresses

      The package was delivered to Mary Dawes, 1417 Oak Street, Reno, Nevada.

Punctuation   **401**

> Mary Dawes
> 1417 Oak Street
> Reno, Nevada 61703

B. In dates, when month, day, and year are given in that order
> On January 2, 1982, the snow was 4 feet deep.
> *but*
> In January 1982, the snow was 4 feet deep.
> *or*
> On 2 January 1982, . . .

C. Before degrees or titles following names
> Professor Tyler, Ph.D.
> Alicia Newton, president

D. In geographical names, when the town and state are given
> Boston, Massachusetts, was the logical choice.

E. In numbers, at three-digit intervals
> 148,613

Dashes

The dash (two consecutively typed hyphens) is used to set off material within a sentence or to signal a sudden break in thought.

> The proposal team — two physicists, a biologist, and a statistician — will compile the final report.

Because the dash adds dramatic emphasis to details, it should be used sparingly.

Ellipses

Ellipses (three periods separated by spaces) are used to indicate words omitted from a quotation. A fourth period indicates a period at the end of a sentence. Three periods centered on the page indicate omission of a paragraph or more.

> "The lack of ductility in glass fibers . . . is one reason microbending occurs. . . ."
>
> . . .

Exclamation Points

Exclamation points are used at the ends of sentences to indicate strong emphasis.

> The error caused a flashback explosion!

Exclamation points are rarely used in technical writing.

Hyphens

1. Hyphens are used between compound modifiers (two or more words joined together to modify or describe another word).
> this well-known person

the heat-treatment process
the twelve-year-old equipment

These compound modifiers are *not* hyphenated when they follow the noun.
This person was well known.
the process of heat treatment

They are also *not* hyphenated when the first modifier ends in -ly.
the nearly completed construction

2. Hyphens are used for compound numbers, fractions, and ratios, if the ratios are used as adjectives and immediately precede the noun.
twenty-three days
four-fifths of a pound
a five-to-one score

3. Hyphens are used for words beginning with *all-*; *great-*; *ex-* (if it means "past"); and *self-*; and prefixes followed by proper names.
all-purpose   self-made
great-uncle   pro-Canadian
ex-president

Parentheses

Parentheses are used to set off supplementary material within a sentence.
The hand auger (a soil-sampling tool) makes testing soil compaction easy.
Remember that parentheses tend to minimize the importance of details.

Periods

1. Periods are used to end a sentence.
Solar-heated swimming pools are expensive but feasible.

2. Periods are used after abbreviations and as decimal points.
Ch.E.
Inc.
14.65%

The period is often omitted after technical abbreviations and acronyms used as the names of organizations.
Btu
NASA

Question Marks

Question marks are used to end direct questions.
What were the results?

Question marks are *not* used after indirect questions.

Wrong: He asked if we had the results?

Correct: He asked if we had the results.

Spelling **403**

Quotation Marks

1. Quotation marks are used to surround another person's exact words.
   The speaker said, "I am not an expert on that question."
   A. Periods and commas are placed *inside* quotation marks.
   B. Colons and semicolons are placed *outside* quotation marks.
   C. Dashes, question marks, and exclamation points are placed *inside* quotation marks if they are part of the quotation but *outside* if they refer to the entire sentence.
   D. Single quotes (i.e., '   ') are used to enclose a quotation within a quotation.

2. Quotation marks are used around minor titles (e.g., articles, reports) and any publication included as part of a larger work (book chapters, paintings, and poems).
   I have enclosed the report, "Design of a Solar-Heated Swimming Pool."

3. Quotation marks indicate words used in a special sense (e.g., irony or slang).
   The pool is "cheap."
   A great big "howdy" to you all.
   These words are rarely used in technical writing.

Semicolons

1. Semicolons may separate two sentences closely related in content.
   The data were entered; the output was recorded.
   The experiment was successful; therefore, the results are significant.

2. Semicolons are used to separate items in a series when the items contain internal commas.
   The convention could be held in Chicago, Illinois; Boston, Massachusetts; or Los Angeles, California.

SPELLING

The following rules are generally applicable. However, check your dictionary when in doubt about specific spellings.

Long Vowels, Short Vowels

1. A long vowel is pronounced according to its name (e.g., plāne, hōpe). An e at the end of a word makes a vowel long.
   plāne   plăn
   hōpe   hŏp

2. In order to retain the long vowel, the final e is dropped before a suffix beginning with a vowel.
   plāning   hōping

3. In order to retain the short vowel, the final consonant is doubled before a suffix beginning with a vowel.
planning
hopping

Plurals

1. Form plurals of most nouns by adding s.

meter   meters

typist   typists

2. Form plurals of numbers by adding s.

1980s   ones

3. Form plurals of some nouns ending in f or fe by changing the f or fe to ve before adding s.

knife   knives

life   lives

self   selves

4. Form plurals of nouns ending in s, ch, sh, or x by adding es.

box   boxes

loss   losses

rich   riches

5. Form plurals of nouns ending in y by changing the y to i and adding es.

city   cities

reply   replies

6. Form plurals of *some* words ending in o preceded by a consonant by adding es.

tomato   tomatoes

potato   potatoes

mosquito   mosquitoes or mosquitos

Prefixes

Add prefixes to root words without dropping letters.
dissatisfied
unnecessary

Similar Words

Do not confuse the following words:

I *accept.*   All *except* the following . . .

Rain *affects* plants.   The *effect* of the rain . . .

angel   An *angle* of 90° . . .

choose (present tense)   chose (past tense)

Dessert *complements* a meal.   I *compliment* you on your work.

Spelling

device (noun)
its (possession)
later (time)
The *loose* feathers . . .
I will *precede* you.
principal (chief)
Not more *than* four . . .
their (possession)
to (direction)
The *weather* is stormy.
whose (possession)
your (possession)

devise (verb)
it's (contraction for *it is*)
latter (last item in a series)
You *lose* a turn.
Proceed with the tests.
principle (a rule)
Then turn on the machine.
there (direction)
too (in addition)   two (number)
whether or not
who's (contraction for *who is*)
you're (contraction for *you are*)

Soft Consonants/Hard Consonants

1. The vowels e and i make the consonants c and g soft (c = s sound; g = j sound).

parcel	sergeant
decide	margin

2. Other letters make these vowels hard (c = k sound; g = g sound).

decade	bargain
particle	angle of a triangle

Suffixes

1. Drop the e at the end of a word before a suffix beginning with a vowel but not before a suffix beginning with a consonant.

plane	planing	care	careful
hope	hoped	complete	completely
value	valuable	safe	safety

2. Retain the final e after a soft c or g when adding a suffix beginning with a or o.

knowledge	knowledgeable
notice	noticeable

3. If a one-syllable word or a word accented on the last syllable ends in a single consonant preceded by a single vowel (e.g., plan, excel), double the consonant before adding a suffix beginning with a vowel.

plan	planning
excel	excelled
occur	occurring
omit	omitted
refer	referring

4. Change a final y to i, except before -ing.

happy	happiness
hurry	hurried
hurry	hurrying

5. When the sound is ē, write ie (except after c).

chief	deceive
field	perceive
yield	

For other sounds, write ei.
height
weight

USAGE

Abbreviations

Standard Abbreviations

1. Abbreviate titles when they come immediately before a name, if a standard abbreviation exists.

   Dr. Brown *but not* Pres. Cadwalden

2. Use the abbreviations a.m. and p.m.

   6:40 a.m.
   10:30 p.m.

3. Abbreviate such words as company or incorporated if a company uses them as part of its name.

   Field and Stream, Inc.
   J. and H. Co.

4. Abbreviate the names of states in the inside addresses of letters.

   14 Main Street
   Westwood, MA

5. Abbreviate certain common Latin expressions.

   e.g. (for example)
   etc. (and so forth)
   et al. (and others)
   i.e. (that is)
   vs. or v. (versus)

6. Avoid non-standard abbreviations in the text of a report (i.e., abbreviations not commonly used in your field).

   Street *not* St.
   September *not* Sept.

Usage   **407**

Technical Abbreviations

You can abbreviate technical units of measurement if they are used often in your report. If your readers may be unfamiliar with the abbreviation, write the words out in full the first time you use them and provide the abbreviation in parentheses, e.g., British thermal unit (Btu). Thereafter, you can use just the abbreviation.

1. Use the singular form of the abbreviation for both singular and plural terms.

   1 ft

   10 ft

2. Use lowercase letters except when a term is derived from a proper noun.

   cu ft   *but*   Btu

   This rule may vary with the particular style manual used. For parallelism, determine the usage in your field and maintain that usage throughout your documents.

3. Use periods only after technical abbreviations that are also words.

   cu in.

4. Spell out short common words.

   acre   rod

   per   ton

5. Use internal spaces when a word in the abbreviation is expressed by more than one letter.

   cu in.

A list of commonly used technical abbreviations follows:[1]

absolute	abs	degrees Celsius	
alternating current	ac	(centigrade)	°C
ampere	A	degrees Fahrenheit	°F
angstrom	Å	direct current	dc
atomic weight	at wt	dozen	doz
Brinell hardness		fluid ounce	fl oz
number	Bhn	foot per minute	ft/min
British thermal unit	Btu	frequency modulation	fm
calorie	cal	gallon	gal
centimeter	cm	gram	g
centipoise	cp	hertz (= cps)	Hz
cubic centimeter	cm^3	horsepower	hp
cubic foot	ft^3	hour	h
cycles per second		inch	in.
(= hertz)	cps	kilogram	kg
decibel	dB	kilometer	km

---

[1]Technical abbreviations may differ with the particular style manual used. Check the style manual for your field when in doubt about the accepted abbreviation of a term.

kilowatt	kW	number	no.
kilowatt-hour	kWh	octane	oct
liter	l	ounce	oz
logarithm	log	pounds per square inch	psi
meter	m	quart	qt
microwatt	μW	revolutions per minute	rpm
miles per hour	mph	square	sq
milligram	mg	temperature	temp
milliliter	ml	volt	V
millimeter	mm	watt	W
minute	min	week	wk
mole	mol	yard	yd
month	mo	year	yr

Due to/Because of

1. "Due to" is used only after a form of the verb *to be*.
   The failure was due to a breakdown.
2. "Because of" is used in all other cases.
   Because of a breakdown, we failed to complete the experiment.

Parallelism

1. Parallel parts of a sentence should be expressed in parallel grammatical form.

   Wrong: Planting a windbreak involves *planning* the design, *planting* the trees, and *management* of the stock.

   Correct: Planting a windbreak involves *planning* the design, *planting* the trees, and *managing* the stock.

2. Headings of equal weight should be expressed in parallel form.

   Wrong: 1.0 Design Considerations
          1.1 Criteria
          1.2 Procedure
          1.3 Evaluating the design

   Correct: 1.0 Design Considerations
           1.1 Criteria
           1.2 Procedure
           1.3 Evaluation

3. Lists should be expressed in parallel form.

   Wrong: We performed the following operations:
          1. Experimentation with computer models
          2. Analyzing the data

Usage   **409**

    3. Recording of the findings
    4. Recommendations for alternatives

Correct: We performed the following operations:
    1. Experimenting with computer models
    2. Analyzing the data
    3. Recording the findings
    4. Recommending alternatives

Shifts

Avoid abrupt changes in expression:

1. Mood: Verbs should be expressed in parallel form.

   Wrong: Plant the trees and then they must *be* correctly *maintained.*

   Correct: Plant the trees and *maintain* them correctly.

2. Number: Pronouns should be either singular or plural.

   Wrong: One must choose the trees carefully before *they* purchase them.

   Correct: One must choose the trees carefully before *one* purchases them.

3. Voice: Use either the active or the passive voice consistently.

   Wrong: They *designed* the windbreak and then the trees *were planted.*

   Correct: They *designed* the windbreak and then *planted* the trees.

4. Tense: Verbs should express parallel time (e.g., present, future).

   Wrong: They *designed* the windbreak and then *plant* the trees.

   Correct: They *designed* the windbreak and then *planted* the trees.

Use of Numbers

1. Generally, exact numbers from 1 to 9 are written out; 10 and above are given as figures.
   eight
   2,750

2. Exact measurements are given as figures. Approximate measurements are written out.
   2 feet by 4 feet
   7 acres
   162 miles
   approximately ten kilometers

3. Decimals, fractions, and percentages are given as figures.
   2.763
   62%
   4¼

4. Addresses, ages, dates, exact dollars and cents, figure numbers, mileage, page numbers, and times with a.m. or p.m. are given as figures.

14 Baron Lane   Figure 4
45 years old   26 miles
$162 or 162 dollars   12:25 p.m.
   but
approximately one hundred and sixty dollars

5. Never begin a sentence with a figure.

Wrong: 13 boxes were delivered.

Correct: Thirteen boxes were delivered.

6. If you express one number in figures, maintain parallelism by expressing adjacent or nearby numbers in figures, even when below ten.

The order included 13 boxes, 6 packets, and 20 sacks.

7. Times other than those used with a.m. and p.m., and numbers used as street names below 100 are written out.

eight-thirty or six o'clock
120 West Sixteenth Street

8. When two numbers are used together in a sentence, write out the first or the shorter one and use a figure for the second or the longer one.

The order included sixteen 2-foot by 4-foot rectangular traps.

The worker replaced 200 sixty-watt lightbulbs.

9. Contracts and other legal documents give numbers in words and in figures.

The bidding price is four thousand three hundred eighty dollars ($4,380) on five hundred (500) pounds of structural steel.

Appendix:
Selected Bibliography on
Technical Writing

GENERAL

Houp, Kenneth W., and Thomas E. Pearsall. *Reporting Technical Information.* New York: Macmillan, 1984.

Lannon, John. *Technical Writing.* Boston: Little, Brown, 1984.

Mills, Gordon H., and John A. Walter. *Technical Writing.* New York: Holt, Rinehart, and Winston, 1978.

Stratton, Charles. *Technical Writing: Process and Product.* New York: Holt, Rinehart, and Winston, 1984.

HANDBOOKS

Brusaw, Charles, Gerald Alred, and Walter Oliu. *Handbook of Technical Writing.* New York: St. Martin's Press, 1982.

Jordan, Stello, Joseph Kleinman, and Lee H. Shimberg. *Handbook of Technical Writing Practices.* 2 vols. New York: Wiley-Interscience, 1971.

Knapper, Arno F., and Loda I. Newcomb. *Style Manual for Written Communication.* Columbus, Ohio: Grid Publishing Co., 1983.

TECHNICAL REPORT WRITING

Mathes, J. C., and Dwight W. Stevenson. *Designing Technical Reports: Writing for Audiences in Organizations.* Indianapolis: Bobbs-Merrill, 1976.

411

Souther, James W., and Myron L. White. *Technical Report Writing.* New York: Wiley, 1977.

Weiss, Edmond H. *The Writing System for Engineers and Scientists.* Englewood Cliffs, N.J.: Prentice-Hall, 1982.

VISUAL AIDS

Lefferts, Robert. *How to Prepare Charts and Graphs for Effective Reports.* New York: Barnes and Noble, 1981.

MacGregor, A. J. *Graphics Simplified: How to Plan and Prepare Effective Charts, Graphs, Illustrations, and Other Visual Aids.* Toronto: University of Toronto Press, 1979.

Turnbull, Arthur T., and Russell N. Baird. *The Graphics of Communication.* New York: Holt, Rinehart, and Winston, 1980.

Index

Index

Abbreviations, 87, 406–408
Abstracts, 301, 379–383
  descriptive, 381–382
  informative, 382
  revision of, 382–383
Abstracting journals, 51–52, 61
Acronyms, 88
Active voice, 90–93, 111–112
Agreement, pronoun and subject–verb, 396
Analogy, 143, 144, 148
Apostrophe, 398–399
Appendix, 346, 387
Application, letter of, 209–210, 223–236
Audience, 2, 5, 11, 93–94, 99
  characterization of, 14–16, 19, 22, 25
  classification of, 13–14, 19, 22, 25
  complex, 12
  educational background of, 14, 25, 87, 88
  effects of, 11–12
  identification of, 13, 19, 22, 25
  limited, 12, 17
  "mock," 5
  primary, 13, 14, 25, 74
  reader attributes of, 14, 16, 25

  role-playing, 5, 6, 99–100
  secondary, 13-14, 25, 74
  technical experience of, 14–15, 19, 25, 85, 86–87, 88
Audience analysis, 12–22, 19, 25
  elements of, 13–16, 19
  methods for, 16–22
  unwritten, 16, 19
  written, 17–22

Bar graph, 317–324
Basis, for partitioning, 140–143, 159
Bibliographies, 46, 59
  computerized, 53
Bibliography, 386, 387
  working, 55, 62
Block form, for letters, 356–359
  modified, 356, 359
Brackets, 399
Brainstorming, 27, 28–29

Capitalization, 395–396
Card catalog, library, 42–44, 58–59

415

Career objectives, 215, 219, 220, 221
Cause and effect
    in definitions, 126, 130
    as pattern of development, 76
    in sentences, 86, 106
Cautions, in instructions, 162, 163, 164,
        166–168, 172
Charts
    egocentric-audience-analysis, 18, 19, 20
    cluster-of-audience-interest, 18, 19, 21
    flow, 32, 33, 80, 162–163, 331–333
    milestone, 248
    organizational, 331, 334
Checklists, 17–18, 27, 37–40, 68–69, 81–
        82, 94, 112–113
Classification
    in definitions, 126, 127, 129
    in information gathering, 28, 31
    in item descriptions, 142–143, 150
    in lists of content for reports, 73–74
    in messages, 195
    of problems in proposals, 243
    in resumes, 217, 218
    of steps in process descriptions, 164–165,
        174
    in table of contents, 371–373
    of tasks in progress reports, 268–269
Cluster-of-audience-interest chart, 18, 19,
        21
Clustering devices, 31, 33, 36–37
Coherence, 102, 103
Colon, 399–400
Comma, 121–122, 400–401
Comma splice, 398
Communication context, analyzing, 11–25
    for definitions, 123
    for final reports, 285–286
    for item descriptions, 139
    for letters of application, 223
    for process descriptions, 160–161, 171–
        172
    for progress reports, 263
    for proposals, 239–241
    for resumes, 210
    for technical letters and memorandums,
        191, 198–199
Comparison
    in definitions, 124, 125, 130

as pattern of development, 76
standard reporting form for, 288–289
Compatibility, point of, 200–201, 203
Composing process, 2, 3–6
    strategies for, 2, 6, 9–117
Computerized information retrieval, 53–55,
        60–62
Connections, in documents, 102–105. *See
    also* Transitions
Connotation, 108
Consonants, 405
Content, arranging, 72–83, 125–126. *See
    also* Format; Structure
    for definitions, 125–126
    for final reports, 286–294
    for item descriptions, 144, 146–147
    for letters of application, 228–229
    for process descriptions, 164–168, 174–
        177
    for progress reports, 267–270
    for proposals, 244–249
    for resumes, 217–218
    for technical letters and memorandums,
        192–195, 199–202
Content, selecting, 72, 99–100
    for definitions, 123–125
    for final reports, 286
    for item descriptions, 144–145
    for letters of application, 225–228
    for process descriptions, 164, 172
    for progress reports, 264–267
    for proposals, 244–249
    for resumes, 215–216
    for technical letters and memorandums,
        192, 199
Continuums, 33, 35
Contrast, 76, 124, 125
Covers, report, 367–368

Dangling modifiers, 91–92, 112, 397
Dashes, 401
Data bases, 53–55, 61
Deduction, 74–75, 77, 193
Definitions, 121–136, 137, 138
    circularity in, 127, 129
    example, 127, 129

Index   **417**

expanded, 122, 126, 128, 130, 132–135, 138
formal, one-sentence, 121–122, 126–l27, 130–132
informal, 121–122, 127, 129
in item descriptions, 144, 147, 148, 151, 152, 153
operational, 122
as pattern of development, 76, 125
placement of, 122
in process descriptions, 162, 163, 173, 174, 177
purpose of, 121–122
Denotation, 108
Description, 76, 124, 125
item, 137–157
process, 159–189
Design report, 287–288
Development, patterns of, 75–78, 125
Diagrams, 32, 34, 80, 140–141, 147–148, 153, 162, 167, 335–336, 338
cutaway, 152, 338
exploded, 337
parts, 144
schematic, 335–336
Documentation, forms of, 385–386
Draft, revising, 98–117
Drawings, 335–338

Editing, 6. *See also* Revising
Egocentric-audience-analysis chart, 18, 19, 20
Ellipsis, 401
Emphasis, 75, 78–80, 99, 100–101, 106–107, 170
Equations, 352
Exclamation point, 401
Executive summary, 299–301

Feasibility study, 288
Figures, as visual aids, 311
Final reports, 284–307
introduction, 290–291
standards in, 291–293
types of, 284, 287–289, 291
Flow chart, 32, 33, 80, 162–163, 172, 247

Footnote page, 384–386
Format, 126, 350–365. *See also* Content, arranging
Free writing, 30–31

Generality, level of, 109
Genus, and differentia, 126, 127, 129, 138
Glossary, 383–384
Government documents, 42, 52–53, 60
Grammar, 396–398
Graphs, 317–330
bar, 317–324
line, 32–33, 35, 324–328
pie, 329–330
Grids
information–gathering, 27, 39
purpose-and-use, 23-25, 72, 74

Headings, 127, 176–177, 218, 248–249, 269–270, 294, 353–355
Hierarchical devices, 31–32
Highlighting, 126, 167–168, 176–177, 195, 249, 269, 294, 350–351
Hyphen, 401–402

Idea-trees, 31–32
Illustrations, list of, 377–379
Independent clause, 106, 107
Indexes, 44, 46–51, 59
Induction, 74–75, 77, 200
Information, gathering, 2, 5, 27–41, 42–69
for definitions, 123, 124
for final reports, 286
for item descriptions, 137–139, 140–144
for letters of application, 224–225
for process descriptions, 161–164, 172
for progress reports, 264
for proposals, 241–244
for resumes, 210–215
for technical letters and memorandums, 191–192, 199, 210–215
Information retrieval, computerized, 53–55, 60–62
Instructions, 159–171

cautions in, 162, 163, 164, 166–168, 170, 172
explanations in, 162, 163, 164, 166–167
format for, 167–168
imperative mood, 168
purposes of, 161
types of, 160
Introductions, 125, 128, 130
Item description, 137–157
bases for, 140–143, 151
classifying parts in, 142–143, 150
part-by-part description in, 144, 146–147
partitioning in, 140–142, 151, 153
as pattern of development, 76, 82
types of, 138–139, 151, 154

Jargon, 88
Job analysis inventory, 225, 226
Journals, abstracting, 51–52, 61
Justification report, 288

Key words, 60, 61

Laboratory report, 288
Language. *See also* Word choice
difficulty of, 87–88
nondiscriminatory, 88–90
Letters
of application, 209–210, 223–236
layout of, 356–360
standard parts of, 361–362
technical, 190–208
of transmittal, 369–370
Library of Congress classification system, 44, 58
Library research, 55–69
Library resources, 42–55, 56, 57, 68
on jobs, 224–225
Line graph, 32, 35, 324–328
Lists, 80, 142, 146, 167, 195, 218, 269, 352. *See also* Nomenclature page
of content, 72–74
of illustrations, 377–379
of questions, 191–192, 199

of symbols, 379
Literature, guides to, 44, 46
Logical progression, 99, 101–105

Maps, 338–339
Memorandums, 190–208
layout of, 363
standard parts of, 363–364
of transmittal, 369–370
Messages, 190–208. *See also* Letters; Memorandums
negative, 198–204
positive, 190–198
standard reporting forms for, 193, 200
Milestone chart, 248
Modifiers
dangling, 91–92, 112, 397
misplaced, 397

Names, 89–90
Narrative. *See* Process descriptions
Nomenclature page, 379
Nondiscriminatory expressions, 89–90
Numbering systems, 355–356
Numbers, 409–410

Operating manual. *See* Instructions
Ordering devices
definition of, 80–81
for definitions, 125
for item descriptions, 147
for process descriptions, 164–167
Organizational chart, 331, 334
Outlines, 80
as relating device, 31–32
skeletal, 101–102
Overview
in definitions, 125
in item descriptions, 144
in process descriptions, 163

Paragraphs, ordering, 76–78
Parallelism, 111–112, 408–409

Index **419**

Parentheses, 121–122, 402
Partitioning, basis for
  in item descriptions, 140–144
  in process descriptions, 159, 164
Passive voice, 90–93, 111–112
Patterns of development
  analytical, 76–78
  chronological, 76–78
  definition of, 75–76
  progressive, 77–78
  relating, 76–78
  spatial, 76–78
Periods, 402
Personal inventory, 210–213, 214
Photographs, 339
Pie graph, 329–330
Prewriting
  definition of, 3, 5
  for definitions, 123–126
  for final reports, 285–294
  for item descriptions, 139–147
  for letters of application, 223–229
  for process descriptions, 160–164
  for progress reports, 263–270
  for proposals, 249
  for resumes, 210–219
  for technical letters and memorandums,
    191–195
Process descriptions, 159–189
  conclusion to, 164
  formatting of, 167–168
  instructions, 159–171
  introduction to, 163, 164
  narratives, 171–181
  purposes of, 161
  types of, 159, 160
  uses of, 159, 161
Progress reports, 262–283
  arranged by task, 268–269
  arranged by time, 267–268
  conclusion of, 267
  formatting of, 269–270
  introduction to, 265
  preliminary assessment of findings in,
    265–267
  purposes of, 263
  types of, 262

uses of, 263
Pronoun reference, 397–398
Proposals, 237–261
  formatting of, 248–249
  request for, 237, 239, 240–241
  to research program, 257–258
  schedule for preparation, 238–239, 245–
    248
  solicited/unsolicited, 237–238, 244–245
  types of, 237
  uses of, 240–241
Punctuation, 398–403
Purposes
  analysis of, 22–25
  of definitions, 121, 123
  of final reports, 285–286
  of item descriptions, 138
  of letters of application, 209
  of process descriptions, 161, 172
  of progress reports, 263
  of proposals, 240–241
  of resumes, 209
  of technical letters and memorandums,
    198–199
  types of, 22
  written/unwritten analyses of, 23–25

Question marks, 402
Quotation marks, 403

Readers. *See* Audience
Redundancy, 110–111
References-cited page, 384–386
Relating devices, 31–37
  clustering, 33–36
  hierarchical, 31–32
  sequencing, 32–33
Repetition
  as connector, 103
  for emphasis, 80
  needless, 106, 110–111
Reporter's formula, 37, 38, 137–138, 140,
  161
Reports, 287–288
  cover for, 367–368

final, 284–307
progress, 262–283
Request for Proposals (RFP), 237, 239, 240–241
Resumes, 209–233
  appearance of, 218, 222
  formatting of, 218
  purposes of, 209
  types, 217, 220, 233
  uses of, 209
Revising, 6, 98–117. *See also* Rewriting
  for amount and kind of detail, 99–100
  for appropriate emphasis, 100–101
  for logical progression, 101–105
  strategies for, 98–113
  for stylistic appropriateness, 105–112
  techniques for, 99–112
Rewriting
  definition of, 2, 4–6
  definitions, 128–129
  final reports, 299
  item descriptions, 149–150
  letters of application, 231
  process descriptions, 170, 179–180
  progress reports, 275
  proposals, 255–256
  resumes, 219–222
  technical letters and memorandums, 197, 204
  techniques for, 5–6
Role playing, 5, 6, 99–100

Samples, as visual aids, 340
Secondary audience, 13–14, 25, 74
Semicolons, 403
Sentence construction, 85–87, 93
  revising, 105–108
Sentences
  avoiding shifts in, 409
  coordination in, 105–106
  incomplete, 398
  partitioning, 144, 164
  run-on, 398
Sequencing devices. *See* Relating devices
Sketches, 335–336. *See also* Diagrams
Skills, hierarchy of, 214

Spelling, rules of, 403–406
Standard reporting forms, 73
  for final reports, 287–289
  for technical letters and memorandums, 193, 200
Standards, 291–293
Structure, 71–82. *See also* Content, arranging; Ordering devices; Patterns of development
Style, planning
  appropriateness of, 105–111
  for process descriptions, 168, 177
  for progress reports, 270
  for proposals, 249
  for resumes, 218–219
  strategies for, 84–94
  for technical letters and memorandums, 195, 202
Suffixes, 405–406
Supplements, report, 366–391
  prefatory, 367–383
  supplemental, 383–397
Symbols, list of, 379

Table of contents, 65, 370–377, 381–383
Tables, 311, 316–317
Tasks/time schedule, 238–239, 245–248
Technical letters and memorandums, 190–208
  closing of, 194, 201
  formatting, 195
  introduction to, 193–194, 200
  negative message in, 198–204
  positive message in, 190–198
  purpose of, 191, 198–199
  types of, 190
  uses of, 191, 198–199
Title page, of reports, 368–369
Titles, personal, 90
Tone, 13, 93–94
  of progress reports, 270
  of proposals, 249
  revising for appropriate, 112
  of technical letters and memorandums, 195, 202
Topic sentence, 65, 76–78

Index   **421**

Transitions, 103–105, 144

Usage, 406–410
Uses
  of definitions, 123
  elements of analyzing, 22
  of final reports, 285–286
  of item descriptions, 138–139
  of letters of application, 209
  of process descriptions, 161, 172
  of progress reports, 263
  of proposals, 240–241
  of resumes, 209
  of technical letters and memorandums,
    191, 198–199
  types of, 22
  written/unwritten analysis of, 23

Verbs, 110
Visual aids, 311–349
  designating, numbering, and titling,
    345–346
  highlighting, 351
  identifying, 346

  in item descriptions, 144
  locating, 346
  pictorial, 311, 335–340
  in process descriptions, 163, 167, 168
  in progress reports, 264
  in proposals, 247–248
  purposes of, 311–315
  symbolic, 311, 316–334
  types of, 316–340
Voice, 90–93, 111–113
  revising, 111–112
Vowels, 403–404

Word choice, 87–90, 93
  denotation/connotation, 108
  economy of, 109–111
  level of generality, 109
  nondiscriminatory, 88–90
  precision in, 108–109
  revising, 108–112
Word order. *See* Sentence construction
Wordiness, 109–111
Writer's block, 4
Writing techniques, 5